Climate Change

Climate Change

THE SCIENCE OF GLOBAL WARMING AND OUR ENERGY FUTURE

EDMOND A. MATHEZ

American Museum of Natural History

Student Companion by Jason E. Smerdon, Lamont-Doherty Earth Observatory, Columbia University

COLUMBIA UNIVERSITY PRESS   NEW YORK

COLUMBIA UNIVERSITY PRESS
Publishers Since 1893
New York   Chichester, West Sussex

Library of Congress Cataloging-in-Publication Data
Mathez, Edmond A.
    Climate change : the science of global warming and our energy future / Edmond A.
    Mathez.
        p. cm.
    Includes bibliographical references and index.
    ISBN 978-0-231-14642-5 (cloth : alk. paper)
    ISBN 978-0-231-51818-5 (e-book)
    1.   Climatic changes.   2.   Global warming.   I.   Title.

    QC981.8.C5M378   2009
    551.6—DC22

                                                                      2008034132

To the memory of Muriel Mathez, who at ninety
and near death found happiness in helping to edit this book.

CONTENTS

FOREWORD

WE LIVE IN THE ERA of global warming. That much is certain. Atmospheric levels of heat-trapping greenhouse gases have increased due to human activities, and a cornucopia of climate changes is now apparent. Global average temperature increased about 1°C (1.8°F) over the past century, and sea level has risen more than 15 centimeters (6 inches). Glaciers are melting worldwide, the major ice sheets in Greenland and Antarctica are fraying at their peripheries, and the imagined Northwest Passage may soon open permanently. Rainstorms are intensifying in many areas, and summer heat waves have become more intense while winters are less severe. At the same time, areas of drought are becoming more extensive, possibly contributing to a significant recent increase in food prices. Species like birds and butterflies that move fast are literally fleeing the warming. Those that cannot fly or live in restricted ecosystems, like polar bears, may be stranded in a hostile environment and gradually fade out of existence.

Whether governments and the public will act sufficiently fast to stem the warming is perhaps the greatest, and most uncertain, part of the picture because global warming presents unique challenges. For political leaders, the issue is most difficult. The worst consequences of warming occur decades after a given level of atmospheric greenhouse gases is reached. By the time the changes are apparent, more gases have been emitted and more changes, some potentially disastrous, could be in the pipeline, irreversibly. With further climate changes building surreptitiously, how can the risks be made clear to a public already assaulted by too many messages about too many problems, many of them quite threatening? For the average person, the climate is an arcane system based on abstract physical concepts. To experts in the physics of radiation or ocean circulation or in Earth's long climate history, the evidence is compelling. Otherwise, the scientific arguments can seem too complex and too dull to be worth the effort seemingly required to comprehend them.

There has been a crying need for a readable account of the vast scope of climate change, one that is accessible to a broad range of readers willing to devote a modest amount of time to understanding the science and its implications. As an educator teaching introductory environmental courses at a major university, I have for years been puzzled by the lack of a text adequate to introduce my students simultaneously to nuances of the science of the global-warming problem and the wide-ranging possibilities for its solution. *Climate Change: The Science of Global Warming and Our Energy Future* by Edmond A. Mathez fills the bill and comes along none too soon. I was particularly pleased to read this book because it provides a companion to the eponymous exhibition, co-curated by Ed Mathez and me, that opened at the American Museum of Natural History in October 2008.

The book's presentation is comprehensive and direct, using simple yet appropriate metaphors to explain otherwise inaccessible details of Earth's climate. This is particularly true of Mathez's descriptions of how the atmosphere works: If Earth is rotating west to east, why do the main air currents above North America also move west to east? Why doesn't the rotation of Earth create the equivalent of an east-to-west breeze? The reader will find equally adept descriptions of the cycling of carbon among rocks, oceans, living things, and the atmosphere, where it persists in the form of carbon dioxide, the major human-made, heat-trapping, greenhouse gas.

This book is equally clear at explaining the vast uncertainties while placing them in the appropriate context. For example, the Greenland Ice Sheet is at risk of melting away completely if Earth warms a few degrees more, but there is conflicting evidence about whether the process would play out over hundreds or thousands of years. A few hundred years would equate with disastrous rates of sea-level rise and the need for prompt action to reduce emissions; a few thousand would give us ample time to think, learn more, and hone our polices. But in a handy "Perspectives" section at the end of each chapter, Mathez puts the various arguments in a context useful to those concerned about whether, when, and how much to act. Although the sea-level rise may occur slowly, the irreversibility could mean that its inevitability would be locked in by emissions over the next several decades.

Finally, there is the question of how to walk back off the limb that we have blithely climbed during the two hundred years of industrialization, before it breaks completely. The main source of carbon dioxide is the burning of fossil fuels: coal, oil, and natural gas. It will require massive decreases in fossil-fuel use and carbon dioxide emissions to stabilize the climate and, eventually, return it to its earlier, safer condition. In *Climate Change*, Mathez is both clear-eyed about the complexities of solving the problem and optimistic about the human capacity to invent and innovate a pathway out of it. He argues compellingly that the key is to grapple with the growing demand for electricity. Increased efficiency is critical, but beyond that, there will be no silver bullet and every solution proposed so far—including solar cells, wind power, and nuclear energy—has both benefits and drawbacks.

The pathway for achieving this goal is largely uncharted. But we know that the focused attention of governments will be required for a long period of time. Developed and developing countries will have to find grounds for cooperative action to reduce emissions worldwide. In the course of acting to stem emissions of the gases, much more will be learned about the climate system and even more about managing an increasingly complex planet. We will have success and failures. The developed countries, having emitted most of the greenhouse gases now in the atmosphere, will have to learn to deal with the rapidly growing developing countries like China on the basis of equality: growing equality of economic power, growing political leverage, and equality of risk from climate change.

But to those who are daunted by the prospect, I like to point to my own experience as a child early in the nuclear age. I spent my childhood diving under desks in school during atomic-bomb drills. We were unsure for many years whether the United States and the erstwhile Soviet Union would annihilate each other. But these two antagonistic countries found a mutual interest in self-preservation. Similarly, we cannot be sure of our success in the fight against global warming, but if I had to bet, I would say that human ingenuity and the instinct for self-preservation will once again carry the day.

Michael Oppenheimer
Princeton University

PREFACE

THIS BOOK SETS OUT the scientific basis for our understanding of climate change. As such, it is a synthesis of a rather extensive field of knowledge that includes the science of Earth's climate, how and why climate is changing, and the consequences of those changes. Climate science involves study of the workings of the atmosphere and ocean and of the interactions between the two and the other components of Earth, such as its flora. It also rests on the study of past climate, so paleoclimatology, as the discipline is known, is an essential part of this book. Furthermore, one cannot talk about climate change without talking about energy because future climate depends directly on our ability to control emissions from fossil-fuel burning and to apply alternative technologies to satisfy the world's insatiable appetite for energy. There is much to be said about energy issues, and writers more knowledgeable of the subject than I have written extensively about them. Nevertheless, as I began to study energy issues, I quickly realized that our energy future is dominated by one overriding issue: how we are going to provide for the world's electricity needs. Accordingly, I have chosen to focus on this matter in the final chapter of this book.

I wrote this book with two audiences in mind. The first comprises undergraduate university students in beginning courses on climate or in more general courses in which climate is an important component. This book paints a wide-ranging picture of an entire field, the hope being that this broad approach will provide the necessary background and at the same time excite students' interests in learning more about the scientific underpinnings of our current understanding of climate and climate change. The second audience includes general readers who may or may not have a background in science but who seek to understand the issue of climate change in some detail. Here I have in mind individuals who are moved to understand the relevant political and policy debates, teachers who need to acquire an understanding and perspective of climate and issues associated with climate change in order to support their teaching, and individuals who need an accessible reference to climate issues as part of their work in other fields—for example, in the financial industry.

I am also thinking of visitors to the exhibition "Climate Change: Threat to Life and Our Energy Future," which was developed by the American Museum of Natural History in collaboration with several other institutions and opened in New York in October 2008. Along with Michael Oppenheimer of Princeton University, I served as the curator of this exhibit. I proposed it to the museum community in January 2006, motivated by a combination of factors. First, I had been sensing a palatable alarm in the scientific community about the growing risk posed to the health of society by climate change, an alarm that I knew was largely unknown to the public. Second, I had been sharing the frustration with many of my colleagues about the generally uninformed nature of the public discussion. The news media, for example, had been treating climate change as if its occurrence were controversial, and the U.S. government had completely abrogated its leadership role. At the same time, many intelligent people outside the sciences—in particular those connected to the businesses of energy, insurance, and finance—had begun to realize that they had to pay attention to climate change if for no other reason than to understand its potential impact on their worlds. So was born the exhibition, and in fact this book as well: I decided to write it as a means of learning the field in sufficient detail to guide the development of the exhibit sensibly and as an opportunity to tell the deeper science behind the exhibit than is possible to tell in the exhibit itself.

In writing the book, I have relied mainly on the primary scientific literature. However, I have limited my citations on each subject to a relatively few recent references, the intent being to provide a bibliography that allows the reader practical entry into the broader literature. It is worth emphasizing, therefore, that the science is based on an enormous body of work involving thousands of people, not all of whose work is cited here. Finally, I must add that parts of this book were rewritten and expanded from the three chapters on the climate system that appeared in *The Earth Machine: The Science of a Dynamic Planet*, a book that I and my colleague James D. Webster wrote in support on an earlier exhibit.

I thank colleagues who helped to make this book what it is. I am particularly indebted to Gavin Schmidt (NASA Goddard Institute of Space Science), who was unfortunate enough to be subjected to an early version of a manuscript but helped to exorcise its numerous and brainless errors; to Jason Smerdon (Columbia University), who provided many useful comments that served to improve a later version significantly and who wrote the student companion that appears at the end of this book; and to Steve Soter (American Museum of Natural History), whose comments helped especially in organization and clarification. I also thank Michael Bender

(Princeton University), Sidney Hemming (Columbia University), Marty Hoffert (New York University), Stephanie Pfirman (Barnard College), and David Walker (Columbia University) for insightful technical reviews of specific chapters, and Michael Oppenheimer (Princeton University) for writing the foreword. Finally, I must acknowledge Patrick Fitzgerald, my editor at Columbia University Press, whose tireless and enthusiastic efforts basically made this book possible; Irene Pavitt, senior manuscript editor at Columbia University Press, who played a major role in its production; Milenda Lee, for her stunning design; and Barbara Balestra, Nanette Nicholson, and Njoki Gitahi (all at the American Museum of Natural History) for helping me put the book together.

Edmond A. Mathez
New York, July 2008

ABBREVIATIONS

AABW	Antarctic Bottom Water
AC	alternating current
CCD	carbonate compensation depth
CFCs	chlorofluorocarbons
CO_2	carbon dioxide
DC	direct current
DML	Dronning Maud Land, Antarctica
D-O EVENTS	Dansgaard-Oeschger events
EDML	European DML core
ENSO	El Niño–Southern Oscillation
GISP2	Greenland Ice Sheet Project 2
GRIP	Greenland Ice Core Project
HCFCs	hydrochlorofluorocarbons
HVAC	high-voltage AC
HVDC	high-voltage DC
IACZ	Indo-Australian Convergence Zone
IGCC	integrated gas combined cycle
IPCC	Intergovernmental Panel on Climate Change
IR	infrared
ITCZ	Intertropical Convergence Zone
NADW	North Atlantic Deep Water
NAO	North Atlantic Oscillation
$^{18}O/^{16}O$	oxygen-18 to oxygen-16 ratio
PC	pulverized coal
PETM	Paleocene–Eocene Thermal Maximum
PV	photovoltaic
UV	ultraviolet

Climate Change

Satellite view looking east from Patagonia over southern Argentina and the South Atlantic Ocean

Climate is a dynamic "system" ultimately driven by the energy of the Sun, but resulting from the dynamic interactions among Earth's atmosphere, ocean, biomass, rock, and ice—all of which, except for ice, are represented in this image. The lighter ocean colors are due to phytoplankton blooms. (Satellite imagery courtesy of GeoEye/NASA. Copyright 2008. All rights reserved, http://visibleearth.nasa.gov/view_rec.php?id=1551)

1 CLIMATE IN CONTEXT

"RAIN, HEAVY AT TIMES, will begin in late morning and continue into the evening hours as a cold front sweeps across the area. . . ." Ah, the weather forecast—what would we do without it? There is no shortage of conversation about the weather, which, after all, touches our daily lives. For some, the weather is pretty important—especially if their harvest depends on it. For others, it is more tangential. I'm thinking of myself here—most days I just want to know about my trek in and out of New York City, where I work. Will rain or snow make it impossible or just more miserable than usual? And climate? What would a climate forecast be like? "The next decade will bring persistent showers and mild temperatures in January to March, and extensive periods of no rainfall at all throughout the summer months." Hmm . . . that seems a bit remote from my immediate worry of getting to work.

Weather and Climate

The musings here illustrate the essential difference between weather and climate, the topic of this book. It also illustrates one of the conundrums in reducing carbon dioxide (CO_2) emissions from the use of fossil fuels, the main culprit in global warming. *Weather* refers to conditions in the atmosphere at any one time. The now familiar radar images on television show that local weather systems develop and dissipate rapidly over the course of hours to a day. On a continentwide scale, weather systems form and decay over days to perhaps a week. A persistent weather system, such as a warm spell, may last for a couple weeks, especially in midlatitudes where the tracks of the systems are commonly determined by the position of the polar jet stream, as chapter 2 explains.

Climate, in contrast, can be thought of as the "average weather" for a particular region over some time. *Region* here can mean the entire globe, as in global warming; it can mean a large land area, say, eastern North America; or it can even mean a small one, as in the "microclimate" of a wine-producing valley near Bordeaux, France. Over all these scales, however, climate implicitly refers to the long term. One conundrum, therefore, is that it is difficult to marshal either the individual or the collective will to make the changes necessary to avoid negative impacts of global warming because they generally do not affect our immediate lives.

Although we have become adept at forecasting weather over hours to days, predictions beyond that become progressively more uncertain with distance into the future. Weather is inherently chaotic. Strictly speaking, the term *chaotic* means that small differences in initial conditions result in large differences in how a system will eventually develop. In other words, to predict weather accurately, one would have to know the temperature, humidity, barometric pressure, wind velocity, precipitation, and other characteristics of a weather system everywhere across an affected region; the more information at hand, the farther out in time a reliable forecast becomes possible.

Being an average condition, climate is more stable and displays distinctive patterns of change on distinctive timescales. Examples include annual changes such as monsoons, which are shifts in winds that bring seasonal rains to a number of regions in the tropics and subtropics. They also include fluctuations that occur only every several years, the most notable of which is El Niño, referring to the periodic warming of the equatorial eastern Pacific Ocean that commonly influences climate across much of the globe.

The Climate System

Climate is a dynamic system resulting from the combined interactions of various parts of Earth with one another and with the Sun. The parts are the *atmosphere*; the ocean (the *hydrosphere*); glaciers, terrestrial ice sheets, and sea ice (collectively known as the *cryosphere*); the living biomass (the *biosphere*); and even the solid Earth (the *lithosphere*) (figure 1.1). Think of it as your body, with all of its parts interacting in a pulsating whole. And like your body, the climate system is not just a set of physicals interactions. It is also a dynamic chemical system, with matter flowing through its various parts.

The *atmosphere*, being the medium that we live in, is the part of the climate system that affects us most directly. It plays a major role in transporting heat around the planet. Because Earth is a sphere, the Sun's heat is more intense near the equator than near the poles. This uneven distribution generates winds that carry heat from the equator to the poles and from the surface to the upper atmosphere. The atmosphere is not isolated from the ocean, however. The ocean circulates, in part driven by the winds and guided by the positions of continents, and thereby also transports

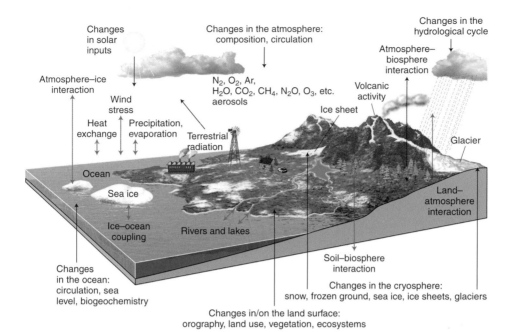

FIGURE 1.1

The dynamic climate system

The diagram shows the climate system's various components and indicates the chemical and physical interactions among them (*double arrows*) and the various processes acting on them (*single arrows*). N_2 = nitrogen; O_2 = oxygen; Ar = argon; CH_4 = methane; N_2O = nitrous oxide; O_3 = ozone; *orography* refers to the mountain landscape.

heat toward the poles. Indeed, the ocean holds far more heat than the atmosphere, but it flows much more slowly.

As for the chemical interactions, the most important are the exchanges of carbon among the atmosphere, ocean, and biosphere (here we can also include the dead biomass held mainly in soils). In fact, we can think of each of these spheres as reservoirs where nearly all the carbon on or near Earth's surface is stored. This description leads to the concept of the *carbon cycle*, referring to the flow of carbon among the various reservoirs. In time periods of months to decades, photosynthesis by plants and decay of organic materials affect the amount of CO_2 in the atmosphere, but over longer times it is the ocean that exerts the dominant control on atmospheric CO_2 content because the amount of carbon in the ocean is more than 50 times that in the atmosphere (or in the entire living biomass). If we think of the climate system as something like our body, the atmosphere and ocean are its main organs, and the carbon cycle is the circulation system that connects them and other organs. Because these elements are so central, they serve as the focus of the first several chapters of this book.

Most of the carbon (more than 99.9 percent) on Earth exists not in the ocean, atmosphere, or biosphere (the "surface" reservoirs), but in a deep reservoir in the form of rocks—that is, the *lithosphere*. The lithosphere is part of the climate system mainly because carbon flows between it and the surface carbon reservoirs, but this flow is far slower than the flow of carbon among the surface reservoirs. Over millions of years, a close balance has apparently persisted between the amounts of carbon flowing from the surface to the rock reservoirs via removal of CO_2 from the atmosphere and ocean by the formation of carbonate- and other carbon-bearing rocks and the return of CO_2 to the atmosphere by the breakdown of those rocks at the high temperatures and pressures of the deep Earth. In fact, this long-term balance appears to have acted as a natural, planetary thermostat, maintaining conditions on Earth's surface that are conducive to the evolution and survival of life since nearly the beginning.

The different parts of the climate system also interact through *feedbacks*, or phenomena that amplify or diminish the forces that act to change climate. An example helps to envision them. As the Arctic warms due to buildup of greenhouse gases, sea ice melts. As sea ice melts, there is less bright ice to reflect solar energy back to space, so more energy is absorbed by the dark ocean. The greater absorption of energy in turn further warms the ocean and overlying atmosphere and causes even more ice to melt. In this way, greenhouse-gas warming is amplified. This feedback in part accounts for why the Arctic is generally more sensitive to global warming than is the rest of the planet. Feedbacks can be complex and operate in unpredictable ways, and they are one reason that projecting future climate is fraught with uncertainty.

The climate system is complicated in other ways, one of which is that the various climate phenomena operate on different timescales (table 1.1). Some of these phenomena and their associated timescales are familiar—for example, the daily variations of warm days and cool nights, and the annual passage of the seasons. Other phenomena occur on longer or irregular intervals, and still others occur at timescales beyond the human experience and are consequently difficult to imagine. Our knowledge of the latter may also be incomplete because the evidence for them is buried (commonly and literally) in the geological record.

Climate Change: Separating Facts from Fears

What we do know from the available records, both geological and observational, is that the climate is changing. Hardly a day goes by without some mention of it in the news. Earth's climate is warming; CO_2 and other greenhouse gases have been building up in the atmosphere mainly as a consequence of the burning of fossil fuel; and the scientific evidence is now overwhelming that this buildup is causing the warming. These statements are the *facts* of climate change.

TABLE 1.1 **DIFFERENT TIMESCALES OF WEATHER AND CLIMATE PHENOMENA**

Timescale	Phenomena
Daily	Warm days and cool nights due to solar heating and Earth's rotation
3–7 Days	Weather events, such as passage of fronts
Months	Eastward-propagating weather disturbances across the tropical Indian and Pacific oceans
Yearly	Warm summers, cool winters, and shifts in zones of precipitation due to tilt of Earth's spin axis and its orbit around the Sun
	Monsoons, notably in the Indian subcontinent, where the cause is summer heating of landmasses that draws moist winds off oceans
23–36 Months	Periodic wind and temperature oscillations in the equatorial stratosphere due to internal atmosphere dynamics
2–7 Years	El Niño events, in which changes in equatorial Pacific Ocean currents and winds result in dramatic shifts in rainfall in equatorial regions globally and in lesser shifts in the climate of some temperate regions
1–3 Decades	Generally ill-defined oscillations, such as the North Atlantic Oscillation (NAO), an oscillation in atmospheric pressure that influences the positions of storm tracks across the ocean and affects the climates of western Europe, the Mediterranean basin, and eastern North America
Centuries	Irregular fluctuations that have led to multicentury cold or warm periods, such as the Medieval Warm Period and the Little Ice Age, the causes of which are uncertain but may be related to one or more natural phenomena, such as variations in solar irradiance
10,000–100,000 Years	Regular variations in orbital parameters (the slow oscillations in Earth's tilt relative to the orbital plane, precession, and eccentricity of its orbit around the Sun) affecting the amount of energy reaching the Northern Hemisphere. Responsible for the approximately 100,000-year glacial cycles of the past million years
Millions of Years	Changes in the positions of continents, solar luminosity, and composition of the atmosphere, all of which affect climate globally

Source: J. R. Christy, D. J. Seidel, and S. C. Sherwood, "What Kinds of Atmospheric Temperature Variations Can the Current Observing Systems Detect and What Are Their Strengths and Limitations, Both Spatially and Temporally?" in *Temperature Trends in the Lower Atmosphere: Steps for Understanding and Reconciling Differences*, edited by T. R. Karl, S. J. Hassol, C. D. Miller, and W. L. Murray (Washington, D.C.: Climate Change Science Program, Subcommittee on Global Change Research, 2006), 29–46.

Less certain are how much the climate will warm in response to growing emissions and to what extent the warming will change the world around us. Should the warming be substantial, it may have huge, negative impacts on biodiversity, ecosystems, agriculture, the global economy, and the health of human societies everywhere. These possible results are the *fears* of climate change.

It is important to separate the facts from the fears because although the facts give us insight, the fears reflect uncertainty. We will need knowledge and ingenuity to respond to global warming. To gain them we must start with the facts.

OBSERVATIONS OF CLIMATE CHANGE: THE FACTS

In addition to CO_2, the greenhouse gases include methane (CH_4), ozone (O_3), and water vapor (H_2O). These gases reside mostly in the *troposphere*, the lower 10 to 15 kilometers (30,000 to 50,000 feet) of the atmosphere where the weather occurs. Here the greenhouse gases absorb heat radiated from Earth's surface and thus act as a giant, insulating blanket.

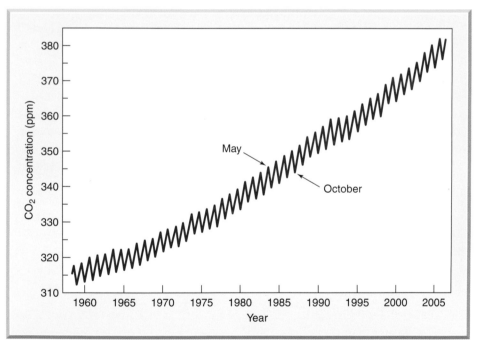

FIGURE 1.2

The Keeling curve

The curve shows the increasing CO_2 content of the atmosphere since 1958, as measured at the top of Mauna Loa volcano, Hawaii. The seasonal variation reflects the life cycle of Northern Hemisphere plants. (After Scripps Institution of Oceanography CO_2 Program, http://scrippsco2.ucsd .edu, with permission)

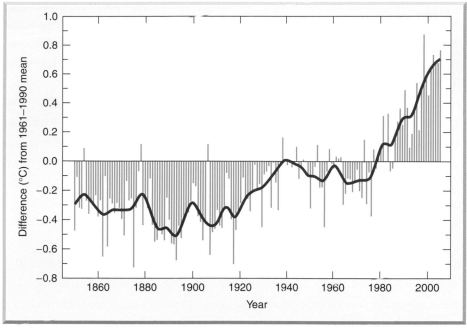

FIGURE 1.3

The change in the average global land-surface air temperature relative to the 1961–1990 mean temperature

Annual averages are represented by the vertical bars (*blue*); the curve (*red*) is the running 10-year mean. The recent warming is notable because it is global and rapid. (After Trenbreth et al. 2007: fig. 3.1)

Greenhouse gases have been building up since the beginning of the industrial age, but only since 1958 has the CO_2 content of the atmosphere been measured directly, beginning first on the top of Hawaii's Mauna Loa volcano (figure 1.2).[1] The remarkable Mauna Loa record shows two interesting features. First, CO_2 concentrations exhibit regular annual fluctuations, reflecting the growth in the springtime and the death in the fall of Northern Hemisphere plants. Second, CO_2 has been climbing on a steady, unbroken path over the years. In 1958, the average CO_2 content of the atmosphere was 315 parts per million (ppm) by volume (that is, 315 ppm = 0.0315 percent); by 2008, it had reached about 385 ppm and is rising at a rate of about 2 ppm per year. Both that rate of increase and that amount of CO_2 now in the atmosphere are greater than at any time in the past 800,000 years,[2] the time over which a continuous record of atmospheric CO_2 contents exists. Furthermore, a number of observations make quite clear that the CO_2 is originating mainly from the burning of fossil fuels.

At the same time, global mean surface air temperature has been rising, too (figure 1.3). The warming began around 1910 and has proceeded in two distinct intervals, the first from 1910 to 1940, and the second beginning in the late 1970s and continuing today. In the intervening interval, global temperature changed little, possibly because of an increase in the amount of pollutants being injected into the atmosphere then. Global mean land-surface air temperature has risen about 1°C (1.8°F) in 100 years, and over the past three decades the rate of increase has accelerated to 0.27°C (0.49°F) per decade.[3]

What is causing the warming? The evidence is overwhelming that it is a result of the rising levels of greenhouse gases in the atmosphere. Several lines of evidence lead to this conclusion. First, there is the basic physics—greenhouse gases absorb the radiant heat, or infrared (IR) energy, being emitted from Earth's surface. As was recognized more than a century ago, the climate *should* warm because now more of that energy is being trapped in the atmosphere rather than escaping to space.

Second, there are no other known natural forces external to the climate system, such as changes in the amount of solar irradiance (that is, the amount of sunlight) reaching Earth, that might account for any but a small fraction of the warming. In the particular case of solar irradiance, there is indirect evidence that irradiance changes with time and that it may account for some of the cool and warm spells in the past, but except for the 11-year sunspot cycle, which represents only a minuscule fluctuation in irradiance, there have been no detectable changes in solar output since the advent of precise measurements by satellite in 1978.[4]

Internal variations in the climate system—that is, fluctuations occurring on timescales of decades and less and resulting mainly from the system's dynamic nature—may conceivably account for warming, at least regionally. These variations include phenomena such as El Niño–Southern Oscillation (ENSO) and the North Atlantic Oscillation (NAO), both of which are ocean–atmosphere interactions resulting in large-scale redistributions of heat (discussed more fully in chapters 2 and 3). However, a variety of *observations* argue against internal variations as being responsible for the warming.

First, warming has been occurring more or less everywhere—it is a global not regional phenomenon, as would be expected if the warming were due to internal variability.[5] Second, the lower atmosphere below about 10 kilometers (33,000 feet) has also been warming, while at the same time parts of the upper atmosphere have been cooling, as expected from the basic theory of greenhouse-gas warming. Third, both the annual average maximum (daytime) temperature and the annual average minimum (nighttime) temperature have increased, but the nighttime temperature has increased more than the daytime temperature. This observation is consistent with what one would expect from increased insulation by greenhouse gases, as explained chapter 5. Fourth, the oceans have been warming by far more than can be accounted for by natural, internal variations in the climate system.[6]

Those are the essential facts in a nutshell: Earth's climate is warming; CO_2 and other greenhouse gases are increasing in the atmosphere; and the scientific evidence overwhelmingly points to the greenhouse-gas buildup in the atmosphere as the cause of the warming.

POTENTIAL CONSEQUENCES OF CLIMATE CHANGE: THE FEARS

The fears concern how much the planet will warm and what the consequences might be, but there is much uncertainty about this future. Three possible consequences that may severely impact society illustrate both the fears and their associated uncertainties.

One concern associated with global warming is the possibility of significant sea-level rise. Sea level is currently rising at a rate of 2.6 ± 0.04 millimeters (0.1 inch) per year,[7] equivalent to 26 centimeters (10 inches) per century. The 2007 report by the United Nations Intergovernmental Panel on Climate Change (IPCC), a massive document assessing the current scientific understanding of climate change, estimated that by the year 2100, sea level may be between 20 and 60 centimeters (8 and 24 inches) higher than it is today.[8] The main contributions to this rising level are the melting of glaciers and thermal expansion of the oceans (that is, warm water is less dense than cold water and therefore occupies more space). Not included in the IPCC estimate is the loss of ice from the Greenland and West Antarctic ice caps (the relevant studies had not yet been published by the time the IPCC report was written). The loss of ice from these ice caps should cause sea level to rise even more, but because we do not know by how much or how quickly the ice will disappear, we do not know by how much or how quickly sea level will rise.

The stakes, nonetheless, are high. Worldwide, two-thirds of cities with populations of more than 5 million people are vulnerable to the effects of rising sea level (the most serious of which are flooding during storms and coastal erosion). In China, for instance, a sea-level rise of just 0.5 meter (20 inches) might affect most regions that are within 10 meters (33 feet) of present sea level, currently home to 144 million people, or 11 percent of China's population.

A second concern is that warming will increase the incidence of harsh and extended droughts, which may significantly affect agriculture and the water supplies of millions of people. A number of regions are particularly vulnerable to drought, including southwestern North America, Southeast Asia, and the Sahel of Africa (the southern borderland of the Sahara Desert). About 1,000 years ago, southwestern North America, for example, experienced a number of "megadroughts," each lasting several decades over a 300-year interval of relative warmth.[9] Such megadroughts have not been seen since. The megadroughts then and the multiyear droughts that have plagued these areas in the recent past appear to be related to conditions in the tropical oceans, but exactly how those conditions influence rainfall patterns is not completely understood. The theory is that warming increases the probability of the occurrence of megadroughts; the fear is that such droughts will occur and have severe economic consequences.

A third concern is that warming will also increase the incidence of extreme weather events. Indeed, these events have been appearing with more frequency over the past half-century.[10] Two recent examples are the heat wave in Europe in 2003 and Hurricane Katrina, which destroyed much of New Orleans in September 2005. Eighteen hundred lives were lost to Katrina, and the storm cost more than $100 billion. In Europe, the summer of 2003 was the hottest in more than 600 years, rainfall was 50 percent below normal, and 22,000 to 45,000 people died in a two-week period as a consequence of the heat.[11]

Extreme events are by definition unusual and may occur in the absence of greenhouse-gas warming, so whether global warming has anything to do with hurricanes such as Katrina is a matter of current debate. Nonetheless, global warming makes the improbable more probable; the European heat wave and Hurricane Katrina illustrate the devastation that such extreme events can bring.

THINKING ABOUT THE FUTURE IN THE FACE OF UNCERTAINTY

We do not know by how much or how rapidly sea level will rise, and we do not know if or when megadroughts and severe weather events will strike. We cannot even state the odds that such events will occur. Yes, we are for the most part, ignorant. But that is exactly the point. We are smart enough to know that we are putting ourselves at risk, but we cannot gauge the risks.

Most of us buy insurance to mitigate risk, such as the personal financial risk associated with one's house burning down. In the case of climate change, we can buy insurance, in a sense, by trying to minimize the change. But there is a big difference in the latter case: unlikely as it might be, if climate change brings human society to its knees, we are out of luck because we will not be able to buy a new planet. In other words, we can seek only to reduce the risk. This is an important point: efforts to limit climate change and to mitigate its impacts are exercises in risk management, and understanding the problem in that light helps to guide our response.

Two characteristics of the climate system also shape our response. First, the climate system possesses *inertia*: it takes time for the system to reach a new balance in response to the forces that have acted to change it. In other words, even if greenhouse emissions were to be immediately capped at today's levels, warming would continue for several decades. By one estimate, there is currently more than 0.6°C (1.1°F) worth of warming already locked in.[12]

Second, climate can reach *tipping points*, or large, abrupt shifts in climate in response to factors that gradually cause climate to change, such as the buildup of greenhouse gases in the atmosphere. The geological record is replete with instances of abrupt and dramatic shifts in climate.

A prime example is the Younger Dryas (named after a tundra wildflower), an interval of extreme cold that abruptly descended on the northern Atlantic Ocean and neighboring regions beginning about 12,900 years ago and that ended just as abruptly 1,300 years later. In Greenland, the end of the Younger Dryas was marked by an increase in temperature of 7°C (13°F) and a doubling of the rate of snow accumulation in just a few decades and perhaps only a few years. The abrupt beginning and end of the Younger Dryas did not come about because of any sudden, external change. Rather, the ultimate cause was the gradual warming following the most recent glacial period. This is the notion of a tipping point—the climate system or any complex system crosses a threshold to a new state at a rate that is determined by the system's internal dynamics and that is faster than the cause.[13] The existence of tipping points and our inability to predict them form together one element of the risk we are taking. We are forcing climate to change, and we do not know how it will react.

On a related note, having described the geological evidence for abrupt, large, and rapid swings in climate, I have been surprised by many students' question why, considering that climate has changed so dramatically in the past in response to natural forces, we should concern ourselves with the human-induced changes. The answers are simple. First, human-induced changes may very well be substantial. Second, complex societies were not around to experience the huge shifts of the past. The climate of the past 11,600 years, known to geologists as the Holocene, has been stable by the standards of the past million years, and complex societies have been around for only about the past 6,000 of those years. The climate system, in short, has within it the possibility of bringing about dramatic changes that modern societies have not experienced but that may seriously challenge their abilities to adapt.

The Story

So, what does this book add to the discussion? As noted, chapters 1 through 4 deal with the climate system. In chapter 5, I then look at the systematic way that the scientific community thinks about climate change, the greenhouse effect, the various factors causing climate to change, and feedbacks and other features of the climate system. Chapter 6 explores the fascinating story of the climate of the past, or *paleoclimate*, which is recorded in tree rings, cave deposits, corals, lake sediments, deep-sea sediments, and even ice. Paleoclimate not only is fascinating, but gives us essential insight into how the climate system works today and how it will change in response to greenhouse-gas emissions. The story focuses on the past 800,000 years, but we shall also visit more distant times to seek additional insight.

Chapter 7 describes the rapid increase in global temperature over the past century, some of the changes that we are beginning to see as a consequence of that warming, and some changes that we can expect based on our understanding of how the climate system works. The Arctic is especially sensitive to warming and at the same time has an outsized influence on global climate. Accordingly, chapter 8 investigates the melting of the Arctic ice cap and permafrost, or perennially frozen ground, and the resulting changes in the tundra. The sensitivity to warming raises another important concern discussed in chapter 8: the fate of the Greenland and West Antarctic ice caps and the extent to which they will contribute to rising sea level.

Chapter 9 seeks to explain climate models. The models offer a way to understand what the future may bring based on fundamental physical laws and an understanding of how the climate system works. The future fundamentally depends, however, on how greenhouse-gas emissions change. Because greenhouse-gas emissions are largely a consequence of energy production, we cannot talk about climate change without talking about energy. Chapter 10 deals with the central question of our energy future—how to satisfy the world's insatiable appetite for electricity while keeping emissions in check.

That is the story. It is complex: it suggests that we face a difficult future, but it also implies that we can avoid the dire consequences of climate change by intelligent action and innovation.

The crescent Moon as seen through Earth's thin upper atmosphere

The atmosphere is the medium of climate. It insulates Earth and keeps it warm, transports heat from the tropics to the poles and from the surface to the heights, and transfers water from the oceans to land by precipitation. A crew member of the International Space Station took this photograph of a crescent Moon from about 360 kilometers (225 miles) above Earth. The cloud deck is about 6 kilometers (3.7 miles) high. (Image Science and Analysis Laboratory, NASA Johnson Space Center, http://eol.jsc.nasa.gov/, photograph ISS008-E-8951)

2 THE CHARACTER OF THE ATMOSPHERE

THE ATMOSPHERE[1] IS THE protective blanket that makes life possible. It is the air we breathe; it protects us from the Sun's deadly ultraviolet (UV) radiation; and it regulates temperature so that liquid water exists now and has always existed somewhere on Earth. Without its atmosphere, Earth's surface would be frozen, and life would not exist here.

Despite its importance, the atmosphere constitutes only a minuscule part of Earth. The total mass of the atmosphere is 5.14×10^{18} kilograms, which is a mere three-thousandth of the mass of the ocean (1.39×10^{21} kilograms) and one-millionth that of the solid Earth (5.98×10^{24} kilograms). If not from the surface, then at least from space the true extent of Earth's thin skin of atmosphere is apparent (see the satellite image introducing chapter 1).

The atmosphere is a moving part, apparent to anyone who steps outside. The movement arises from the fact that more solar energy falls on the equatorial regions than on the polar regions. In redistributing the heat, the atmosphere's motions help to drive the ocean's surface currents, which also transport heat poleward. Furthermore, the atmosphere connects the different parts of the climate system. It is the medium through which the oceans and the land surfaces communicate, as winds transport water evaporated from the ocean to land, where it may fall as precipitation. The atmosphere is also one path by which carbon moves between the ocean and land (chapter 4).

How the Sun's heat shapes and moves the atmosphere is the subject of this chapter. The chapter details the parts of the atmosphere and how and why it circulates on a global level. It also explains some of the important regional circulation patterns and their influence on climate. It ends with the story of ozone and how human activities affect the farthest reaches of the atmosphere, which in turn affects life back on Earth.

The Structure and Composition of the Atmosphere

Reaching 500 kilometers (300 miles) into the sky, the atmosphere is layered—compositionally, dynamically, and thermally. This feature is fundamental in that it determines both the character of the climate system and, more generally, the conditions on Earth's surface.

THE LAYERS

The layer closest to Earth is called the *troposphere* (figure 2.1).[2] The troposphere accounts for about 80 percent of the total mass of the atmosphere. This is where the weather occurs and where most of the planet-insulating greenhouse gases are con-

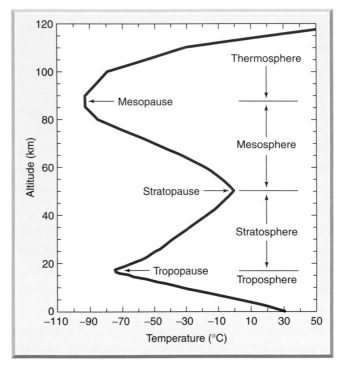

FIGURE 2.1

The layered structure of the atmosphere

The troposphere makes up about 80 percent of the total mass of the atmosphere, is where the weather occurs, and holds most of the planet-insulating greenhouse gases. Its temperature decreases, on average, 6.5°C (11.7°F) per kilometer (3,280 feet) in altitude because it is heated from below, loses energy from above, and becomes less dense with altitude. The overlying stratosphere is relatively warm due to the presence of ozone, which absorbs both ultraviolet radiation from above and infrared radiation from below. With essentially no greenhouse gases, the mesosphere is the coldest part of the atmosphere. The vanishingly thin thermosphere is heated by solar radiation and by its interactions with the solar wind. (After Hartmann 1994, with permission)

centrated. The top of the troposphere, known as the *tropopause*, separates it from the overlying stratosphere. This boundary is at an altitude of 16 to 18 kilometers (52,000 to 59,000 feet) in the tropics, but drops to 8 to 10 kilometers (26,000 to 33,000 feet) in the polar regions. The drop is gentle from the equator to about 40°N and 40°S latitudes, but there the tropopause suddenly plunges about 2 kilometers, owing to the circulation patterns in the troposphere, before continuing its gentle drop to the poles.

The tropopause exists because of the stability of the overlying *stratosphere*. The stratosphere extends to about 50 kilometers (30 miles) and is notable because it contains most of the ozone in the atmosphere. The oxygen-based ozone molecule absorbs UV light, which is lethal, so without ozone, life would probably not exist on land or in water shallower than a couple meters, the depth to which UV light penetrates. In fact, according to the fossil record, life did not spread from water to land until about 400 million years ago, presumably with the development of the ozone layer.

Above the stratosphere is the *mesosphere*, which reaches to about 85 kilometers (50 miles), and above that is the *thermosphere*, which constitutes only a minute fraction of the atmosphere but extends out some 500 kilometers (300 miles).

TEMPERATURE PROFILE

Without any incoming heat, the temperature of the atmosphere would simply decrease smoothly with altitude as the density of air decreased. The layered temperature structure exists, however, because of how the atmosphere is heated. About half of the incoming solar radiation is absorbed by the ocean and land surfaces. A fraction of this energy is emitted as infrared (IR) radiation, the form of radiation we usually refer to as "heat," to warm the troposphere from below (chapter 5). The troposphere also radiates energy from its top. In addition, motions of the atmosphere transport water vapor and heat away from the surface and stir it into the troposphere. Mainly from these processes, the troposphere reaches an overall "radiative-convective equilibrium" whereby temperature decreases with increasing elevation at an average rate of 6.5°C (11.7°F) per kilometer, the obvious manifestation of which is snow-capped mountain peaks.[3]

The temperature of the tropopause reaches about −60°C (−76°F), but then temperature increases again in the lower stratosphere. The stratosphere is a relatively warm blanket layer because its ozone absorbs both solar UV radiation from above and IR radiation from below. In the overlying mesosphere, the temperature again decreases with increasing altitude, reaching a minimum of about −100°C (−148°F) at an altitude of 85 kilometers (53 miles). Finally, temperature increases with altitude in the thermosphere, in part due to the absorption of solar radiation and in part from bombardment of the outer atmosphere by the solar wind, the continuous stream of protons and electrons given off by the Sun.[4] During periods of strong solar activity, the temperature at the top of the thermosphere may reach 1,500°C (2,700°F). But this does not mean that it is very "hot." The amount of heat is minuscule because there are so few molecules of gas at this altitude, and a body there would feel very cold.

CONSEQUENCES OF A STABLE ATMOSPHERIC STRUCTURE

The structure of the atmosphere is essential for life, basically because it keeps the planet from drying out. The very cold tropopause and the even colder mesopause (the top of the mesosphere) trap water vapor below because any that rises to these heights forms ice particles that settle out, keeping the upper atmosphere dry. Water vapor that does reach the upper atmosphere is dissociated by solar radiation to create free hydrogen (H_2). The H_2 molecule has so little mass that at high altitudes it is not held to Earth by gravity and so may be lost to space. Had this happened to an appreciable degree for an appreciable time, Earth may conceivably have completely dehydrated, in which case the surface would now be devoid of both water and life. In fact, this process may have happened to our neighboring planet Mars.

The structure of the atmosphere and its stability also has consequences for weather and pollution. Weather originates mostly from the thermal motions in the troposphere, and that is where the weather mostly stays. Although the air of the troposphere may mix into the base of the stratosphere, the mixing of stratospheric air into the rest of the atmosphere is slow. For this reason, when pollutants do reach the stratosphere, they tend to stay there and are quickly transported around the globe. As we will see (chapter 5), this outcome happens in a spectacular way in the case of volcanically produced sulfate aerosols, which circle the globe in weeks and stay in the stratosphere for several years.

THE COMPOSITION OF AIR

Three gases—nitrogen (N_2, 78.1 percent), oxygen (O_2, 20.9 percent), and argon (Ar, 0.93 percent)—make up 99.96 percent of dry air by volume. The remaining gases are water vapor (H_2O), carbon dioxide (CO_2), methane (CH_4), ozone (O_3), and nitrous oxide (N_2O). The amount of water vapor in the atmosphere is highly variable, but on a hot day may reach several percent. In the context of climate, CO_2 is particularly important because it is a greenhouse gas, meaning a gas that absorbs IR radiation emitted from Earth's surface (chapter 5). Indeed, the buildup of CO_2 in the atmosphere is the principal reason that climate is warming. The CO_2 content of the atmosphere reached 385 parts per million (ppm) by volume (or 0.0385 percent) by 2008, a concentration that perhaps disguises its importance as a greenhouse gas. The concentration is currently rising at the rate of about 2 ppm per year. Other greenhouse gases—such as methane, ozone, and nitrous oxide—are also on the rise but exist in much lower abundances.

The most important greenhouse gas in terms of the IR radiation it absorbs is water vapor. We refer to the amount of water vapor in the air as *humidity*. Most of this vapor represents water evaporated from the ocean. It condenses to form clouds made of liquid droplets or ice particles, which eventually fall as rain or snow. From the point of view of climate, clouds are important because they trap heat from below and reflect sunlight from above (chapter 5). Primarily because the amount of water that air can hold depends on temperature, humidity decreases rapidly with altitude, and the cold air of the polar regions is much drier than the warm air of the tropics.

Finally, even though we perceive air to be weightless, it has mass. To be exact, a column of air 1 centimeter on a side at its base and extending from sea level to the outer limit of the atmosphere weighs 1,034 grams (14.7 pounds per square inch). This is the "air pressure" around us. The weight of air causes air pressure to decrease progressively with altitude (figure 2.2). Air pressure also changes slightly from one place to another, depending primarily on whether the local air mass is rising, in

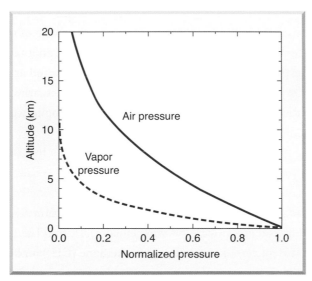

FIGURE 2.2

The decreases in air and vapor pressure with altitude

Air pressure is highest at the surface because of the weight of the overlying air, and vapor pressure decreases with altitude as air temperature falls. Air and vapor pressure are normalized to their respective values at the surface. (After Hartmann 1994, with permission)

which case the local atmospheric pressure at the surface is slightly lower than normal or is sinking, in which case it is slightly higher than normal.

The Circulation of the Atmosphere

Mariners are familiar with the oceanwide wind systems, which have been pushing ships back and forth across the oceans for hundreds of years. The easterlies (winds that blow from east to west), also known as the *trade winds*, carried them westward. These winds are driven by the difference in atmospheric pressure between the so-called subtropical highs at about the 30° latitudes and a band of low pressure on the equator. In the early days of sailing, the return trip was made possible by the westerlies (winds out of the west). These winds blow between about the 30 and 60° latitudes and are driven by the pressure difference between the subtropical highs and the bands of low pressure known as the *subpolar lows*. As suggested by the easterlies and westerlies, the winds are strongest parallel to latitude, not in the north–south (longitudinal) direction. The reason for this distinction has to do with Earth's rotation and is known as the Coriolis effect. The Coriolis forces interact with the north–south circulation patterns transporting warm air to high latitudes to give us the global wind patterns we see.

THE CORIOLIS EFFECT

The *Coriolis effect* is the deflection of the winds by Earth's rotation and is named for the nineteenth-century French mathematician Gaspard-Gustave de Coriolis (1792–1843), who first explained how Earth's rotation influences the movement of fluid. In particular, the Coriolis effect is a deflection in the path of a moving object, including air and water masses, to the right in the Northern Hemisphere and to the left in the Southern Hemisphere.

To understand the Coriolis effect, imagine standing still on the equator and holding a ball. Both you and the ball have an eastward velocity because, even through you are standing still, you are anchored to the eastward-rotating Earth. Now, think of a point somewhat north of the equator. That point is also moving eastward but not quite as fast as you are moving on your equatorial perch because the rotating circle on which it is anchored is smaller. Next, imagine that you throw the ball directly north. The ball still maintains the eastward velocity it had at the equator, but the Earth beneath it is not moving as fast. Thus from the perspective of the Earth below, the ball appears to be moving north but also drifting eastward, or to the right if looking toward the pole. How the air is deflected to create the trade winds is determined by the interaction of the Coriolis effect and a phenomenon called Hadley circulation.

HADLEY CELLS

The English mathematician George Hadley (1685–1768) was the first to explain the persistent trade winds. In 1735, he pointed out that more solar heat reaches the surface at the equator than at the poles because the Earth is spherical, which causes warm equatorial air to flow toward the poles and cool air toward the equator as a means of evening out the heat. The *convective circulations*—referring to the rising of warm, relatively light (low-density) air and the sinking of cool, higher-density air—also known as *Hadley cells*, start with the rise of warm air near the equator to create an equatorial zone of low pressure, the upward limb of the Hadley cells. The cells' downward limbs, which exist at about the 20 to 30° latitudes, create two zones of high pressure—the subtropical highs mentioned earlier. Because the sinking air is dry, these regions are also where most of the world's great deserts are located. Some of the downward-flowing air is directed poleward, but most flows back toward the equator, completing the circulation. At the same time, the Coriolis effect deflects the Hadley cells. As a result, the Hadley cells' upper, poleward limbs are deflected eastward, and the lower, returning limbs (the trade winds) flow westward (figure 2.3).

In reality, the Hadley cells are more complex. In winter, there is a dominant Hadley cell, and it is displaced to the summer hemisphere. For example, during the Northern Hemisphere winter, the upward limb is not at the equator but in the Southern Hemisphere at about 10°S, and the downward limb is at about 20°N latitude. The opposite is true during the Southern Hemisphere winter. In other words, the northern and southern Hadley cells oscillate in strength through the seasons, and the equatorial low-pressure and subtropical high-pressure zones likewise undulate back and forth. This behavior reflects the fact that the latitudinal zone where Earth is heated most shifts with the seasons.

Finally, the upward limb of Hadley circulation, which is also where the Northern Hemisphere and Southern Hemisphere trade winds meet near the equator, is known as the Intertropical Convergence Zone (ITCZ). Here, the large-scale upward motion of air results in low surface air pressure, high evaporation rate of water from the ocean, high rainfall, and transfer to the atmosphere of substantial amounts of heat from the surface (chapter 5). We revisit the ITCZ later (chapter 3) because it is characterized by a set of large-scale east–west circulation patterns that are distinct from Hadley circulation and that figure prominently in the important phenomenon known as El Niño–Southern Oscillation (ENSO).

FERREL AND POLAR CELLS

The Coriolis effect breaks up simple Hadley circulation to create a set of midlatitude, weaker, and much less stable circulation patterns between about the 30 and 60°

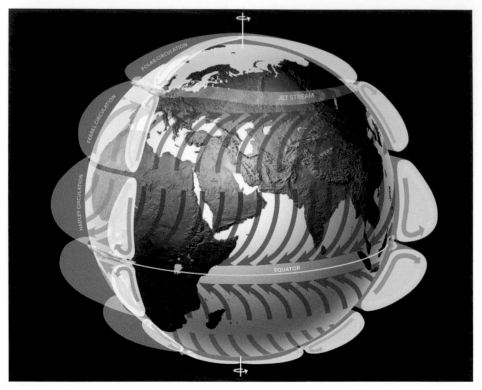

FIGURE 2.3

Idealized surface and global wind patterns

Warm air rises near the equator to create a zone of low pressure and the upward limbs of the Hadley cells. Downward Hadley flow creates two zones of high pressure at about 20 to 30°N and S latitudes, where most of the world's great deserts are. The Coriolis effect deflects the Hadley cells and breaks up simple Hadley cell circulation to create the Ferrel cells, which represent eddy circulations and result in westerly surface winds. The Polar cells are caused by the descent of dry air near the poles. (American Museum of Natural History)

latitudes. These patterns are termed *Ferrel cells*. In the Ferrel cells, cold-air masses rise at about the 50° latitude and sink at about the 30 to 40° latitudes. These cells are caused by eddy circulations, or small swirls in the opposite direction from the main flow, that transport heat and moisture poleward and result in southwesterly Northern (and northwesterly Southern) Hemisphere surface winds (see figure 2.3). This circulation gives rise to a parade of transient eddies that progress from west to east and are responsible for weather disturbances (and are thus familiar features of weather maps).

Near the poles, cold, dry air descends in vortices to create high-pressure regions from which the downward-flowing air diverges toward the equator. These high-pressure regions are the *Polar cells*. The Coriolis force directs the air westward in a

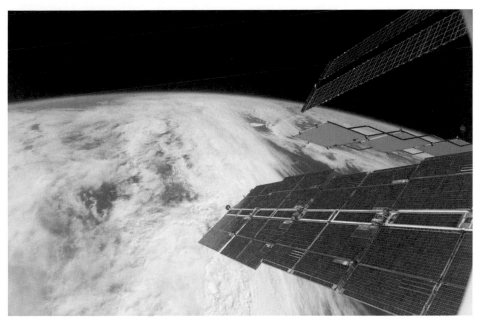

FIGURE 2.4

The Northern Hemisphere polar jet stream

The view looks east over Cape Breton Island, Canada. The undulating paths of the polar jet stream influence weather, especially in the winter, between about the 45 and 60°N latitudes. (Image Science and Analysis Laboratory, NASA Johnson Space Center, http://eol.jsc.nasa.gov/, photograph ISS008-E-12443)

surface wind system called the *polar easterlies*, which encounter the Ferrel cell westerlies along the polar front, creating a zone of unstable and severe weather.

This brings us to the *jet streams* (figure 2.4). The jet streams are high-speed, horizontal rivers of wind some hundreds of kilometers wide and several kilometers thick. The polar jet streams, with peak winds of about 300 kilometers (186 miles) per hour, undulate back and forth near the polar fronts at an altitude of about 10 kilometers (32,000 feet); they are in essence the upper tropospheric manifestations of the polar fronts.[5] The slow undulations affect weather patterns between about the 45 and the 60° latitudes and are mostly what we hear about in weather forecasts. Another set are the subtropical jet streams at the 30° latitudes. These locations correspond to the downward limbs of the Hadley cells, which account for the sudden drops in the tropopause mentioned earlier. Wind speeds up to 540 kilometers (330 miles) per hour have been recorded in the subtropical jet stream, but because they are at an altitude of about 12 kilometers (40,000 feet), they have less influence on subtropical weather than the polar jet streams have on weather at higher latitudes.

Some Climate Phenomena

The previous sections describe the "general circulation" of the atmosphere—the global circulation that results from the uneven heating of Earth by the Sun—but there are also local fluctuations in circulation that lead to important climate phenomena. They include, for example, the seasonal monsoons and more subtle decadal fluctuations, such as the North Atlantic Oscillation (NAO). These natural rhythms of climate operate over different scales of time and space. They are likely to be affected by global warming, and they help to illustrate how difficult it is to sort out human-induced from natural changes in climate.

THE MONSOONS

The term *monsoon* derives from an Arabic word referring to the "season of winds," and nowhere do people live more by a monsoon's seasonal rhythm than in India, especially in the western part of the country, where the monsoon usually starts abruptly around mid-June and ends in September. The summer monsoon accounts for about 90 percent of the region's yearly rainfall.

The Asian monsoon comes about because during the Northern Hemisphere summer, the Sun warms the land surface of the Tibetan Plateau and Indian subcontinent, in turn warming the air, which rises. The upwelling draws warm, moist air off the western Indian Ocean to yield the high summer rainfall over India (figure 2.5). The summer winds also drive ocean currents that flow north along the coast of the Horn of Africa and east around the tip of India. In the winter, the air over the Himalayas is cool and dry. It flows off the mountains, bringing with it winter drought conditions to India. These winds also blow out across the ocean, reversing the ocean currents.

Monsoons bring seasonal rains to other regions, notably sub-Saharan Africa, northern Australia, southwestern North America, and southern Brazil. The African monsoon, which is caused by the seasonal shift in the ITCZ, is particularly important because it profoundly influences the ecosystem and the livelihoods of peoples of the Sahel, the southern borderland of the Sahara Desert stretching from Senegal in the west to Sudan in the east.

What will the effects of global warming be on the monsoons? There is no clear answer at this point, only that monsoons are sensitive to changing climate. In general, one might expect that enhanced warming over land areas will intensify the monsoons or lengthen their seasons. The past several decades have seen rapid global warming (chapter 7), and at the same time the incidence of heavy rainfall in specific

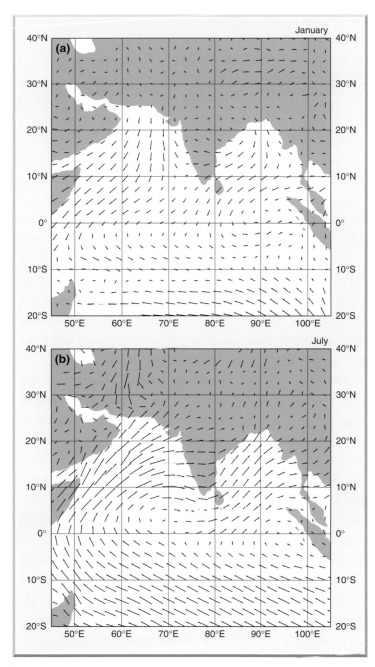

FIGURE 2.5

Contrasting wind patterns during the winter and summer monsoons over southwestern Asia and the northern Indian Ocean

The monsoon occurs because during the Northern Hemisphere winter (*a*), the cold, dry air of the Himalayas tumbles off the mountains and out over the ocean to bring dry conditions to India. In the summer (*b*), the Sun warms the land surface of the Tibetan Plateau and Indian subcontinent. The warm air over this region rises, drawing moist air from the western Indian Ocean into India and reversing the ocean currents. (After Hartmann 1994)

episodes (but without change in total rainfall) in India has increased, bringing with it more landslides, floods, and crop damage.[6] Climate records also link the strength of the Asian monsoon to large-scale changes in climate;[7] a similar linkage has been established in the case of the African monsoon.[8]

THE NORTH ATLANTIC OSCILLATION

The NAO is a wintertime oscillation of atmospheric pressure between the polar and the subtropical North Atlantic. These oscillations occur over the course of a few weeks, but there are also longer-term, decadal oscillations, which are the focus here.[9] The NAO is of some import because in the short term it influences the weather and in the long term the climates of Europe, western Asia, parts of the Mediterranean basin, and certain regions of eastern North America.

The so-called positive phase of the NAO describes the situation of high atmospheric pressure centered over the Azores and low pressure near Iceland (figure 2.6). The pressure difference, which is most pronounced in the winter, results in strong winds transporting relatively warm, moist air from the Atlantic Ocean to northern Europe, Scandinavia, and Asia, bringing relatively warm and wet winters to these regions. Unaffected by these winds, southern Europe and the Mediterranean basin remain relatively dry. The strong winds also create circulation patterns that draw cold, dry, polar air into northeastern Canada and western Greenland and warm, moist air from the subtropical western Atlantic to the east coast of North America, causing the winters there to be unusually warm and wet too.

In the NAO's negative phase, both the Azores high and the Iceland low are much weaker and displaced farther south.[10] The lower pressure gradient between these two centers means fewer and weaker storms that tend to follow a more east–west track, bringing moisture to southern Europe and the Mediterranean region. Meanwhile, northern Europe and the east coast of North America are cooler.

As noted, the NAO appears to fluctuate over several decades and influences climate on that timescale. The important consequences have been changes in marine and terrestrial ecosystems, water supply, energy consumption, and agriculture. For example, by bringing warmth to the North Sea, the positive phase of the NAO has had a negative effect on the already suffering cod fishery there.[11] In the Mediterranean region, it has reduced the supply of drinking water and the harvests of olives and grapes.

The reason for the NAO's decadal oscillations is not clear. One possibility is that these oscillations simply reflect random fluctuations in atmospheric circulation. Another is that the shift to a dominantly positive phase of NAO is somehow a consequence of global warming.[12]

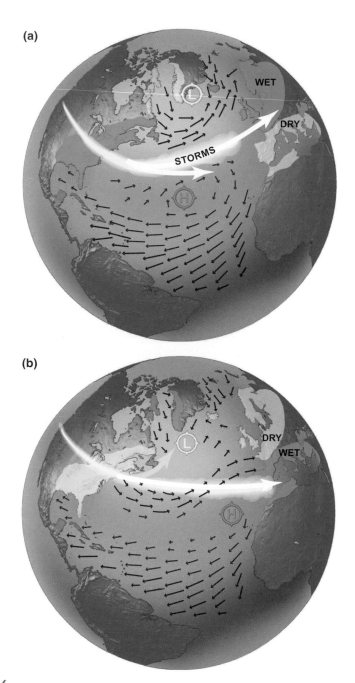

FIGURE 2.6

The North Atlantic Oscillation

In the NAO's positive phase (*a*), winter storms cross the Atlantic Ocean on a northerly track, bringing relatively warm and wet winters to northern Europe and eastern North America and dry winters to the Mediterranean basin. In its negative phase (*b*), the storm tracks are farther south, bringing cooler and drier winters to northern Europe and eastern North America and wetter winters to the Mediterranean region. (After Martin Visbeck, IFM-GEOMAR at the University of Kiel, http://www.ldeo.columbia.edu/NAO/, with permission)

Ozone: A Stratosphere Story

Compared with the happenings in the troposphere, those in the stratosphere bear only minimally on climate. As we have seen, however, the atmosphere is stratified because of the relative warmth of the stratosphere owing to the ozone that exists there. The ozone molecule consists of three atoms of oxygen (O_3).[13] The gas is not mysterious—it can be produced by an electric spark, so its distinctive, pungent odor is likely familiar to anyone who has played with electric trains or used arc welders.

STRATOSPHERIC VERSUS TROPOSPHERIC OZONE

The extremely low concentration of ozone in the atmosphere (less than 10 ppm) disguises the fact that it either protects or harms life, depending on where it resides. About 90 percent of the ozone exists in a layer 20 to 25 kilometers (12 to 15 miles) thick in the stratosphere, and it is good here because it absorbs most of the incoming UV radiation. The term *ozone layer* is rather misleading because ozone is not really a "layer," but drifts and swirls around the stratosphere in changing concentrations, similar to the way clouds drift and swirl. Also, an "ozone hole" (which is actually a region of low ozone concentration) develops in a seasonal cycle over Antarctica in the Southern Hemisphere spring. As the ozone layer diminishes, we can expect higher incidences of skin cancer, cataracts, and immune system impairments, as well as damage to crops and other plants.

Ozone also exists in the lower troposphere as a component of the smog that hangs over many cities. Here it is harmful because it is highly reactive and damages living tissue and because it is a greenhouse gas. This distinction raises the obvious question of why the "bad" ozone near the surface does not rise to replace the "good" ozone in the stratosphere. One reason is that because ozone is so reactive, it does not remain in the troposphere long enough to mix. Another is that only 10 percent of the total ozone exists in the troposphere, whereas the remaining 90 percent is in the stratosphere. Thus there is not enough tropospheric ozone to replenish the ozone being destroyed in the stratosphere.

FORMATION AND DESTRUCTION OF STRATOSPHERIC OZONE

This description brings us to the central issue of how stratospheric ozone is created and destroyed. Ultraviolet radiation is the key on both counts. It causes the ordinary oxygen molecule (O_2) to decompose into two highly unstable O atoms. Within a fraction of a second, these atoms combine with O_2 to form O_3. When the O_3 molecule absorbs lower-energy UV radiation, it splits into O_2 and O. The latter may recombine with O_2 to form O_3 again or react with O_3 to produce two O_2 molecules.

The incident UV radiation thus maintains a dynamic balance among O_2, O_3, and O. Ozone's ability to absorb UV radiation explains why it is so important: it protects Earth from much of the Sun's damaging rays.

Certain natural compounds containing nitrogen, hydrogen (originating from the breakdown of atmospheric water vapor), and chlorine (originating from sea salt) get carried into the stratosphere, where they destroy ozone. As the concentrations of these destructive compounds increase and decrease with the changing seasons and incident sunlight, so do the concentrations of ozone. In addition, explosive volcanic eruptions may inject other ozone-destroying chlorine- and sulfur-bearing compounds into the stratosphere.[14] Long before humans began polluting the atmosphere, natural processes regulated the amount of ozone in the stratosphere as ozone was created at about the same rate it was destroyed.

The natural balance started to change with the human manufacture of certain chlorine-, fluorine-, and bromine-containing compounds, known as *halocarbons*, which destroy ozone. Best known among these compounds are the chlorofluorocarbons (CFCs), used primarily in air conditioners. They tend to be extremely stable and do not readily react with other chemicals in the lower atmosphere. In the stratosphere, however, UV radiation breaks them down to release free chlorine (Cl) atoms. According to our present understanding,[15] the chlorine then reacts with ozone to form chlorine monoxide (ClO) and ordinary (molecular) oxygen,

$$Cl + O_3 \rightarrow ClO + O_2$$
atomic chlorine + ozone → chlorine monoxide + molecular oxygen

If each chlorine atom took out only one ozone molecule, CFCs would hardly threaten the ozone layer. However, when chlorine monoxide encounters a single oxygen atom, free chlorine forms again,

$$ClO + O \rightarrow Cl + O_2$$
chlorine monoxide + atomic oxygen → atomic chlorine + molecular oxygen

and thus is available to decompose another ozone molecule. In this way, a single chlorine atom can destroy thousands of ozone molecules.

STRATOSPHERIC CLOUDS AND THE OZONE HOLE

Intimately involved in the process of formation and destruction of ozone are stratospheric clouds. Invisible from the ground, these clouds can be the size of a continent. They form when temperatures dip below −80°C (−112°F) and water vapor condenses to form microscopic ice particles. The particles provide surfaces on which reactions

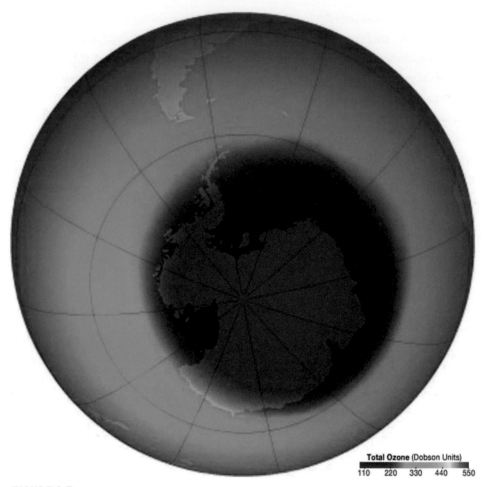

Total Ozone (Dobson Units)
110 220 330 440 550

FIGURE 2.7

The ozone hole

The hole in the ozone layer (*purple*) over the Antarctic as it appeared in October 2006. The Dobson Unit is a measure of the amount of ozone in the total atmospheric column. (NASA, http://ozonewatch.gsfc.nasa.gov/)

occur that form chlorine molecules, Cl_2. Unlike the chlorine atom, Cl_2 is inert—it does not react with anything. Thus when stratospheric clouds form over Antarctica in the winter, Cl_2 builds up, but ozone is not affected. When the spring sunlight strikes the Antarctic stratosphere in late August, however, UV radiation breaks up Cl_2 to chlorine atoms, which then destroy ozone to create the "ozone hole" that now regularly appears there at that time of year (figure 2.7).[16]

Although the use of halocarbons has been curtailed by treaty, another complication has significant bearing on how fast the ozone layer will recover. Chlorine can exist in the stratosphere as the chlorinated nitrite $ClONO_2$. Fortunately, this compound is inert. However, ice particles incorporate nitrogen to make a compound known as nitric acid trihydrate (NAT). The NAT-bearing ice particles enlarge in the stratospheric clouds and eventually settle out of the stratosphere.[17] By this mechanism, the stratosphere becomes "denitrified." The loss of nitrogen means that none is available to form $ClONO_2$, resulting in relatively more chlorine available to destroy ozone.

Because stratospheric clouds form every winter over Antarctica, the region is always effectively denitrified. But in the Arctic, the winters are warmer, and until recently stratospheric clouds formed only rarely. Now, with the increased amount of greenhouse gases in the troposphere, the stratosphere is cooler (chapter 7). The lower temperatures mean that clouds form more frequently, causing denitrification of the Arctic stratosphere,[18] which in turn is causing an ozone hole to open up in the Arctic spring where none existed before the advent of CFCs.

Fortunately, chlorine does not remain in the stratosphere forever. It eventually reacts with compounds such as methane to produce stable hydrochloric acid (HCl), which is also incorporated into ice particles and settles into the troposphere to be washed out by rain. As the use of halocarbons is being phased out, their abundance in the stratosphere has begun to decline, and the ozone layer will eventually repair itself if the world's governments abide by their treaties. Current projections suggest that pre-1980 levels of ozone will return by the latter part of the twenty-first century.[19] Needless to say, substantial issues and questions remain. For example, the environmental impacts of the compounds that are being developed to replace CFCs are not completely known, and some sources of chlorine have yet to be regulated. Also, climate change should result in more water vapor reaching the stratosphere, which may slow ozone recovery.

Perspective

This chapter began by stressing the importance of the compositional, dynamic, and thermal layering of the atmosphere in determining both conditions at the surface of Earth and the character of its climate. Weather is restricted mainly to the troposphere, with the overlying and relatively warm stratosphere representing, in effect, a stable lid on the troposphere. The atmosphere's thermal and dynamic structure is related to the distribution of greenhouse gases. Ozone accounts for the warmth of the stratosphere, but most of the greenhouse gases exist in the troposphere, in part simply because the density of air dwindles with altitude, but also because thermal motions distribute water vapor and greenhouse gases throughout the troposphere.

Global warming is likely affecting the atmosphere's natural rhythms, which occur mainly from the uneven heating of Earth by the Sun. The atmosphere's "general circulation" moves heat from the tropical to the polar latitudes, but its motions are complicated by the Coriolis effect, which results in distinctive circulation patterns for different latitudinal zones: the Hadley, Ferrel, and Polar circulations. More local patterns lead to important climate phenomena. The seasonal monsoons and the more subtle decadal fluctuation known as the NAO are but two of a number of such phenomena.

It is interesting to think about the atmosphere in the context of the long-term evolution of the planet and of life on it. As mentioned, life did not spread from water to land until about 400 million years ago, presumably in response to the buildup of the ozone layer, which would have reduced the intensity of UV radiation reaching Earth's surface. Also, the ozone layer has played a prominent role in determining the structure of the atmosphere. That the world acknowledged the problem of ozone destruction and then took action to reverse the destruction through the Montreal Protocol[20] and subsequent treaties gives us reason to hope that it can also find the means of reversing global warming.

The atmosphere also connects the different parts of the climate system. In particular, it transports water from ocean to land and is an important pathway for carbon to move between land and ocean. The importance of these connections in determining climate are made apparent in the next two chapters.

The Blue Planet

Seventy percent of Earth's surface is covered by ocean, making it the Blue Planet. The ocean is the "caldron of climate": it holds most of the heat of the climate system, transports heat from the tropics to the poles in concert with the atmosphere, controls the CO_2 content of the atmosphere, and thus has a major influence on surface temperature on timescales of a century and longer. The image is a mosaic based mostly on satellite observations. (Image by Reto Stöckli, render by Robert Simmon, and based on data from the MODIS Science Team, http://visibleearth.nasa.gov/view_rec.php?id=15993)

3 THE WORLD OCEAN

HOW CLIMATE WILL CHANGE in response to the buildup of greenhouse gases in the atmosphere will depend in large part on what happens in the world ocean. The reasons for this connection begin with the large amount of solar energy the ocean absorbs and holds, about 1,000 times more heat than the atmosphere holds. Consequently, the ocean is effective at both distributing that heat around the globe and moderating regional temperature (which is why summers in oceanic climates are not too hot or winters too cold). Moreover, the ocean also holds much more carbon than the atmosphere does. And because the two bodies readily exchange carbon dioxide (CO_2), the ocean exerts the primary control on the CO_2 content of the atmosphere and thus on climate, on timescales ranging from a century to a millennium and longer. For good reason, the ocean has been called the "caldron of climate."[1]

The ocean is immense, especially when considering its depths. Being mostly beyond view and unfamiliar, those depths are mysterious. Vast regions are more than 4,000 meters (13,000 feet) deep—as deep as the highest mountains are high. The ocean bottom is for the most part a broad, flat, featureless expanse known as the *abyssal plains*. The edges of the continents extend into the ocean as the *continental shelves*. In some places, these shelves reach out hundreds of kilometers from the shoreline. The entire North Sea, for example, is really part of the continent and is hardly anywhere more than 90 meters (300 feet) deep. Elsewhere, the shelf extends only a few tens of kilometers from shore—for example, along the west coast of South America. Then what is called the *continental slope* plunges steeply from the shelf edge to the deep ocean. The continental slope is a factor in guiding ocean circulation.

This chapter begins with the basic properties of water, which ultimately explain why the ocean exerts such an important control on the heat and carbon budgets of Earth's climate system. It continues with an exploration of the ocean's structure and circulation, both the surface currents, which are largely tied to the winds, and the circulation of the deep ocean, which can have a profound, long-term influence on climate. Finally, the chapter addresses the coupled interactions between the ocean and the atmosphere, using as an example the all important combination of the phenomena known as El Niño and Southern Oscillation, which affects weather patterns around the world.

The Important Properties of Water

It may seem odd to begin a discussion of the ocean with such an obvious and ubiquitous material as water (H_2O) itself, but water has some very special properties relevant to the ocean's character. Imagine the H_2O molecule as an oxygen ion with two rabbit ears of hydrogen ions (figure 3.1). The oxygen ion has a negative electric charge, whereas the hydrogen ions have positive charges, making a water molecule "dipolar" (it has electrically negative and positive sides). (An ion is a charged particle that forms when a neutral atom or cluster of atoms, known as a molecule, either loses or gains one or more electrons to acquire either a positive [+] or a negative [−] charge.) This dipolarity causes individual water molecules to aggregate, with the positive side of one loosely attached to the negative side of another. This feature makes liquid water relatively stable. The uneven distribution of charge also enables water molecules to latch onto other ions, and for this reason water is an excellent solvent for many substances.

HOW THE OCEAN MODERATES TEMPERATURE

The stability of water means that the amount of energy needed to evaporate water and the amount of energy that must be removed to freeze it are among the highest of any substance. In other words, evaporation absorbs heat, and freezing releases

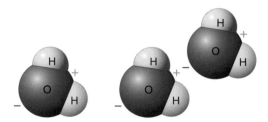

FIGURE 3.1

The water molecule

The water molecule's oxygen side has a small net negative charge, and its hydrogen side has a small net positive charge, allowing it to aggregate with other water molecules and other polar molecules. Because of these characteristics, liquid water is more stable than it would otherwise be and is an excellent solvent for many substances.

it. Furthermore, the *specific heat* of water, which is the amount of energy required to raise a unit mass of water 1°C (1.8°F), is also among the highest of all materials. Water can thus absorb and release relatively large amounts of heat with very little change in temperature. Water's high *thermal capacity* has an important consequence as far as climate is concerned. It results in the ocean being a huge reservoir for heat, holding, as noted earlier, about 1,000 times more heat than the atmosphere. So even though the ocean circulates much more slowly than the atmosphere, it moves a significant amount of heat around the planet.

Water's high specific heat has other consequences: the temperature of the ocean changes more slowly in response to heating or cooling than does the temperature of the atmosphere, and the total range and seasonal changes in ocean temperature are much less than those on land. To appreciate this, consider that the highest and lowest land temperatures, which have been recorded in the Libyan Desert and at the Russian Vostok research station in central Antarctica, respectively, are 58°C (136°F) and −88°C (−126°F), a range of 146°C (262°F). Ocean temperature, in contrast, varies from a maximum of about 36°C (97°F) in the Persian Gulf to −2°C (28°F) near the poles, a range of only 38°C (69°F). Annual sea-surface temperatures vary by no more than 2°C (3.6°F) in the tropics, 8°C (14.4°F) in the middle latitudes, and 4°C (7.2°F) in the polar regions, whereas seasonal temperature variations of more than 50°C (90°F) are not uncommon on continents. That is why the presence of a nearby ocean spares most coastal regions the frigid winters and sweltering summers that some continental interiors experience at similar latitudes. The ocean is a natural thermostat—it releases and takes up heat on timescales of decades to centuries, whereas the atmosphere does so in days to weeks. This difference occurs because of the fundamental character of water.

WHY THE OCEAN IS SALTY

Seawater is only 96.5 percent water by weight. The rest of it is mainly sodium chloride (NaCl) and other dissolved salts (figure 3.2). Why is the ocean salty? The simple answer is that every year rivers dump into the ocean 2.5 billion metric tons of material, including dissolved positively charged ions of sodium (Na^+), magnesium (Mg^{2+}), and calcium (Ca^{2+}) leached from rocks (chapter 4). In contrast, chlorine (Cl^-) and sulfate (SO_4^{2-}), the important negatively charged ions in seawater, have accumulated over geologic time and originated mostly from gases that escaped from Earth's interior during volcanic eruption. Other materials in the ocean include dust from deserts and various pollutants from human activities.

But this description raises another question. Rivers have added far more dissolved matter to the ocean than the ocean actually contains; furthermore, the compositions of the matter dissolved in rivers and of the matter found in ocean water are

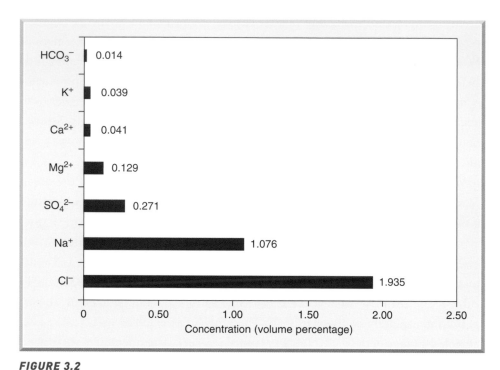

FIGURE 3.2

The composition of seawater

Shown are the ions of chlorine (Cl⁻), sodium (Na⁺), sulfate (SO₄²⁻), magnesium (Mg²⁺), calcium (Ca²⁺), potassium (K⁺), and bicarbonate (HCO₃⁻). (Data from Bigg 1996)

different. For example, the ratio of Na⁺ to Ca²⁺ in seawater is about 26, but in river water the ratio is only 0.4. Why this difference, and why is the ocean not even more salty than it is? The answer is that the dissolved material continually precipitates and accumulates as sediment, and some compounds have a longer residence time in the ocean than others. Thus Na⁺ resides in the ocean much longer than Ca²⁺ because marine animals, such as phytoplankton, remove Ca²⁺ to make calcium carbonate (CaCO₃) shells. When the animals die, the carbonate shells accumulate on the ocean floor to make carbonate mud, which eventually becomes limestone. In contrast, Na⁺ is removed mainly by clay particles. The particles also settle to form sediment, but this process is much slower than carbonate removal. The composition of ocean water has remained remarkably constant over geologic time because the various dissolved components have attained a long-term balance as they cycle through the various parts of Earth.

SEAWATER SALINITY

The term *salinity* refers to the total salt content of water. The salinity of seawater is about 35 parts per thousand (ppt), essentially equivalent to 35 grams (1.2 ounces) of salt per liter of water. The addition of salt to water decreases the freezing point and increases density. Consequently, seawater freezes at −1.8°C (28.8°F) rather than at 0°C (32°F) and has a density of 1.026 grams per cubic centimeter, compared with 1.0 for pure water. (Pure water reaches its maximum density at 3.98°C [39.2°F].) Increasing temperature, however, decreases density. Density and thus salinity and temperature are among the fundamental factors that determine how the ocean circulates and affects climate. For example, in certain polar regions water sinks because it attains a relatively high density, and in the process it removes CO_2 from the atmosphere and stores it in the deep ocean.

Several processes can cause salinity to change. Evaporation removes pure water and thus increases the salinity of the remaining water; precipitation and influx of fresh river water decrease salinity by dilution; freezing removes and melting of ice adds freshwater to increase or decrease salinity. These processes combine to make ocean salinity vary from one place to another (figure 3.3). The most saline parts of the open ocean correspond to the belts of high atmospheric pressure on either side of the equator (the tropical highs, also marked by the global belts of desert

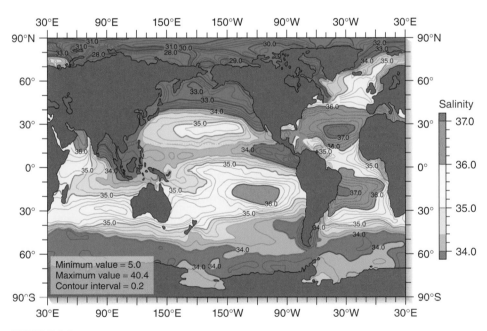

FIGURE 3.3

The average annual salinity of ocean surface water, 2005

The most saline parts (*red*) of the open ocean exist where evaporation is highest and precipitation lowest—at the subtropical belts of high pressure on either side of the equator. Salinity decreases (*blue*) toward the poles because of greater precipitation and lower evaporation in these regions. Salinities are locally low near the mouths of large rivers and high in semienclosed seas in arid regions, such as the Mediterranean. Salinity is given as total salt content of water in parts per thousand. (After National Oceanic and Atmospheric Administration, National Oceanographic Data Center, http://www.nodc.noaa.gov/)

[chapter 2]), where evaporation exceeds precipitation. Salinity is lower at the equator and, aside from the tropical highs, tends to decrease toward the poles because of greater precipitation and lower evaporation in these regions. Low salinities also exist locally near the mouths of large rivers, whereas very high salinities develop in semienclosed seas in arid regions—the Persian Gulf, the Red Sea, and the Mediterranean Sea being prime examples.

The Ocean's Layered Structure

The average temperature of the ocean is 3.6°C (38°F), but if you swim in it, it is pretty obvious that the surface part of it is much warmer than that.[2] Indeed, because the density of water depends on both temperature and salinity, the ocean is thermally and compositionally stratified (figure 3.4).

The ocean has three major layers. A thin, warm surface layer, termed the *mixed layer*, extends to about 20 to 200 meters (66 to 660 feet) in depth (the average is 70 meters [230 feet]). Here temperature changes little as waves and convective overturning keep the waters stirred. Over years to decades, the ocean's thermal capacity is essentially that of the mixed layer because over these timescales the mixed layer mixes little with the colder water below it.

Below the mixed layer is the *thermocline*, a zone in which temperature decreases and salinity increases rapidly with depth. The thermocline's base is about 5°C (41°F) and generally extends to 500 to 900 meters (1,600 to 3,000 feet) depth, but this depth varies from location to location and season to season. In winter at certain high-latitude locations, the thermocline can reach all the way to the ocean bottom, as occurs, for example, in the Norwegian-Greenland Sea during particularly cold spells. Conversely, cold water exists near or at the surface around parts of Antarctica, in which case there is neither thermocline nor mixed layer.

Below the thermocline is the *deep zone*, where salinity and temperature vary only slightly with depth. The deep zone constitutes about 65 percent of ocean water and at its deepest and coldest is about 2°C (35.6°F) (figure 3.5).

The Ocean's Surface Currents

Ocean surface currents are driven mainly by the wind, so like the wind they are affected by the Coriolis force and form distinctive patterns dominated by *subtropical gyres*, or semicircular current systems, on either side of the equator (figure 3.6). The generally easterly (out of the east or west-directed) trade winds drive the westward-flowing arms of the subtropical ocean gyres near the equator. This flow, in turn, generates eastward-flowing equatorial countercurrents both at and below the surface. Near the poles, easterly winds drive smaller gyres, but in the Southern Ocean

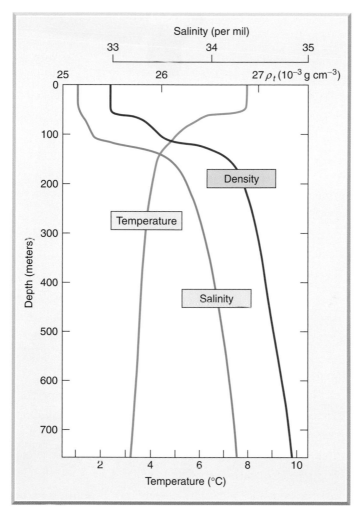

FIGURE 3.4

Vertical profiles of density, temperature, and salinity through the upper several hundred meters of the ocean

The mixed layer extends to a depth of about 170 meters (560 feet) in this profile and mixes little with the thermocline beneath it. Salinity is in parts per thousand; density $= (\rho_t - 1{,}000) \times 10^{-3}$ g/cm^3. (After Denman and Miyake 1973)

strong westerly winds drag water around the entire globe in the Antarctic Circumpolar Current.

THE WESTERN BOUNDARY CURRENTS

Of course, the positions of the continents also determine current patterns. In particular, the continents deflect the westward-flowing arms of the subtropical gyres, transforming them into strong, poleward *western boundary currents* flowing generally

FIGURE 3.5

A conductivity, temperature, and depth measurement device

A basic instrument in oceanographic research, the CTD device measures temperature, salinity (by measuring electrical conductivity of the seawater), and depth as it is lowered into the ocean, thus providing a continuous profile of these properties with depth. The vertical tubes are water bottles that may be closed at specific depths. When the device is returned to the ship, the water it has captured is typically analyzed for oxygen, CO_2, nutrients, and other substances. The photograph was taken on the Woods Hole Oceanographic Institution research vessel *Atlantis* during an expedition to the northeastern Pacific Ocean in 1997. (Photograph by the author)

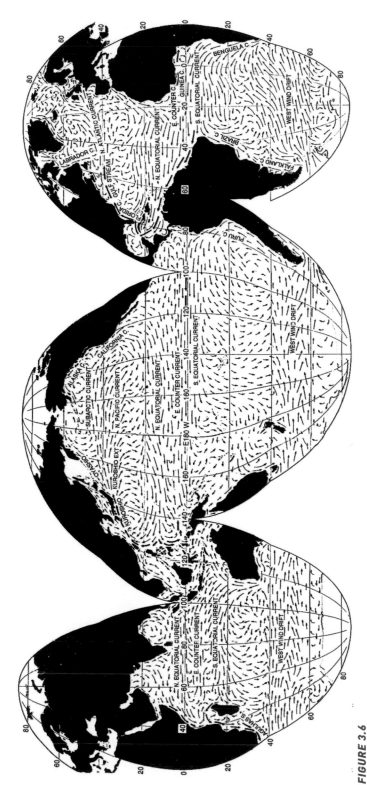

FIGURE 3.6

The global ocean surface currents

The currents are driven mainly by the winds and, like the winds, are affected by Coriolis forces. The continents deflect the currents into strong, poleward western boundary currents, the most important of which from the point of view of climate is the Gulf Stream. (After Hartmann 1994, with permission)

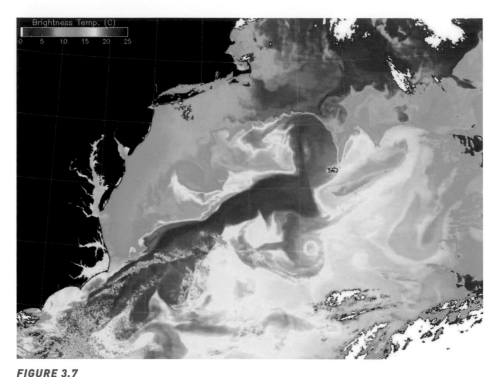

FIGURE 3.7

False-color satellite image of the Gulf Stream along the eastern United States and its eddies

The flow (*reds and yellows*) amounts to about 30 million cubic meters (39 million cubic yards) of water per second. The Gulf Stream and other boundary currents transport an enormous amount of heat to the high latitudes. (Liam Gumley, MODIS Atmosphere Team, University of Wisconsin–Madison Cooperative Institute for Meteorological Satellite Studies, http://visibleearth.nasa.gov/view_rec.php?id=1722)

parallel to the coastlines. Important from the point of view of climate is the well-known Gulf Stream in the North Atlantic Ocean (figure 3.7), but other oceans also have their western boundary currents. In the North Pacific, the Kuroshio Current flows past Japan; in the South Atlantic, the Brazil Current flows along the coast of South America. The western South Pacific and Indian oceans have the analogous East Australian and Agulhas (Mozambique) currents, respectively. These currents are fast. For example, the Gulf Stream flows at a rate of nearly 2 meters (6.5 feet) per second, and it takes only about a month for warm water from the tropics to reach the midlatitudes.

The western boundary currents are important for several reasons. First, they carry a great deal of heat—especially the Gulf Stream,[3] which not only is fast but has an enormous mean flow of about 30 million cubic meters (39 million cubic yards) of water per second. Mention should be made of the common misconception that the Gulf Stream is primarily responsible for northern Europe's relatively warm climate. It is true that the Gulf Stream delivers a great deal of heat to the North Atlantic, but

North Atlantic surface water retains a substantial amount of summer heat anyway, and winds out of the west and southwest transport warm surface water from lower latitudes toward Europe,[4] so it is really the entire warm ocean surface that keeps western Europe relatively warm.

The western boundary currents are also important in that they remove CO_2 from the atmosphere. Cold water dissolves more CO_2 than does warm water. Therefore, the warm waters carried poleward by the western boundary currents initially contain relatively little CO_2, but as the waters cool, they take up CO_2 from the atmosphere.

The eastern boundary currents arise for the same reason that the western boundary currents do. They include, for example, the California and Peru currents, which flow along the west coasts of North and South America, and the Benguela Current, which flows along the west coast of southern Africa. All these currents flow from high latitudes, carry cold water toward the equator, and then turn westward away from the coasts.

EKMAN TRANSPORT

It is fortunate for those of us who like to eat fish that ocean currents do not flow only in the direction of the wind. This phenomenon was first observed by the Norwegian explorer Fridtjof Nansen (1861–1930), who with his comrade F. H. Johansen in their ship the *Fram* spent the winter of 1893/1894 frozen in the Arctic pack ice. Their purposes were to study the Arctic Ocean currents and to get to the North Pole (they never reached it, despite mounting a mad dash with dogs and sledges when they realized the currents would not take them there). As the ice carried them along, Nansen noted that instead of drifting with the wind, the ice pack drifted at an angle 20 to 40 degrees to the right of the wind direction. Based on Nansen's observation, the Swedish oceanographer Vagn Ekman (1874–1954) developed a model showing how, due to the Coriolis effect, ocean surface water should move 45 degrees to the right of the wind direction in the Northern Hemisphere and 45 degrees to the left of it in the Southern Hemisphere.[5]

What does all this have to do with fish, though? An important consequence of *Ekman transport*, as this phenomenon is known, is the upwelling of water along coastlines. Atmospheric circulation in subtropical regions produces winds parallel to the west coasts of South America and Africa. By Ekman transport, the winds carry the upper few tens of meters of the ocean surface layer away from shore, where it is replaced by colder water from depths of 100 to 200 meters (300 to 650 feet). The upwelling water is rich in nitrates and phosphates and thus supports vigorous growth of plankton and therefore of fish—hence, for example, the anchovy and sardine fisheries off Peru and Ecuador.

Ekman transport can also drive currents toward shore, in which case water piles up and then sinks. It also operates in the open ocean, producing, for example, the cold tongue of water extending from South America into the equatorial Pacific.

Global Flows of Water Through the Ocean and Atmosphere

Water circulates through all the major oceans in what has been termed the *global ocean conveyor system*.[6] More esoterically, this phenomenon is known as *thermohaline circulation* because differences in water density drive the circulation, and, as we have seen, water density depends on temperature and salinity. The ocean conveyor starts with downwelling in the North Atlantic Ocean and the Southern Ocean around Antarctica. This water flows to and mixes with Pacific Ocean water, which returns as a shallow current to replace the downwelling water (figure 3.8). The current exerts a stabilizing influence on global climate over hundreds to thousands of

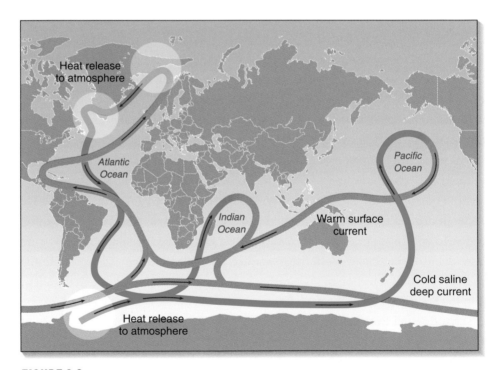

FIGURE 3.8

The global ocean conveyor system

In this system, also known as *thermohaline circulation*, the current starts with the downwelling of cold, dense water in the North Atlantic and Southern oceans around Antarctica. This deep water flows to and wells up in the Pacific Ocean and then returns as a shallow current to replace the downwelling Atlantic Ocean water. This current exerts a major influence on global climate over hundreds to thousands of years. (After Intergovernmental Panel on Climate Change 2001: fig. 4.2)

years, but changes in the current are paradoxically implicated in abrupt swings in climate, especially in the North Atlantic.

THE OCEAN CONVEYOR SYSTEM

The ocean conveyor works like this: warm, near-surface water in the Atlantic Ocean at about 35°N forms with a salinity of 34 to 36.5 ppt and flows northward at a depth of about 800 meters (2,600 feet). In the Norwegian-Greenland and Labrador seas during the winter, the northward-flowing water rises to the surface as surface waters are displaced by winds. Upon reaching the surface, the new water, which is initially at a relatively warm 10°C (50°F), cools rapidly to about 2°C (35.6°F) as it loses heat to the atmosphere. The cooling increases the water's density, so it sinks to become what is known as North Atlantic Deep Water (NADW). Now at a temperature of 2 to 4°C (35.6 to 39.2°F) and with a salinity of 34.8 to 35.1 ppt, NADW forms a deep, dense current that flows south, through the South Atlantic, and all the way to the Southern Ocean.

In the vicinity of about 60°S, NADW flows over Antarctic Bottom Water (AABW). The AABW originates mainly by downwelling in the Weddell Sea, where the densest water in the ocean forms due to the cold and increased salinity brought about as winter sea ice forms. It spreads as tongues into the Pacific, Indian, and Atlantic oceans, but in the Atlantic it mixes with and becomes entrained in the much stronger southward flow of NADW, which eventually spreads out through the Indian Ocean and into the southern Pacific Ocean.

The cold bottom water may make it as far as 30°N latitude in the Pacific Ocean, but eventually it mixes with the water above it. Pacific Ocean water returns to the Atlantic Ocean as a warmer, less salty current at shallow to intermediate depths through the Indian Ocean and around the southern tip of Africa. The amount of water flowing in the global ocean conveyor system is about 15 times that flowing in all the world's rivers, and the deep water mixes through the world ocean in a period of about 1,000 years.

The northward flow of water in the North Atlantic, also known as *meridional overturning circulation*, represents an enormous transport of heat—in fact, about one-quarter of the total global meridional (that is, directed north to south) heat transport. This explains why winters in the North Atlantic, despite the high latitude, are relatively mild, and it illustrates how the ocean can have a significant moderating effect on regional climate. Geological records of past climate indicate that on a number of occasions the thermohaline circulation has slowed or stopped and then restarted abruptly (that is, within a few years), implicating it in some of the rapid and dramatic shifts in climate experienced in the past by Greenland and northern Europe (chapter 6).

OCEAN SALINITY VARIATIONS
AND THE GLOBAL HYDROLOGIC CYCLE

The global ocean conveyor system is intimately coupled to water transport through the atmosphere. Over the whole of the North Atlantic, there is more evaporation than precipitation (table 3.1), so the air leaving the Atlantic is more moist than the air entering it, which makes Atlantic water saltier than North Pacific water (by about 5 percent) (see figure 3.3). This extra saltiness is why deep water forms in the North Atlantic rather than in the North Pacific. The Pacific, in contrast, experiences slightly more precipitation than evaporation.[7] The global ocean conveyor acts to offset this imbalance by redistributing the salt throughout the world ocean.

The Mediterranean Sea is also implicated in this affair. Mediterranean water is relatively saline because of a high evaporation rate. (The salinity is 37 to 38 ppt, which is why the lazy swimmer floats so easily in it.) Freshwater flows into the Mediterranean from several major rivers (most important the Nile), and less-saline Atlantic Ocean surface water flows in through the Strait of Gibraltar. But beneath the incoming surface flow, the more saline Mediterranean water flows out through the strait in the reverse direction. In the Atlantic, this saline water sinks, mixes, and spreads for thousands of kilometers. The water from the Mediterranean increases the salinity of the North Atlantic by about 6 percent over what it would otherwise be.

TABLE 3.1 **THE WATER BALANCE OF THE CONTINENTS AND OCEANS**

Region	Evapotranspiration (mm/yr)	Precipitation (mm/yr)	Runoff (mm/yr)
Europe	375	657	282
Asia	420	696	276
Africa	582	696	114
Australia	534	803	269
North America	403	645	242
South America	946	1,564	618
Antarctica	28	169	141
All land	480	746	266
Arctic Ocean	53	97	44
Atlantic Ocean	1,133	761	−372
Indian Ocean	1,294	1,043	−251
Pacific Ocean	1,202	1,292	90
All oceans	1,176	1,066	−110

Source: D. L. Hartmann, *Global Physical Climatology* (San Diego: Academic Press, 1994).

As water flows around the globe, land experiences more precipitation than evaporation, but over the ocean the opposite is true (see table 3.1). The different continents receive significantly different amounts of water as precipitation and lose different proportions of water by runoff (the water that flows back to the ocean via rivers) and through the combined processes of evaporation and plant transpiration, termed *evapotranspiration*. The flow of water through the ocean, atmosphere, and land is known as the hydrologic cycle. The amount of water that moves through the hydrologic cycle every year is equivalent to about a 1-meter-deep (3-foot) layer of water over the entire Earth.

The Ocean–Atmosphere Interaction of El Niño–Southern Oscillation

Every several years, a dramatic and rapid warming occurs in the waters off the coast of Peru and Ecuador. Within a month, water temperature increases by 2 to 4°C (3.6 to 7.2°F). Anchovies disappear; sardines move south; and birds, fur seals, and other animals whose livelihoods depend on the fish die. Heavy rains inundate the normally arid coastal regions of northern Peru and Ecuador and cause flooding, but droughts strike the Andes of southern Peru and more distant regions such as northeastern Brazil. The warming of the Pacific usually takes place around Christmas and persists into May or June. Local fisherman refer to it as El Niño, Spanish for "Christ Child."

El Niño is an ocean phenomenon, but it is intimately connected to the atmospheric phenomenon known as the Southern Oscillation, which is a swing in barometric pressure (the weight of air around us) between the eastern and western parts of the equatorial Pacific Ocean. Because the two are coupled, they are now generally referred to together as *El Niño–Southern Oscillation* (ENSO).

As we have come to learn, ENSO is hardly limited to the Pacific Ocean and surrounding regions. Rather, it affects climate around the globe. It commonly creates severe drought in Australia, Indonesia, and southern Africa;[8] weakens the Asian monsoons; and brings mild but stormier winter weather to North America. It even affects the North Atlantic, where hurricanes are markedly less common and less intense in El Niño years than in "normal" years. The effects on ecosystems and agriculture can be dramatic (figure 3.9). For example, the unusually strong 1997/1998 event resulted in the death of 16 percent of the world's tropical corals.[9]

El Niño is the consequence of a change of atmosphere and ocean circulation across the entire equatorial Pacific and is a dramatic example of how the two interact. It occurs at irregular intervals of 2 to 7 years, appears on average about every 4 years, and has been a feature of the climate system for at least the past 130,000 years.[10] Next to the seasons, ENSO is the most important recurrent change in climate.

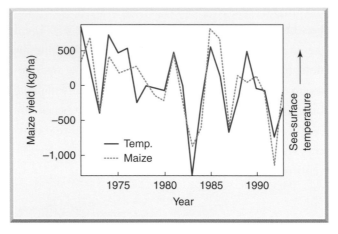

FIGURE 3.9

The correlation between annual maize yield in Zimbabwe and sea-surface temperature in the eastern equatorial Pacific

The maize yield in Zimbabwe is a close reflection of rainfall. Sea-surface temperature is the average for February and March, the height of the growing season in Zimbabwe, for the region in the equatorial Pacific bounded by 90 to 150°W longitudes and 5°N to 5°S latitudes. The correlation demonstrates the effect that El Niño–Southern Oscillation has on distant parts of the world. The maize yield is in kilograms per hectare relative to national average. (Adapted by permission from Macmillan Publishers Ltd: Nature, Cane, Eshel, and Buckland 1994, copyright 1994)

ENSO DYNAMICS

How does ENSO work? In "normal" years, sometimes referred to as La Niña (the Girl) years, especially when the winds are strong, the easterly trade winds of the Northern and Southern Hemispheres converge along the equator to blow west across the equatorial Pacific Ocean, driving the westward-flowing equatorial current (figure 3.10). The easterly winds cause Ekman transport away from the equator and upwelling of cold water in a narrow band along it, while coastal winds cause upwelling along the west coast of South America. A strong, eastward-flowing, equatorial "undercurrent" provides the water to replace these upwellings. The westward-flowing surface water piles up in the western Pacific, causing sea level to be about 60 to 70 centimeters (23 to 27 inches) higher there than in the eastern Pacific. The upwelling brings the thermocline essentially to the surface in the east, and in the west the pileup of surface water causes the thermocline to be much deeper, typically about 150 meters (490 feet) below the surface.

In the meantime, ocean and land surfaces of the western Pacific heat the atmosphere. The warm, moist air rises to form strong thunderstorms in what is known as the Indo-Australian Convergence Zone (IACZ). The convergence occurs because the rising air, which causes the barometric pressure of the IACZ to be low, is replaced by easterly winds from the Pacific and westerly winds from Australasia (figure 3.11). The rising air feeds a returning, eastward-directed flow of air near the

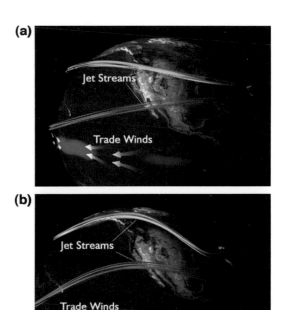

(a)

Jet Streams

Trade Winds

(b)

Jet Streams

Trade Winds

FIGURE 3.10

Shifts in sea-surface temperatures, winds, and positions of the jet streams in normal years and El Niño years

In normal (La Niña) years (*a*), the winds blow west across the equatorial Pacific Ocean, causing upwelling of cold water as surface waters are continuously displaced through Ekman transport. In El Niño years (*b*), the winds weaken, and warm western Pacific water flows back toward the east. El Niño and La Niña events affect the jet streams and Walker circulation, thereby changing weather patterns in distant parts of the globe. Most prominently, El Niño years bring droughts to parts of Southeast Asia, South America, and Africa (*dark brown*). (Krishna Ramanujan, NASA Goddard Space Flight Center, http://www.nasa.gov/vision/earth/lookingatearth/elnino_split.html)

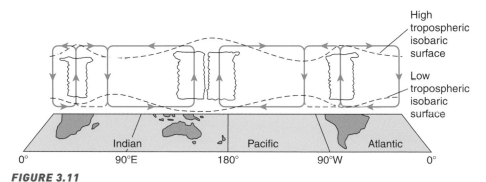

High tropospheric isobaric surface

Low tropospheric isobaric surface

Indian | Pacific | Atlantic

0° | 90°E | 180° | 90°W | 0°

FIGURE 3.11

Convective atmospheric circulation along the equator, also known as Walker circulation

The circulating cells shift eastward in El Niño events, causing changes in precipitation patterns throughout the tropics. The isobaric surface is a surface of constant atmospheric pressure. (After Webster 1983, with permission)

top of the troposphere that eventually sinks in the eastern Pacific, creating a zone of high barometric pressure. This equatorial convective cycling of the atmosphere has become known as *Walker circulation*, for British physicist Sir Gilbert Walker (1868–1958), who studied records of the Indian monsoon while stationed in India and recognized the relationship between the strength of the monsoon and the changing wind patterns in the Pacific Ocean.[11]

It is worth emphasizing the strong, positive feedback between the ocean and the atmosphere in this coupled circulation pattern:[12] the strong easterly surface winds maintain a warm western Pacific Ocean; the warm waters heat the overlying air and cause it to rise; the rising air results in low barometric pressure in the west and high pressure in the east; the consequent pressure gradient causes strong winds to blow west across the eastern equatorial Pacific; and the strong winds keep the warm water in the western Pacific. The ENSO phenomenon represents one of several coupled atmosphere–ocean circulation patterns that influence climate on timescales of years to several decades.

In an El Niño year, this coupled ocean–atmosphere circulation pattern of the equatorial Pacific collapses (see figure 3.10). The easterly trade winds of the eastern Pacific weaken, warm western Pacific water flows back toward the east, and sea level flattens out—in essence, the water that typically piles up in the west sloshes back in an enormous (but imperceptible to us) "wave." Upwelling ceases along the coast of South America and along the equator, the thermocline deepens to tens of meters in the east, and the equatorial undercurrent stops. The warmer water in the east now heats the air over the eastern Pacific, causing the air to rise and bringing the heavy rains to coastal South America.

In the meantime, in the western Pacific the IACZ becomes less localized and shifts eastward. The regular Walker circulation of normal years breaks down, and descending dry air commonly falls over Indonesia and Australia to create drought conditions there. This change in Walker circulation is also the mechanism through which ENSO influences the Asian monsoon. The shift to El Niño also causes a shift in the trade winds (see figure 3.10), changing weather patterns in other, distant parts of the globe.

The equatorial Pacific oscillates between its El Niño and normal states probably because the depth of the thermocline in the east, which influences sea-surface temperature for a given wind velocity, does not change exactly in phase with changes in the winds.

ENSO PATTERNS

The swings in barometric pressure that accompany changes from one state to the other provide a record of ENSO that is known as the Southern Oscillation Index, the difference in barometric pressure between the IACZ and the eastern Pacific (figure 3.12).

The historical record of the Southern Oscillation Index reveals several important features. First, although ENSO events occurred on average about every four years from 1935 to 1995 and in that sense are quasi-periodic, their occurrence is highly irregular. For example, from 1978 to 1986 there was only one ENSO event, but from 1990 to 1995 there was either one prolonged event or several that followed each other in close succession. Second, some ENSO events are intense, while others are not. The 1997/1998 ENSO was perhaps the fiercest in the past 1,000 years, and the 1982 event was also unusually intense. Others, however, are barely noticeable. Third, ENSO states alternate with what we have been calling "normal states," or La Niña events.

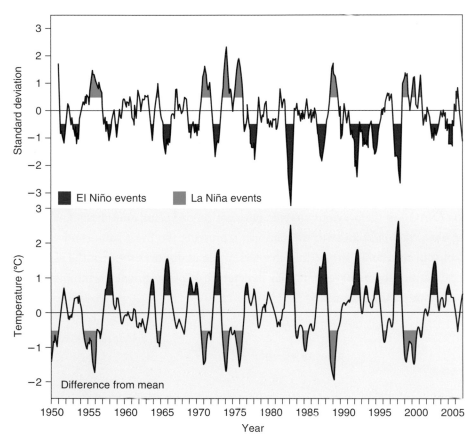

FIGURE 3.12

The Southern Oscillation Index and corresponding variations in eastern equatorial sea-surface temperature

The SOI is the deviation from the mean difference in atmospheric pressure between Tahiti and Darwin, Australia. Sea-surface temperature is the monthly difference from the mean in the region bounded by 5°N to 5°S latitudes and 120 to 170°W longitudes. Note the 1982/1983 and 1997/1998 events, which are thought to have been the strongest of the twentieth century. (After McPhaden, Zebiak, and Glantz 2006, with permission)

In fact, La Niña and El Niño events can simply be viewed as two alternating states of ocean–wind circulation patterns, with neither one nor the other being abnormal.

Needless to say, what ultimately causes ENSO has engendered and continues to engender debate; it also raises the question of what will happen to ENSO as climate warms. This question is fairly important because sea-surface temperature in the eastern tropical Pacific Ocean appears to affect climate, especially rainfall, in parts of western North America (chapter 7). A clear picture of how warming will affect ENSO or how ENSO will affect climate in a warmer world has yet to emerge.

Perspective

This chapter has sought to show how the ocean is an integral part of the climate system for two primary reasons. First, the ocean stores far more heat than the atmosphere, so the two in concert move heat around the planet, primarily from the tropics to the poles. Second, the ocean also holds far more carbon than the atmosphere. More important, because the two exchange CO_2, the ocean also affects global climate by controlling the CO_2 content in the atmosphere over periods of a century and longer. In essence, the ocean acts as a global thermostat by being a large reservoir for both heat and carbon.

The tight physical and chemical coupling of the ocean and the atmosphere should now be evident. The present climate system displays this coupling by a variety of features. We encountered two features, the monsoons and North Atlantic Oscillation (NAO), in the previous chapter, but another one stands out, ENSO, which has a global reach and, next to the seasons, represents the most important recurrent change in climate. In succeeding chapters, we shall discover some of the roles these ocean–atmosphere interactions play in determining how global warming will affect conditions on the continents.

A valley in Kashmir, high in the Himalayas, India

Erosion and weathering of rocks removes CO_2 from the atmosphere by dissolving it in water. The carbon ends up in the ocean, where it is used by organisms to make shell. When the organisms die, the shells accumulate on the ocean bottom and eventually form rocks. This process is part of the carbon cycle, referring to the flow of carbon among the ocean, atmosphere, biosphere, and lithosphere, all of which are carbon-storage reservoirs. The carbon cycle is one of several biogeochemical cycles that determine the character of Earth's environment, including its climate. (Photograph by the author)

4 THE CARBON CYCLE AND HOW IT INFLUENCES CLIMATE

THE GROWING AMOUNT of carbon dioxide (CO_2) in the atmosphere is the primary cause for global warming (chapters 7 and 9), so it is important to understand the factors that naturally control its abundance. Atmospheric CO_2 levels have been measured continuously only since 1958. In that year, Charles Keeling (1928–2005), who would go on to become a professor at Scripps Institution of Oceanography, began measuring CO_2 at a newly established observatory on the top of Hawaii's Mauna Loa volcano.[1] The change in CO_2 content with time, widely referred to as the *Keeling curve*, shows two interesting features (figure 4.1). First, every year CO_2 concentration reaches a maximum in May and then decreases until October, when it begins to rise again. The reason for this change is that in the Northern Hemisphere early summer plants begin to grow rapidly, drawing down CO_2, but in the fall they become dormant or die and decay, returning CO_2 to the atmosphere. This process is like the natural rhythm of breathing, but on a much grander scale.

The second feature of the Keeling curve is equally obvious. Atmospheric CO_2 content has been rising at a rapid clip since Keeling started measuring it. In 1958, the average CO_2 content of the atmosphere was 315 parts per million (ppm) by volume; in 2006, it was 380 ppm and increasing by about 2 ppm per year. This rate of increase is far greater than any common natural phenomenon can produce and is unprecedented in the climate record. Emissions of CO_2 and other greenhouse gases as a consequence of human activity—burning of fossil fuel, cement production, and other activities, together commonly referred to as *anthropogenic emissions*—account for this rise (chapter 10).

But there is one curious statistic: only about one-third of the carbon emitted from all human activities and one-half of that emitted from fossil-fuel burning and cement production since 1800 are in the atmosphere. Where has the rest gone? The answer is that it has been removed by the ocean and the biosphere and is now stored there.

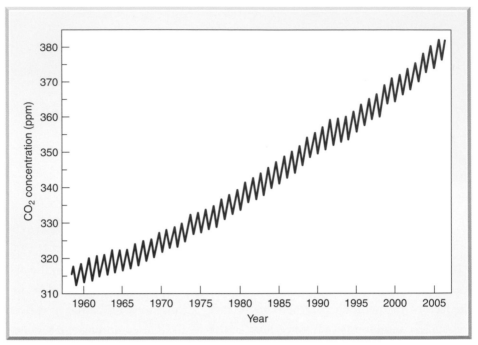

FIGURE 4.1

The Keeling curve

The CO_2 content of the atmosphere has been rising since the measurement at Mauna Loa volcano began in 1958, when it averaged 315 ppm. In 2006, CO_2 content was 380 ppm and rising by about 2 ppm per year. The annual fluctuation of CO_2 level reflects the natural cycle of plant growth and death in the Northern Hemisphere. (After Scripps Institution of Oceanography CO_2 Program, http://scrippsco2.ucsd.edu, with permission)

This brings us to the subject of this chapter, the *carbon cycle*, which refers to the flow of carbon among the various global "reservoirs" that store carbon. The chapter explores how the ocean and biosphere control the CO_2 content of the atmosphere and how the ocean is particularly affected by the increase in atmospheric CO_2.

Reservoirs of Carbon

To understand the carbon cycle, we must first characterize the carbon reservoirs. One might suppose that most carbon on Earth is present in living organisms or perhaps in the form of coal or oil. That, however, is emphatically not the case. More than 99.9 percent of the carbon on or near the surface of Earth is sequestered in rocks, mostly limestones, carbon-rich shales, and coal and oil deposits. The rock "reservoir" amounts to some 50 million billion metric tons (50 million gigatons, or 10^{15} metric tons) (table 4.1).

In contrast, *all* the surface reservoirs combined contain only about 42,000 gigatons of carbon. By *surface reservoirs*, we mean the atmosphere, where carbon exists

TABLE 4.1 **SIZES OF CARBON RESERVOIRS**

Rock reservoir	50×10^6 metric gigatons
Limestone	40×10^6
Organic carbon in sedimentary rocks	10×10^6
Fossil fuels	4.7×10^3
Marine carbonate sediments	2.5×10^3
World ocean	39×10^3
Bicarbonate ion	37×10^3
Carbonate ion	1.3×10^3
Dissolved CO_2	0.74×10^3
Organic carbon in soils and terrestrial sediments	1.6×10^3
Organic carbon in permafrost	0.9×10^3
Atmospheric CO_2	0.76×10^3
Living biomass	0.6×10^3

Sources: L. R. Kump, J. F. Kasting, and R.G. Crane, *The Earth System*, 2d ed. (Upper Saddle River, N.J.: Prentice Hall, 2004); S. A. Zimov, E. A. G. Schuur, and F. S. Chapin III, "Permafrost and the Global Carbon Budget," *Science* 312 (2006): 1612–1613.

mainly as CO_2; soil, where carbon occurs in a variety of organic compounds; the ocean, where it is present mainly as the dissolved bicarbonate ion (HCO_3^-); and, of course, living biomass. The largest by far of these surface reservoirs is the ocean, which holds 39,000 gigatons, a fact that, as we shall see, has an important bearing on climate. The other reservoirs are minuscule in comparison; for example, all the biomass contains a piddling 600 gigatons of carbon, even less than the atmosphere!

The Carbon Cycle

It is convenient to think of the carbon cycle as two cycles rather than one. The long-term carbon cycle operates over hundreds of thousands to millions of years, but carbon also cycles through surface reservoirs in timescales measured only in months to years to centuries. The short-term carbon cycle bears directly on the state of the present climate and the way it changes over human timescales, but we also have to understand the long-term carbon cycle to appreciate our present predicament.

THE LONG-TERM CARBON CYCLE

The *long-term carbon cycle* involves the movement of carbon between the solid Earth and the ocean and atmosphere. The weathering of rock, which is especially rapid in mountainous regions where erosion rates are high, removes CO_2 from the atmosphere. This happens because CO_2 reacts with surface water and with the silicate and carbonate minerals of rocks. The result is the formation of calcium (Ca), magnesium (Mg), bicarbonate (HCO_3^-), and silica (SiO_3, SiO_4) in solution in the water (H_2O), which may be represented by a simplified reaction:

$$4CO_2 + 6H_2O + CaSiO_3 + MgSiO_3 \rightarrow Ca^{++}$$
$$+ Mg^{++} + 4HCO_3^- + 2H_4SiO_4$$

atmospheric carbon dioxide + water + calcium- and
magnesium-bearing silicate minerals → dissolved
calcium and magnesium cations + bicarbonate ions + silicic acid

The dissolved ions wash into rivers that eventually end up in the ocean, where organisms use the calcium and bicarbonate ions to make shells:[2]

$$Ca^{++} + 2HCO_3^- \rightarrow CaCO_3 + CO_2 + H_2O$$

calcium + bicarbonate ions dissolved in seawater → calcium carbonate
minerals calcite or aragonite + carbon dioxide + water

FIGURE 4.2
Coquina
Coquina is a form of limestone composed of fossilized shell debris. The shells and the cement that holds them together are made of the calcium carbonate mineral calcite. This sample is on display in the Gottesman Hall of Planet Earth at the American Museum of Natural History, New York. (Photograph by D. Finnin, American Museum of Natural History)

FIGURE 4.3
Chalk

Chalk, a variety of limestone, makes up the White Cliffs of Dover, in southeastern England. The chalk in the cliffs formed about 70 million years ago by the accumulation of the calcium carbonate skeletal remains, known as coccoliths (see figure 4.9), of single-celled, microscopic algae. (Photograph by Maki Itoh, http://www.makikoitoh.com, with permission)

The dissolved silica in the ocean precipitates to opal in the shells of microscopic plants:

$$H_4SiO_4 \rightarrow SiO_2 + 2H_2O$$
dissolved silica → opal + water

As shelled organisms die, their carbonate shells accumulate on the ocean floor to form limestone (figures 4.2 and 4.3). New layers of sediment cover older layers, and with burial the sediments slowly transform to rock. The rocks are sometimes buried to depths of thousands of meters, a process that may take millions of years, in which case the heat and pressure of the deep Earth may break down the carbonate minerals to give off CO_2. The CO_2 slowly percolates out of the crust and back into the atmosphere, completing the cycle. Alternatively, the sedimentary rocks may be dragged down into Earth's mantle in subduction zones, where one of Earth's lithospheric plates is thrust beneath another. Here the carbonates similarly break down, and the CO_2 finds its way back to the surface as part of the gas emanating from

FIGURE 4.4
Coal

Coal is the remains of decayed plants buried in swamps and marshes and then subjected to high temperature and pressure. This metamorphism from dead plants to coal, which typically occurs over 1 million years or more, drives off water and most hydrocarbons, leaving coal as the carbon-rich residue. This sample is on display in the Gottesman Hall of Planet Earth at the American Museum of Natural History, New York. (Photograph by D. Finnin, American Museum of Natural History)

erupting volcanoes. In either case, a reaction for the return of carbon from the deep Earth to the surface reservoirs is

$$SiO_2 + CaCO_3 \rightarrow CaSiO_3 + CO_2$$
silicate minerals + carbonate minerals → calcium
silicate minerals + carbon dioxide

A variant of the cycle is the removal of CO_2 from the atmosphere by plants as they engage in photosynthesis and then the direct burial of dead plant matter. This process commonly occurs in swamps, where burial of the plant matter may eventually produce coal beds (figure 4.4), or in river deltas, where plant matter may be buried in mud to form carbonaceous shale. The reaction for photosynthesis, which is driven by sunlight, may be represented by

$$CO_2 + H_2O \rightarrow [CH_2O] + O_2$$
carbon dioxide + water → carbohydrates, starches, and
other organic compounds of plants + oxygen

The oxidation of carbon-rich sedimentary rocks as they are exposed to air by erosion or as they are metamorphosed during burial returns CO_2 to the atmosphere and completes the cycle:

$$[CH_2O] + O_2 \rightarrow CO_2 + H_2O$$

organic compounds + oxygen → carbon dioxide + water

By these mechanisms, carbon is cycled between the lithosphere, on the one hand, and the ocean and the atmosphere, on the other, on timescales of hundreds of thousands to millions of years and more.

Now we have to return to the notion of carbon reservoirs. The enormous mismatch in the amounts of carbon held in the rock and in the surface reservoirs has a very important implication—that over millions of years, there must have existed a close balance in the exchange of carbon between them. The amount of carbon that is being put into the rock reservoir through the formation of carbonate-bearing and organic-carbon-bearing rocks must over the long term be balanced by the amount returned by degassing as the rocks get buried or subducted and thermally decompose. Indeed, this long-term balance may be why Earth has not experienced a runaway greenhouse, in which so much CO_2 accumulates in the atmosphere that the planet heats up to the extent that it cannot support life. In other words, the long-term carbon cycle appears to have acted as a natural thermostat, maintaining conditions on Earth's surface conducive to the evolution and survival of life since nearly the beginning.

However weighty this thought might be, the important point in the context of climate change is that fossil-fuel consumption and cement production represent a rapid transfer of carbon from the rock reservoir to the surface reservoirs. This street goes only one way as far as the climate system is concerned because carbon returns to the rock reservoir only by natural processes that operate much more slowly than the rate at which we are removing carbon from this reservoir. This, broadly, is the fundamental problem: by burning fossil fuel, we are removing carbon from the long-term reservoir and putting it into the short-term, surface reservoirs, but natural processes cannot return it to the long-term reservoir quickly enough.

By examining the geological record, we can gain a sense of how long it takes for a large excess of carbon to be drawn down into the rock reservoir. About 55 million years ago, a sudden influx of CO_2 into the atmosphere–ocean system sent temperatures soaring at a time known as the Paleocene–Eocene Thermal Maximum (PETM) (chapter 6). The recovery of the climate system and return to cooler conditions took 40,000 to 60,000 years, which may be the shortest period of time that the long-term carbon cycle requires to correct sudden perturbations in the climate system.

THE SHORT-TERM CARBON CYCLE

The *short-term carbon cycle* refers to the circulation of carbon among the surface reservoirs—the ocean, atmosphere, soil (land), and biosphere (figure 4.5). It does not involve rocks at all, and the cycling can be rapid, taking only from months to decades to a millennium. Because the circulation is rapid, the short-term cycle is immediately relevant to climate over the next several decades to millennia. In the land-based part of the short-term carbon cycle, photosynthesis removes carbon from the atmosphere (as described earlier). About half the terrestrial organic material from photosynthesis is returned to the atmosphere by respiration of plants and animals, including ourselves:

$$[CH_2O] + O_2 \rightarrow CO_2 + H_2O$$
organic matter + oxygen → carbon dioxide + water

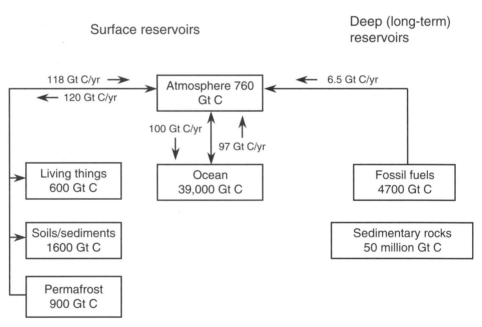

FIGURE 4.5

The global short-term carbon cycle, showing the surface carbon reservoirs, the rates of flow of carbon among them, and the flow of carbon to the surface from the long-term reservoirs

The burning of fossil fuels represents a net transfer of carbon from the long-term rock reservoir to the short-term surface reservoirs. The transfer is irreversible on timescales relevant to humanity unless we find ways of putting carbon back into the rock reservoir. Gt C = billion metric tons of carbon. (Data from table 4.1 and NASA)

whereas the remainder ends up in soil by accumulation of dead debris. Microbes in the soil engage in aerobic respiration to produce CO_2 as well. Or, if they live in deeper, oxygen-poor levels of the soil, microbes may produce methane (CH_4):

$$2[CH_2O] \rightarrow CO_2 + CH_4$$
organic matter → carbon dioxide + methane

The resulting methane enters the atmosphere but eventually combines with oxygen (oxidizes) to form more CO_2. The methane content of the atmosphere is a measure of the terrestrial methanogenesis rate and, indirectly, of the extent of wetlands in mainly tropical and temperate climates.

The average carbon atom resides in the atmosphere for about a decade,[3] which is about how long it takes for the atmosphere to respond to changes in the rates of carbon flowing in and out of the biosphere. The natural flows of carbon in and out of the surface reservoirs balance themselves, but adding to the natural flow is a steady trickle of carbon in the form of anthropogenic CO_2 (see figure 4.5).

THE OCEAN CARBON PUMPS

Although the atmosphere's interactions with the land-based biosphere have an immediate (that is, months to decades) effect, they are not really the main story. On the longer timescale of several decades to a millennium and more, the ocean is the primary regulator of CO_2 in the atmosphere. Indeed, the ocean has taken up about one-half of all the carbon originating from fossil-fuel emission and cement production for the period 1800 to 1994 and about one-third of it for the two decades from 1980 to 2000 (table 4.2);[4] climate models suggest that 90 percent of anthropogenic CO_2 will eventually find its way into the ocean.[5]

Several mechanisms are involved in the exchange of carbon between the atmosphere and the ocean. First, photosynthetic organisms known as phytoplankton live in the *photic zone* (the region penetrated by light, extending to about 100 meters [330 feet] below the ocean's surface). Phytoplankton are the lunch of various zooplankton, such as foraminifera, radiolarians, and copepods. The zooplankton produce fecal pellets, which rain down into the ocean depths, where the organic material is consumed by other microorganisms, releasing CO_2 (and nutrients) to deep water. The combined processes of photosynthesis in the shallow ocean and the CO_2 transfer to and enrichment of the deep waters is known as the *biological pump*. The biological pump is particularly effective at high latitudes where surface waters are relatively rich in nutrients[6] and thus contain abundant phytoplankton (figure 4.6).

TABLE 4.2 **WHERE ANTHROPOGENIC CARBON HAS COME FROM AND WHERE IT GOES**

	1800–1994	1980–1999
Emissions		
From fossil-fuel and cement production	244 ± 20	117 ± 5 metric gigatons
From land-use change (for example, deforestation)	100–180	24 ± 12
Stored in		
The atmosphere	165 ± 4	65 ± 1
The ocean	118 ± 19	37 ± 8
The biosphere	61 − 141	39 ± 18

Source: C. L. Sabine, R. A. Feely, N. Gruber, R. M. Key, K. Lee, J. L. Bullister, R. Wanninkhof, C. S. Wong, D. W. R. Wallace, B. Tilbrook, F. J. Millero, T.-H. Peng, A. Kozyr, T. Ono, and A. F. Rios, "The Oceanic Sink for Anthropogenic CO_2," *Science* 305 (2004): 367–371.

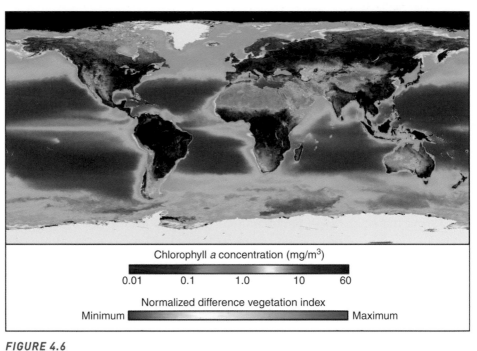

FIGURE 4.6
The distribution of ocean chlorophyll and land vegetation, 1997–2007 composite image

Surface ocean waters at high latitudes (*light green*) are relatively rich in nutrients and thus contain abundant phytoplankton. The chlorophyll concentration scale refers to the ocean; the normalized difference vegetation index is a measure of the density of land vegetation. (Satellite imagery courtesy of GeoEye/NASA. Copyright 2008. All rights reserved, http://oceancolor.gsfc.nasa.gov/)

Another biological pump, sometimes referred to as the *carbonate pump*, involves the removal of carbon from the ocean when foraminifera, mollusks, and other organisms build calcium carbonate ($CaCO_3$) shells, as described earlier. When these marine "calcifiers" die and their shells settle to the deeper ocean, one of two things happens. The shells may reach the bottom, accumulate as sediment, and eventually become part of the rock reservoir, with the carbon removed from the surface reservoirs for millions of years. Alternatively, because carbonate is not stable in the deep ocean, shells that settle below the *carbonate compensation depth* (CCD) (about 4,000 meters [13,000 feet]) dissolve and release CO_2 faster than they accumulate on the bottom as sediment.

It may seem counterintuitive, but the reaction in the ocean that produces carbonate for shells also produces CO_2, as illustrated by the second reaction in the long-term carbon cycle. So, despite the fact that the carbonate pump may transfer carbon to the deep ocean or even to the rock reservoir, in those parts of the ocean where calcifying organisms are active, the ocean actually pumps CO_2 into the atmosphere rather than removing it.

A third pump for carbon is called the *solubility pump*. It turns out that the solubility of CO_2 (and most other gases) in water increases as temperature decreases and pressure increases.[7] For this reason, the deep ocean can hold substantially more CO_2 than can the shallow ocean. The solubility pump refers to the cooling and sinking of surface water at high latitudes, thus removing CO_2 from the atmosphere and transporting it to the deep ocean.

Like the biological pump, the solubility pump operates mainly at high latitudes, where CO_2-rich surface waters tend to sink as they cool, making these regions particularly important in the regulation of atmospheric CO_2. The deep, CO_2-rich waters that form at high latitudes may take many decades to centuries and longer to well up to the surface, depending on location. Upwellings of deep water occur along certain coastlines in equatorial regions (figure 4.7). As the deep water rises and warms, it gives up CO_2 to the atmosphere.

The ocean therefore both removes CO_2 from the atmosphere and adds it back, and over the course of years the transfer back and forth establishes a CO_2 balance between the two.[8] But because the ocean contains so much more carbon than the atmosphere, it is the 800-pound gorilla in this story and exerts the main control in the transfer process.

The Acidification of the Ocean

Because the atmosphere and the ocean are in chemical equilibrium, as the CO_2 content of the atmosphere increases, so does that of the ocean. This equilibrium has

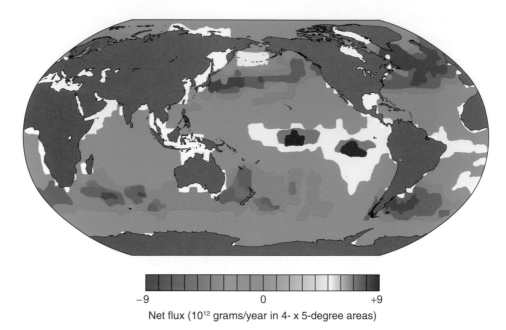

−9   0   +9

Net flux (10^{12} grams/year in 4- x 5-degree areas)

FIGURE 4.7

The annual net exchange of CO_2 between the ocean and the atmosphere

The ocean removes CO_2 from the atmosphere in the North Atlantic Ocean, in western parts of the South Atlantic Ocean and the Pacific Ocean between about 40 and 50°S latitudes, and throughout the Southern Ocean around Antarctica. It gives up CO_2 to the atmosphere mainly in the equatorial Pacific Ocean. (After Takahashi et al. 1997)

what may turn out to be an important consequence: the CO_2 taken up by the ocean goes through a series of reactions that reduce the pH of water (pH is a measure of acidity; the lower the pH the higher the acidity):[9]

$$CO_2(aq) + H_2O \rightarrow H_2CO_3$$
dissolved CO_2 + water → carbonic acid   (1)

$$H_2CO_3 \rightarrow HCO_3^- + H^+$$
carbonic acid → bicarbonate + hydrogen ions   (2)

$$HCO_3^- \rightarrow CO_3^{2-} + H^+$$
bicarbonate → carbonate + hydrogen ions   (3)

In preindustrial times, ocean pH was about 8.2. Now it is 8.05, and if the CO_2 content of the atmosphere doubles, it will decrease to about 7.9.[10] (It is important to understand that the ocean is not actually acidic, which refers to compositions for which pH is less than 7; rather, it is becoming less basic because its pH, although decreasing, remains well higher than 7.) Although the pH change may seem small, the acidification is likely to have a profound influence on ocean biogeochemistry, particularly on calcifying organisms. As pH decreases (H^+ concentration increases), the amount of carbonate (CO_3^{2-}) in seawater decreases (reaction 3), which has the effect of decreasing the stability of calcite and aragonite, the carbonate minerals that constitute the skeletons and shells of calcifying organisms (but this decrease appears to depend on the organism).[11]

Because CO_2 solubility in ocean water increases with decreasing temperature and increasing pressure, there are depths in the ocean above which calcite and aragonite (both of which are composed of calcium carbonate but differ in how the atoms are put together) are stable and below which they are not.[12] Furthermore, these depths will migrate to shallower water depths as ocean acidification proceeds. The specific depths at which carbonates become unstable (saturation depths) differ from one location to another (figure 4.8). In the North Atlantic, for example, the depths exceed 2,500 meters (8,200 feet), but in the North Pacific they rise to within a couple of

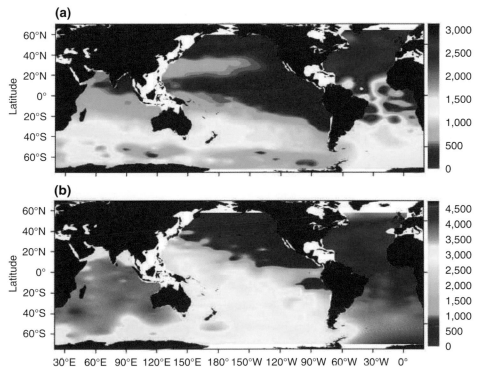

FIGURE 4.8

Aragonite and calcite saturation depths

The saturation depth for aragonite (a) is shallower than the saturation depth for calcite (b) because aragonite is more soluble in seawater. The depths are sensitive to ocean water acidity. The color scales on the right are in meters of water depth. (After Feely et al. 2004, with permission)

hundred meters of the surface. One of the consequences of CO_2 buildup and acidification is that the carbonate-saturation depths will continue to rise closer to the surface. If CO_2 emissions continue to increase, calcite and aragonite may become unstable in the surface waters in the Southern Ocean beginning sometime around the mid-twenty-first century, and no shelled organisms will live there.[13]

The changes occurring in the ocean lead to two important questions, neither of which has a clear answer at this point. Will ocean acidification influence the ocean's ability to absorb CO_2 from the atmosphere and thus affect climate? And how severely will acidification affect marine calcifiers and disrupt marine ecosystems?

EFFECTS ON CLIMATE

Regarding the first question, if less carbon is being converted into shells, there should be a concomitant decrease in CO_2 production in ocean waters (the second reaction in the long-term carbon cycle) and thus an *increase* in the net flow of CO_2 from the atmosphere into the ocean. Calculations suggest, however, that the increased rate of CO_2 removal from the atmosphere will be small compared with anthropogenic CO_2 emissions.[14]

Or there may be feedbacks (chapter 5) associated with the ocean's changing biochemistry that we simply do not know about, some of which may have the opposite effect. For example, it is possible that as CO_2 in the atmosphere increases, the rate of carbon uptake by the ocean (and land) will slow, resulting in a progressive increase in the fraction of anthropogenic CO_2 emissions that remain in the atmosphere, thereby also amplifying warming.[15]

EFFECTS ON MARINE ECOSYSTEMS

Increasing CO_2 content and decreasing pH of the ocean may influence marine organisms in several ways. Most obvious is the negative influence on the ability of calcifying organisms to reproduce and to build and maintain their shells, as noted earlier.[16] Of particular concern is how the ocean's changing composition will affect the ability of corals to build their skeletons, which is necessary to maintain reef structures, not to mention entire reef ecosystems. Corals are sensitive to ocean temperature, pollution, and other factors. For example, the particularly warm 1997/1998 El Niño year destroyed 16 percent of the world's coral reefs, and many more are under serious threat from the combined stresses related to climate change. Several studies suggest that corals will be unable to build their carbonate reef structures as the CO_2 content of the atmosphere approaches 480 ppm, a level that will be reached within 50 years at the current rate of emissions.[17]

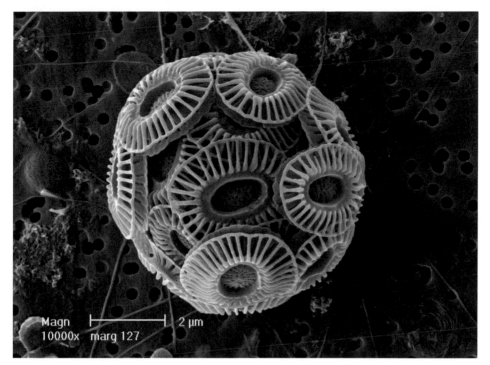

FIGURE 4.9

Scanning electron photomicrograph of the phytoplankton **Emiliania huxleyi**

The organism *Emiliania huxleyi* secretes and is armored with delicate calcium carbonate coverings known as coccoliths. It lives near the ocean surface, where it photosynthesizes. Such shell-building microorganisms are important for removing carbon from the ocean, and how they will fare as ocean pH decreases is uncertain. (Photomicrograph by Jeremy Young, http://www.noc.soton.ac.uk/soes/staff/tt/eh/index.html)

As for phytoplankton, the few existing experiments suggest that increasing ocean CO_2 content and declining pH increase photosynthesis only marginally in most phytoplankton.[18] There is at least one important exception, however—the abundant phytoplankton *Emiliania huxleyi* (figure 4.9). For this organism, both photosynthesis and calcification rates increase markedly with increasing CO_2.[19] According to reaction 3, increasing hydrogen ion (H^+) concentration decreases carbonate ion (CO_3^{2-}) concentration, but at the same time increases bicarbonate ion (HCO_3^-) concentration. The bicarbonate ion is the source of carbon for carbonate formation; at least for *Emiliania huxleyi*, then, the effect of increasing bicarbonate ion concentration apparently more than offsets that of decreasing carbonate ion concentration.

Finally, a reduced pH should increase the concentrations of dissolved copper, zinc, and other toxic metals as well as nutrients dissolved in seawater.[20] There is little research on this topic, so how these competing factors will ultimately affect marine organisms' health is not known.

Perspective: Uncertainties in the Carbon Cycle

The numerous unknowns in how the carbon cycle operates translate into a huge question about how climate will respond to the 36 gigatons (and growing) of CO_2 we are dumping into the atmosphere every year as a consequence of our activities.[21] First, one needs to appreciate the single, overriding uncertainty that concerns all elements of the carbon cycle. The atmosphere contains more CO_2 now (385 ppm) than at any time in at least the past 800,000 years (chapter 6). In the past, atmospheric CO_2 varied regularly and in step with the glacial and interglacial periods (there were 10 during that time). At the ends of the glacial periods, atmospheric CO_2 content rose along with temperature at a rate of perhaps 10 to 20 ppm per 1,000 years. The current rate of increase is 100 times faster than that, and all the natural biogeochemical processes are now operating beyond the long-term bounds of atmosphere composition. Will global biogeochemical processes operate in the same way in this new world as they did in the past? We do not know.

Another uncertainty is how climate change will affect carbon uptake by terrestrial plant ecosystems. In general, elevated CO_2 should enhance many plants' rate of growth. However, growth rate is also sensitive to moisture, availability of nutrients, and other factors, any of which may be the ultimate limit on plant growth. To illustrate, for certain perennial grassland species, the CO_2-enhanced growth rate is transient only because growth rate is eventually limited by the availability of nitrogen in the soil.[22] Also, although climbing temperature causes plants to increase their rates of CO_2 respiration dramatically, this phenomenon also may be transient because plants tend to acclimate to the warmth and eventually thus to reduce respiration.[23]

A related question is what happens to the carbon in soils. Warming and increased CO_2 content should increase the rate at which plant matter is added to soils, but it should also lead to an increase in soil microbial respiration and a consequent reduction of carbon content in the soil.[24] Over decades to centuries, such effects are potentially important as far as atmospheric CO_2 is concerned because of the far greater abundance of carbon in soils than in the atmosphere. The carbon content of and the rates of CO_2 and methane generation from soil depend on temperature, atmospheric CO_2 content, local rainfall, soil moisture, soil oxygen content, fertility, the plant species present, the time since a fire, the nature of the rock substrate, the extent of physical and chemical protection, and the enzymes and inhibitors present. Needless to say, soils vary greatly in terms of this long list of characteristics, which is one reason they are poorly understood.

Yet another question is how climate change will affect the rate at which the ocean removes CO_2 from the atmosphere. In addition to the uncertainties associated with ocean acidification, climate models suggest that increased warming should increase ocean thermal stratification. In other words, the ocean should become more stable, and downwelling should slow, which in turn should reduce the ocean's ability to remove CO_2 by means of the solubility pump. However, the general expectation is that warming will cause the biological pumps to become more active, thus offsetting the reduced activity of the solubility pump.

Although uncertainty persists in how the global carbon cycle will operate in a warmer, more CO_2-rich world, the natural processes of that cycle are clearly incapable of absorbing all 36 gigatons of anthropogenic CO_2 now being injected into the atmosphere every year.

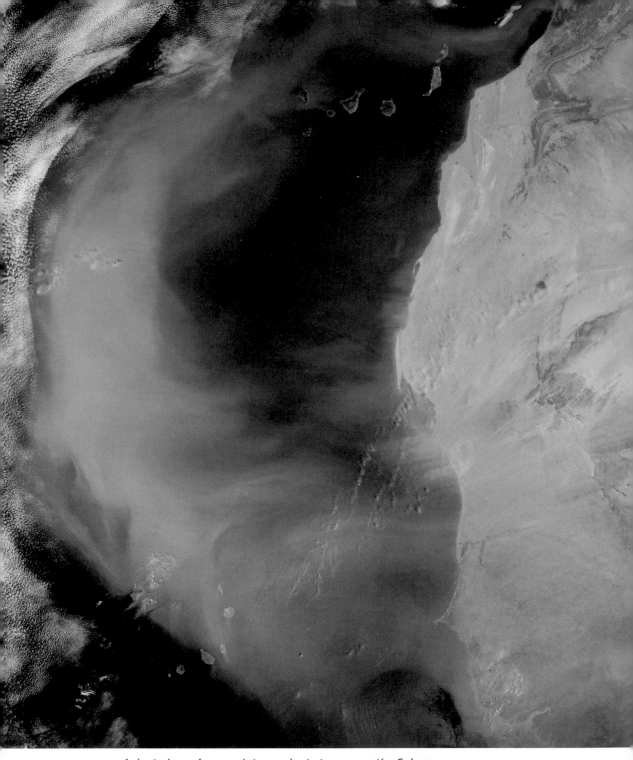

A dust plume from an intense dust storm over the Sahara

Dust shades Earth's surface and keeps it cooler than it would otherwise be. Suspended in the atmosphere, such particles are among the factors that change the balance between Earth's incoming and outgoing radiation. Surface temperature depends on the energy balance; quantifying that balance is one of the endeavors of climate science. This image, taken on March 2, 2003, extends from the coast of Senegal and the Cape Verde Islands in the south to the Canary Islands and coast of Morocco in the north. (Jacques Descloitres, MODIS Rapid Response Team, NASA Goddard Space Flight Center, http://visibleearth.nasa.gov/view_rec.php?id=5078)

5 A SCIENTIFIC FRAMEWORK FOR THINKING ABOUT CLIMATE CHANGE

ENERGY. THAT IS WHAT drives climate, and the balance between the energy Earth receives from the Sun and that which it radiates out to space will ultimately determine our climate future. To understand climate change, we need to quantify the amounts of energy that flow in and out of the Earth system, identify the factors that are causing the balance between incoming and outgoing energy to shift, and determine which of those factors are important and which are not.

The present balance between incoming and outgoing energy is such that the mean global temperature at Earth's surfaces is about 15°C (59°F). If for some reason Earth were to begin trapping more incoming energy than it radiates, the amount of outgoing energy would increase to bring the amounts of incoming and outgoing energy back into balance, and global temperature would be higher. The opposite is also true: were Earth to begin losing more energy than it receives, the climate would cool until energy loss and gain return to a balance.

This chapters explains the scientific framework needed to think systematically about climate change. We begin by exploring the balance between incoming energy and outgoing energy and continue by examining the mechanisms that influence how the balance may change. This description leads us to an explanation of the so-called greenhouse effect, the phenomenon by which greenhouse gases—principally water vapor and carbon dioxide (CO_2), but also several other constituents of the atmosphere that we learned about in chapter 2—trap heat near Earth's surface. It also leads us to explore other factors important in determining energy balance, including aerosols suspended in the atmosphere, land cover, volcanic eruptions, and changes in solar irradiance.

We then examine some of the important ways these factors may interact with one another. The factors that change energy balance tend to shift the entire climate system. As these changes permeate the system, they affect yet other factors that influence energy balance, commonly in complex and unpredictable ways.

Energy Balance

The Sun emits radiation over a range, or spectrum, of energies, represented in part by the rainbow of colors that make up sunlight. Radiation exists in the form of waves, and radiation wavelength is the inverse of energy: the higher the energy, the lower the wavelength. Therefore, we can think of spectra in terms of energy or wavelength.[1]

Solar radiation occupies but a small portion of the electromagnetic spectrum, referring to the whole range of wavelengths at which radiation can exist. On the high-energy ("short"-wavelength) side of the solar spectrum is ultraviolet (UV) radiation; on the low-energy ("long"-wavelength) side is infrared (IR) radiation. Some of the energy radiating from Earth is in the longer-wavelength/lower-energy part of the IR spectrum and is known as *thermal IR radiation*.

INCOMING SOLAR ENERGY

The temperature of the Sun's surface—about 6,000°C (10,800°F)—determines both the spectral character and the amount of energy received by Earth. The amount, known as the *solar constant*, is 1,367 watts per square meter. Averaged over the entire surface of the upper atmosphere, this amount is equivalent to 342 watts per square meter.[2] (A *watt* is a unit of power equal to 1 joule of energy per second, so radiation intensity is the rate of energy flow per square meter.)

As we know from common experience, however, the amount of energy received from the Sun at any one location on Earth depends on how high the Sun is in the sky (figure 5.1). For this reason, the amount of radiation received per year is greater at the equator than at the poles (figure 5.2). The excess heat at the equator drives winds and ocean currents that transport the heat poleward.[3] In addition, because Earth's equator is tilted at 23.45 degrees relative to the orbital plane, *insolation* (the amount of radiation received per unit time, typically given in units of watts per square meter) at high latitudes changes dramatically with the seasons: insolation is high during the long days of summer, but low to nonexistent during winter. The high summer insolation is one of several factors that make polar regions particularly sensitive to global warming (chapter 8).

ALBEDO

The outgoing energy has two components. The first is the solar energy reflected back to space. Of the 342 watts per square meter received from the Sun, about 30 percent is reflected back to space by Earth's surface, clouds, and the atmosphere

FIGURE 5.1

The spring Sun in the Arctic

Even in the summer, the Sun is never far off the horizon in the Arctic. Yet summer insolation is high because the Arctic is bathed in sunlight for all or most of the day. This image was taken in 1949 on Alaska's North Slope. (Photograph by Rear Admiral H. D. Nygren, National Oceanic and Atmospheric Administration, http://www.photolib.noaa.gov/, photograph corp1011)

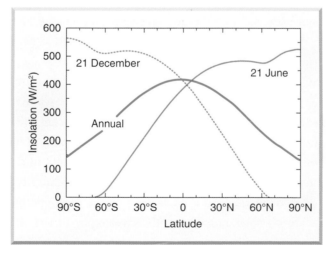

FIGURE 5.2

Variations in total annual and solstice insolations with latitude

The excess heat received at the equator during the course of the year is transported poleward by wind and ocean currents. The high summer insolation experienced by polar regions is one reason that they are so sensitive to global warming. (After Hartmann 1994, with permission)

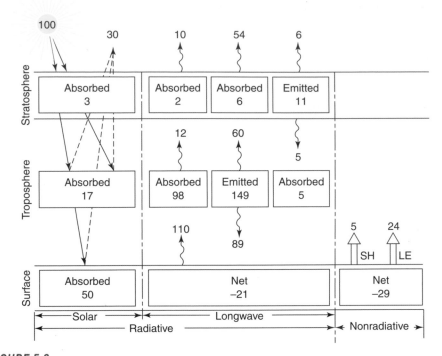

FIGURE 5.3

The flow of energy from the Sun to Earth and between the surface and the atmosphere

Of the total solar radiation that reaches Earth, about 50 percent is absorbed by the surface, 20 percent is absorbed by the troposphere and stratosphere, and 30 percent is reflected back to space. The surface emits infrared (IR) radiation, most of which is absorbed by greenhouse gases in the troposphere. The troposphere reemits IR radiation back to the surface and to the overlying stratosphere, where some is absorbed by ozone, one of the greenhouse gases. Latent heat (LE) and sensible heat (SH) also flow from the surface to the atmosphere. The numbers are the percentages of the received solar radiation of 342 watts per square meter. (After Hartmann 1994, with permission)

(figure 5.3). The fraction of total incident radiation reflected back to space is known as *albedo*. Albedo is highly variable from one place to another, however, because it depends on extents of snow, ice, and cloud cover; proportion of land versus sea; land-surface type; abundance of vegetation; solar zenith angle (the angle between the vertical, or *zenith*, and the line of sight to the Sun); and other factors (table 5.1).

Total albedo also depends on the wavelength of the solar energy received. For example, plants absorb solar radiation with wavelengths of less than 0.7 micron (1 micron is 1/1,000 of a millimeter, or 0.0394 inch), but reflect longer wavelengths (which helps them keep cool). Water is most reflective for blue light, which of course is why Earth appears as the Blue Planet from space. The consequence is that albedo is high near the poles because of the presence of snow, ice, and clouds there and low over tropical regions that are covered mostly by dark ocean and forest. (But

TABLE 5.1 **ALBEDO OF VARIOUS SURFACES**

Surface	Typical Albedo (%)
Deep water, low wind, low latitude	7
Deep water, high wind, high latitude	12
Moist dark soil	10
Moist gray soil	15
Dry soil, desert	30
Wet sand	25
Dry light sand	35
Asphalt pavement	7
Concrete pavement	20
Short green vegetation	17
Dry vegetation	25
Coniferous forest	12
Deciduous forest	17
Forest with snow cover	25
Sea ice, no snow cover	30
Old, melting snow	50
Dry, cold snow	70
Fresh, dry snow	80

Source: D. L. Hartmann, *Global Physical Climatology* (San Diego: Academic Press, 1994).

albedo in the tropics may be locally high where there are abundant clouds or reflective surfaces such as deserts.) Mainly for this reason, there is a net flow of energy from Earth to space poleward of about 40°N and 40°S latitudes and a net flow in the opposite direction on the equator sides of these latitudes.

THERMAL RADIATION FROM EARTH'S SURFACE

The second component of outgoing radiation is, as noted, long-wavelength thermal IR radiation, which is basically radiant heat. It requires a brief explanation. For practical purposes, all matter emits IR radiation, which is also known as *black body radiation* because the total intensity (that is, amount) of emitted radiation depends *only* on the temperature of the surface of an object, not on its composition or anything else.[4] So the more solar energy absorbed by a surface, the warmer that surface becomes and the more IR radiation it emits.

The consequence of this phenomenon is that IR radiation flows back and forth between Earth's surface and the atmosphere (see figure 5.3). In particular, the warm surface radiates on average 390 watts per square meter to the atmosphere.[5] Some of that energy escapes to space, but the atmosphere absorbs most of it and by that mechanism is kept warm. Most of the energy absorbed by the atmosphere is radiated back to Earth's surface, which leads to the surprising fact that the surface radiates more energy than it receives from the Sun. In other words, the troposphere acts as a huge blanket, trapping heat near Earth's surface. The main stuff of the blanket is greenhouse gases.

Two other mechanisms transfer heat from Earth's surface to the atmosphere: the movement of *sensible heat* and *latent heat* (see figure 5.3). Imagine the evaporation of water from, say, a lake. Evaporation requires an input of heat, formally known as the *latent heat of vaporization*. Now imagine that the water vapor rises and condenses to form water droplets somewhere high in the atmosphere. Here it gives up an equal amount of heat.[6] Thus evaporation at the surface and condensation higher up result in the transfer of heat from the surface to the atmosphere and represent the movement of latent heat. Sensible heat, which is far less important, is the transfer of heat due simply to the difference in temperature between the surface and atmosphere.

Radiative Forcing

The notion of energy balance is embodied in a particularly useful concept known as *radiative forcing*: the change, relative to the year 1750, in incoming energy minus outgoing energy in response to a factor that changes energy balance.[7] (The year 1750 is commonly taken as the pre–Industrial Revolution benchmark for investigating changes in the climate system since then.) Thus the statement that the radiative forcing of the greenhouse gas CO_2 is 1.66 watts per square meter means that the CO_2 in the atmosphere has reduced the outgoing radiation by that amount compared with what it was in 1750 (figure 5.4).

The utility of the concept is that it allows us to quantify and assess the relative importance of the various factors that shift the energy balance. It does not, however, tell us anything about how the climate will actually change as a consequence. There are several reasons why this is so, and we get to them at the end of the chapter. First, let us discuss the factors themselves, the *forcing factors*, meaning the influences *external* to, or not part of, the climate system that can cause climate to change. Buildup of greenhouse gases, injection of aerosols into the atmosphere by volcanic eruptions and human activity, changes in land cover that modify surface albedo or affect energy transfer between the surface and atmosphere, changes in solar irradiance—all are examples of external forcing factors.

This description distinguishes between purely natural forcing factors (volcanic eruptions, changes in solar irradiance) and those forcing factors that result from

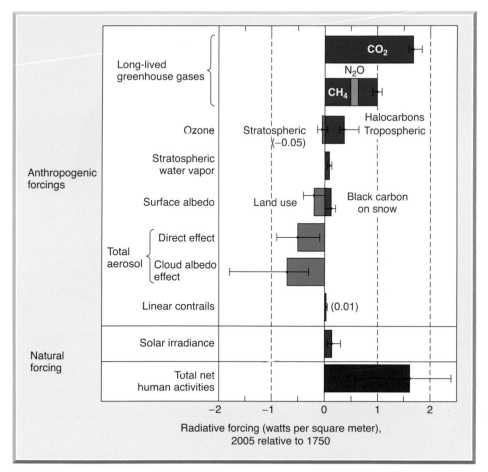

FIGURE 5.4

Changes in global mean radiative forcings, 1750–2005

The long-lived greenhouse gases are the most important of the forcing factors. Aerosols have an opposite, cooling effect, but their influence is highly uncertain. The bar associated with each forcing represents the range of uncertainty. (After Forster et al. 2007:FAQ fig. 2)

human activity (buildup of greenhouse gases, aerosol loading of the atmosphere, changes in land cover) (see figure 5.4). Although changes in land cover can also come about naturally, over the past century they have been almost entirely the result of clearing for agriculture, urban development, and the like.

The Greenhouse Effect

As can be seen in figure 5.4, greenhouse gases are the most important forcing factor. The "greenhouse effect"[8] refers to the absorption of IR radiation by gases in the atmosphere. The main constituents of the atmosphere, nitrogen (N_2) and oxygen (O_2),

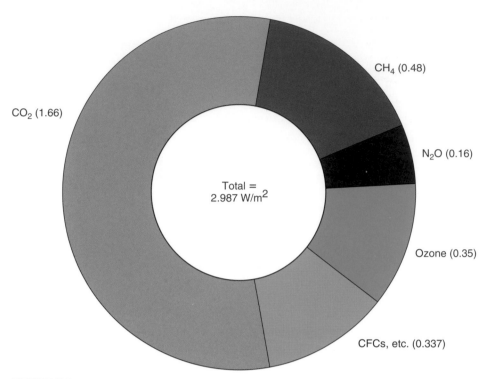

FIGURE 5.5

Contributions of different greenhouse gases to radiative forcing

The numbers are in watts per square meter. CH_4 = methane; N_2O = nitrous oxide; CFCs = chlorofluorocarbons. (Data from Forster et al. 2007)

are transparent to incoming solar and outgoing IR radiation and thus do not contribute at all to greenhouse warming. Rather, the trace gases are the culprits. The infamous CO_2 is an important greenhouse gas, but water vapor is far more so. Water vapor does not appear in figure 5.4 or figure 5.5, however, because its abundance in the atmosphere depends on temperature. In other words, water vapor is not an external forcing factor, but a "feedback" in the climate system. We examine water vapor later along with other feedbacks.

CARBON DIOXIDE AS A GREENHOUSE GAS

The relative importance of the different greenhouse gases (see figure 5.5) depends not only on their abundances, but also on how much they absorb specific wavelengths of energy and on the locations of these spectral absorption "bands" relative to the spectral distribution of incident energy (figure 5.6). For example, CO_2 strongly absorbs energy having a wavelength of about 15 microns (thousandths of a millimeter), which is particularly important because this wavelength coincides with the maximum intensity of IR radiation.[9]

The possibility that CO_2 added to the atmosphere by human activity may cause global warming was first recognized more than 100 years ago. The Swedish chemist Svante Arrhenius (1859–1927), famous for work that serves as the basis for modern physical chemistry, predicted that if atmospheric CO_2 doubled, Earth would become several degrees warmer. Over the past 250 years, atmospheric CO_2 has increased nearly 40 percent, from 275 to about 385 parts per million (ppm) by volume (0.0275 and 0.0385 percent) in 2008. Today, most of this excess CO_2 comes from

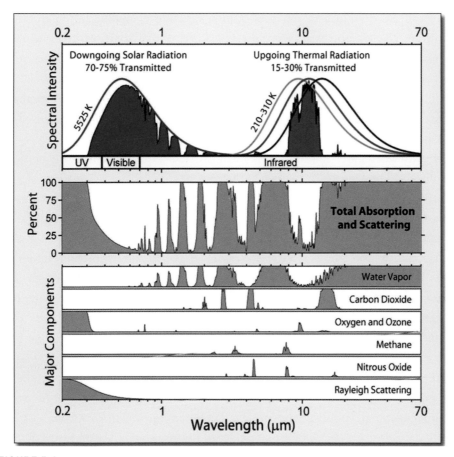

FIGURE 5.6

Spectral characteristics of the Sun and Earth's atmosphere

(*Top*) The normalized radiation of the Sun (*red*) and the upgoing thermal (IR) radiation (*blue*) transmitted by the atmosphere. The solar spectrum corresponds to that produced by a body at a temperature of 5,800°C (10,500°F). Thermal spectral emissions from Earth vary from place to place and with altitude (*black, blue, and pink curves*) but correspond on average to a body at –18°C (0°F). (*Middle*) The total absorption and scattering spectrum of the atmosphere and (*bottom*) the absorption spectra of its various component greenhouse gases. (After Global Warming Art, http://www.globalwarmingart.com/wiki/Image:Atmospheric_Transmission_png, original created by Robert A. Rohde)

the burning of fossil fuels, but cement production, gas flaring, and changes in land use—for example, deforestation—and extensive biomass burning are also important. It is mainly this excess CO_2 that has been implicated in the global warming of the past century.

OTHER GREENHOUSE GASES

Methane (CH_4) is the second most important greenhouse gas (see figure 5.5). The primary natural source of methane is decaying vegetation in soils and wetlands; major sources from human activities include agriculture, landfills, and natural-gas emissions. The 800,000-year record preserved in Antarctic ice shows that the methane content of the atmosphere has varied in a narrow range from about 400 to 700 parts per billion (ppb) by volume (0.00004 to 0.00007 percent) through glacial and interglacial cycles (chapter 6). Now the atmosphere contains 1,775 ppb methane. Curiously, unlike CO_2, the methane content of the atmosphere has remained approximately constant since the late 1990s. No one knows why.

Of somewhat lesser importance is nitrous oxide (N_2O), which forms naturally in soils and the ocean (see figure 5.5). Nitrous oxide is also a by-product of agriculture and fossil-fuel combustion, so its atmospheric concentration has increased from a preindustrial level of about 270 ppb to a present-day concentration of 321 ppb. The manufactured halocarbons, which include a whole host of trace gases such as chlorofluorcarbons (CFCs) (for example, CCl_3F) and hydrochlorofluorocarbons (HCFCs) (for example, $CHClF_2$), are also greenhouse gases. Although their abundances are minuscule, some of these gases are of concern because of their long lifetimes. For example, perfluoroethane (C_2F_6) remains in the atmosphere for many thousands of years, so although it is not a problem now, it will become one if we keep emitting it for hundreds of years.

Tropospheric ozone is also a powerful greenhouse gas (see figure 5.5). Unlike the naturally formed ozone of the stratosphere (the atmosphere layer above the troposphere [chapter 2]), tropospheric ozone is a pollutant formed mainly from reactions involving hydrocarbons, carbon monoxide, and nitrous oxide. As the levels of these pollutants decrease, so presumably will the level of tropospheric ozone.

The greenhouse gases are deservedly notorious in the context of global warming. Perhaps little appreciated, however, is the fact that except for the halocarbons, greenhouse gases have always existed naturally in the atmosphere. More to the point, our existence depends on greenhouse gases because without them the Earth's average surface temperature would be −18°C (0°F) and lifeless. With them, surface temperature has for nearly all of Earth's history remained within a range where liquid water is stable, enabling surface life to exist.

Aerosols

Aerosols are the second most important forcing factor after greenhouse gases (see figure 5.4). The forcing is negative—aerosols tend to cool rather than warm the surface. The magnitude of aerosol forcing is highly uncertain, however, so aerosols deserve some attention.

SOOT TO DUST

Aerosols are particles, at least partly solid and typically less than two microns across, suspended in the air. For several reasons, the effect of aerosols on climate is still poorly known.[10] One reason is that the varieties of aerosols have different properties as far as cooling is concerned (figure 5.7). These varieties include urban haze con-

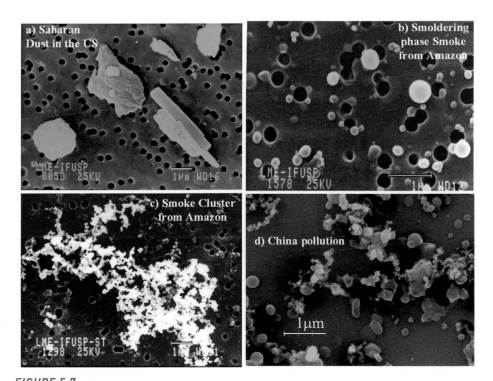

FIGURE 5.7

Scanning electron microscope images of various aerosols

The aerosol particles are resting on filters, and the dark circles are pores in those filters: (*a*) Sahara Desert dust transported by winds and collected in Virginia; (*b*) particles collected in the Amazon during the smoldering (low-temperature) phase after biomass burning and consisting mostly of organic compounds; (*c*) an aggregate smoke particle from biomass burning in the Amazon produced during the flaming (high-temperature) phase of a fire; (*d*) air pollution particles from China. Note the 1-micron scales near the bottom of each photo. (Photographs by Vanderlei Martins, http://climate.gsfc.nasa.gov/, image of the week, May 29, 2005)

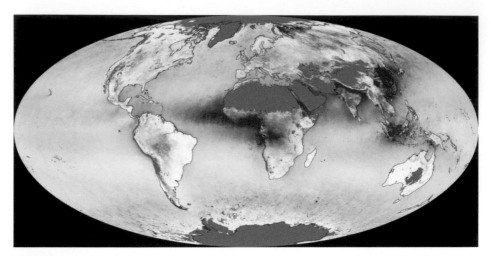

FIGURE 5.8
The global distribution of aerosols, 2006

The depth of color corresponds to the annual mean optical depth at 550 nanometers, a measure of opacity of the atmosphere as measured by satellite. Note the highly uneven global distribution of aerosols, which is one reason their effect on climate is difficult to gauge. The data do not include measurements taken over clouds, bright desert, or snow- or ice-covered surfaces (*gray areas*). (Created by Reto Stöckli, NASA Goddard Space Flight Center, http://climate.gsfc.nasa.gov/viewImage.php?id=199)

sisting of soot and other carbonaceous materials from fossil-fuel burning, sulfate, and other pollutants; sea salt and sulfate originating in the ocean; organic particles of smoke from forest fires; and mineral dusts mainly from desert surfaces, but also from other land surfaces. Most of these aerosols have a cooling effect on the surface by shading it (that is, by increasing the planet's albedo), but at the same time soot and other materials made mostly of black carbon absorb sunlight, thereby warming the atmosphere while cooling Earth's surface.

Another reason for the uncertainty surrounding aerosols is that their concentrations vary enormously from place to place, reflecting the fact that they originate from specific sources (cities, wildfires, deserts, and so on) (figure 5.8). A third reason is that they persist in the atmosphere for short times (typically a week or so), making it rather difficult to determine their integrated, large-scale effect on climate. The 2007 report by the Intergovernmental Panel on Climate Change (IPCC) assembled and attempted to estimate the direct radiative forcings of different aerosols (table 5.2). The large uncertainty in its estimate illustrates how uncertain the effects of aerosols are.

SECONDARY EFFECTS

If the complexities just described are not enough, aerosols also have important secondary effects. The warming of the atmosphere and the cooling of the surface by black carbon reduce the temperature gradient from the surface upward. The reduced temperature gradient, in turn, diminishes evaporation and cloud formation. In contrast, water vapor condenses on aerosols to form clouds, increasing reflectance due to clouds and thereby enhancing surface cooling.[11] The aerosol-laden droplets in clouds are 20 to 30 percent smaller than aerosol-free droplets. This difference has an interesting effect: because smaller droplets tend to stay in the atmosphere longer, less rain falls over areas where the atmosphere is polluted, and more falls over unpolluted areas, such as the ocean.

Two examples illustrate the combined effects of aerosols on cloud formation and weather patterns. In central China, rainfall over hilly areas that are downwind of urban and industrial centers is less when conditions are hazy than when they are not.[12] In addition, Asian monsoons are weakened by "brown clouds," or the pollution from biomass and fossil-fuel burning that plagues the region. To make matters worse, brown clouds also warm the atmosphere and as a consequence may enhance melting of high-altitude Himalayan glaciers. The more rapid melting increases current water supply, but as the glaciers disappear, the water supply over the long term will dwindle.[13]

TABLE 5.2 **MODEL ESTIMATES OF THE DIRECT EFFECTS OF AEROSOLS ON RADIATIVE FORCING**

Aerosol (Dominant Components and Source)	Radiative Forcing (Watts per Square Meter)
Sulfate (sulfuric acid via sulfur dioxide [SO_2], from fossil-fuel burning)	-0.4 ± 0.20
Organic carbon (from fossil-fuel and biomass burning)	-0.05 ± 0.05
Black carbon (from incomplete fossil-fuel combustion)	$+0.20 \pm 0.15$
Smoke (compounds from forest fires)	$+0.03 \pm 0.12$
Nitrates (ammonium, from ammonia and nitrogen oxide [NO_x] emissions)	-0.10 ± 0.10
Anthropogenic dust (agriculture, cement production, and drying)	-0.1 ± 0.20
All aerosols combined, direct effects	-0.50 ± 0.40
Increase in cloud albedo due to aerosols	-0.3 to -1.8

Source: P. Forster, V. Ramaswamy, P. Artaxo, T. Berntsen, R. Betts, D. W. Fahey, J. Haywood, J. Lean, D. C. Lowe, G. Myhre, J. Nganga, R. Prinn, G. Raga, M. Schulz, and R. Van Dorland, "Changes in Atmospheric Constituents and in Radiative Forcing," in *Climate Change 2007: The Physical Science Basis. Contribution of Working Group I to the Fourth Assessment Report of the Intergovernmental Panel on Climate Change*, edited by S. Solomon, D. Qin, M. Manning, Z. Chen, M. Marquis, K. B. Averyt, M. Tignor, and H. L. Miller (Cambridge: Cambridge University Press, 2007), 129–234.

Both satellite and ground-based observations have detected an increase in the amount of solar radiation reaching Earth's surface.[14] This "brightening," which became apparent around 1990, reversed many years of presumed "dimming" and appears to reflect a steady decline in sulfate and black carbon aerosols as pollution-control measures have taken hold.[15] The decline in these pollutants may have the side effect of increasing global warming over what would be the case from greenhouse gases alone, an unfortunate by-product of efforts to curb pollution.[16] The study of aerosols is being advanced by satellite observations, and so far these observations have confirmed aerosols' important role in reducing radiative forcing.[17]

Changes in Land Use

As with aerosols, the effect of human-induced changes in land use on climate is highly uncertain. The changes impinge on climate directly, but they also impinge on it indirectly by affecting practically all components of the environment, such as water, ecosystems, and biodiversity.[18]

For example, cutting down a forest will cause less CO_2 to be removed from the atmosphere if loss of the forest results in a decline in total photosynthetic activity, which is usually the case. In contrast, draining a wetland may diminish methane emissions to the atmosphere by reducing the rate of plant decomposition. Changing land use may also modify albedo, aerosol production, surface hydrology, vegetation transpiration characteristics, and vegetation structure. All these factors influence the amounts of heat and water transferred back and forth between land and atmosphere.

The sources of uncertainties regarding the effects of land-use change on climate arise from incomplete understanding or characterization of these processes, huge regional variations in land-use changes, imprecise characterization of the nature of land cover, and the veracity of the models used to estimate the impacts. To illustrate the importance of regional variations, one study based on model simulations asserted a rise in average temperature of more than 2°C (3.6°F) over the Amazon due to tropical deforestation, but almost no change in temperature averaged over the entire globe because the effects varied so much from one place to another.[19]

The industrial age has seen substantial changes in land use. In 1750, cropland and pastureland occupied 6 to 7 percent of global land surface, but in 1990 they covered 35 to 39 percent.[20] Much of the expansion in farmland up to 1950 was in temperate regions; although the expansion there has since stabilized, it has been replaced by rapid expansion in the tropics. We see this expansion most profoundly in the Amazon basin, where 16 percent of the original forest has been cleared, with about one-half to two-thirds of that cleared land now devoted to agriculture.

The 2007 IPCC report compiled the results of studies that modeled the effects of land-use change on radiative forcing in connection with changing albedo and

CO_2 emissions. In keeping with the large uncertainties of these direct but opposite influences, the radiative forcing from higher CO_2 emissions ranged from +0.20 to +0.57 watt per square meter, whereas that due to changes in albedo ranged from −0.08 to −0.66 watt per square meter. In other words, the higher CO_2 emissions from changing land use have had a slight warming effect, but the resulting changes in albedo have had a slight cooling effect.

Natural Forcings

Up to now, the discussion has focused on forcing factors arising from human activities. There are, in addition, several purely natural forcing factors. Chief among them are volcanic eruptions, which alter climate by injecting aerosols into the atmosphere, and changes in solar irradiance, which come about from the changing activity of the Sun itself.

To avoid confusion, it is worth mentioning that the amount of solar radiation reaching Earth may also change as the result of the combined fluctuations in the ways Earth orbits the Sun. There are the slow oscillations in the degree of Earth's tilt relative to the orbital plane, in its precession, and in the eccentricity of its orbit around the Sun. These fluctuations combine to produce what are known as *Milankovitch cycles*, which largely account for the periods of alternating glacial and interglacial conditions that Earth has experienced at various times. Milankovitch cycles can lead to major shifts in climate, but are not contributing to climate change today. Because they have had such a major effect on past climates, they are examined in that context in chapter 6.

VOLCANIC ERUPTIONS

Volcanoes influence climate when especially large, explosive eruptions inject sulfur dioxide (SO_2) into the stratosphere. Within weeks, the sulfur dioxide reacts with water vapor to form an aerosol composed of particles of sulfuric acid. This stratospheric aerosol circles the globe in two or three weeks (figure 5.9). It absorbs solar radiation from above and IR radiation from below, thereby warming the stratosphere but cooling the underlying atmosphere as well as Earth's surface.[21] The aerosol also destroys stratospheric ozone (chapter 2) and may influence weather patterns.[22] Because stratospheric aerosol takes two or so years to settle out, the impact on climate lasts well beyond the proximate time of the eruption. In contrast, sulfuric acid formed in the troposphere washes out with rain in weeks. Consequently, non-explosive eruptions, such as those of Hawaii's Kilauea volcano, do not influence climate unless they are unusually voluminous, in which case the large thermal plume can transport sulfur dioxide and other gases into the stratosphere.

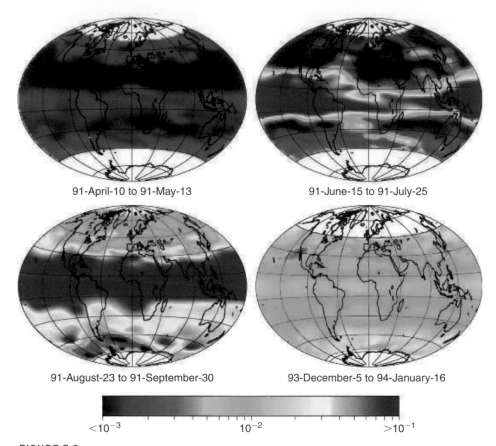

91-April-10 to 91-May-13   91-June-15 to 91-July-25

91-August-23 to 91-September-30   93-December-5 to 94-January-16

$<10^{-3}$ 10^{-2} $>10^{-1}$

FIGURE 5.9

Stratospheric aerosol before and on several occasions after the eruption of Mount Pinatubo, June 15, 1991

The colors illustrate optical depth (opacity), which is related to aerosol abundance (*reds are high and blues low*). Note in the upper-right image that the aerosol circled the globe in little more than a month. The two lower images, representing later dates, show how the aerosol eventually spread throughout the entire stratosphere. (NASA Langley Research Center, Aerosol Research Branch, http://visibleearth.nasa.gov/view_rec.php?id=1803)

The first major climate-influencing eruption to be carefully observed with modern instruments occurred on June 15, 1991, when Mount Pinatubo in the Philippines erupted in a giant explosion. It shot some 15 million metric tons of sulfur dioxide into the stratosphere. The calculated mean global radiative forcing was about –3 watts per square meter for several months after the eruption, with a corresponding mean global surface cooling of 0.3°C (0.5°F).[23] This forcing is large compared with forcings from human activities (see figure 5.4), but the eruption perturbed climate for only two to three years.

Pinatubo was significant, but much larger eruptions have occurred with correspondingly larger consequences. One was the dramatic eruption in 1815 of Mount

Tambora in Indonesia. The Tambora eruption appears to have been the largest of the past 20,000 years, ejecting 125 cubic kilometers (30 cubic miles) of pumice and 175 cubic kilometers (40 cubic miles) of ash and killing more than 90,000 people.[24] The year following the eruption, 1816, became known as the "year without a summer" or more wryly as "eighteen hundred and froze to death."

The precise effect of the Tambora eruption on climate is not easily gauged. The decades from about 1790 to 1830 were unusually cold, and within these cold decades the years 1812 to 1818 were among the coldest. Other large eruptions during the same period—Soufriere Hills on the Caribbean Island of Montserrat in 1812; Mayon in the Philippines in 1814; Colima in Mexico, also in 1814; and Beerenberg in the North Atlantic in 1818—probably contributed to the cold conditions as well.[25] Sparse climate records also add to the uncertainty.

The Tambora event is instructive in that it apparently led to a wildly variable climate response.[26] The summer of 1816 in northeastern North America was very cold (and dry). Hard frosts occurred in June, July, and August in New England, contributing to widespread crop failure and hardship. Elsewhere on the continent, summer conditions were not so unusual. In western North America and around the North Pacific, the weather was rather typical. But on the Colorado Plateau, tree rings indicate high growth rates, implying cooler and wetter-than-normal conditions, and in the central part of the United States, from the Gulf Coast to as far north as Illinois, it was unusually warm.

As for Europe, the summer of 1816 was the coldest in Great Britain and the second coldest (after 1814) in central Europe since 1750, and western Europe was exceptionally cold and wet; but in Scandinavia, conditions were near normal. On the Indian subcontinent, the summer monsoons of 1815 to 1818 were particularly weak, locally causing famine. Japan and most of China experienced extremes in weather during this period, producing famine and other hardships. The limited data from South America and southern Africa suggest that 1816 and 1817 were only slightly cooler than normal. Perhaps the consequences of a similar eruption today would not be so dire because of our ability to distribute food and other essential needs effectively around the globe.

CHANGES IN SOLAR IRRADIANCE

The Sun goes through a well defined, 11-year cycle of changes in the numbers of sunspots, or dark regions where irradiance is low, and faculae, or bright regions where irradiance is high (figure 5.10). This activity results in a total irradiance change of about 0.15 percent between cycle maxima and cycle minima. Although this change is too small and the cycles are too short to have any appreciable influence on Earth's

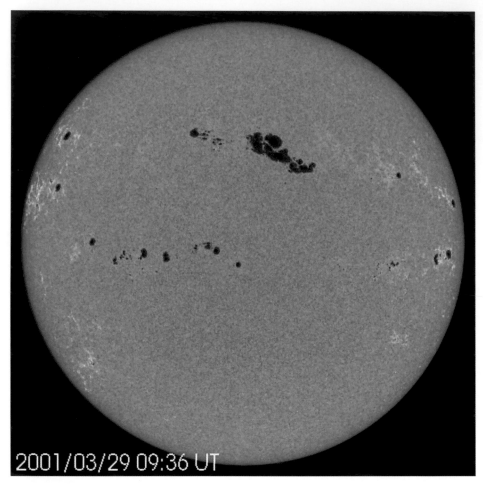

2001/03/29 09:36 UT

FIGURE 5.10

Sunspots observed during a cycle of high sunspot activity, March 30, 2001

The sunspots are evident because they are slightly cooler than the rest of the Sun's surface, which is more than 6,000°C (10,800°F). They form when intense magnetic fields protrude from the interior. This particular sunspot group is unusually large, with an area more than 13 times that of Earth's surface. (SOHO/MDI Consortium, http://soho.nascom.nasa.gov/gallery/images/bigspotfd.html)

climate, there is the question of whether solar irradiance may exhibit larger variations over longer time spans that may have climatic effects.

Solar irradiance has been measured continuously by satellite only since 1978, during which time the Sun has displayed no statistically significant, long-term change.[27] There is some evidence, however, that longer-term variations may have triggered periods of unusual cold or warmth. One of these variations is the Dalton Minimum, a time of relatively low solar irradiance that approximately corresponded to the cold period from 1790 to 1830. The cold is generally attributed to the low solar output, although, as we have seen, the several large volcanic eruptions of the period also contributed to the cold. The Medieval Warm Period, which lasted from approximately 850 to 1200, and the seventeenth-century Maunder Minimum, another period of unusual cold, may also have been due to variations in solar irradiance.[28]

In evaluating the current state of knowledge, the 2007 IPCC report estimated that solar irradiance has increased an amount corresponding to a radiative forcing of + 0.12 watt per square meter since 1750 (see figure 5.4), considerably less than that from anthropogenic greenhouse gases.[29] Although this estimate has been debated, a recent analysis that separated and evaluated the anthropogenic, volcanic, ENSO, and solar irradiance impacts on climate concluded that changes in solar output account for only about 10 percent of the warming since 1905 and essentially none of the rapid warming of the late twentieth century.[30]

Important Features of the Climate System

As noted earlier, although the concept of radiative forcing is useful for quantifying the relative importance of the various forcing factors, the magnitude of a particular forcing does not tell us much about how the climate will actually respond to the forcing. In this section, we explore the various features of the climate system that complicate the task of understanding the climate's response to forcing factors.

FEEDBACKS

Feedbacks are one such complex feature because they affect the climate's sensitivity to forcings. Sensitivity in this context is a measure of how the climate system responds to a forcing change. The feedbacks turn out to be quite important and represent major uncertainties in understanding and modeling how climate will change in response to both natural and anthropogenic forcings (chapter 9).

A feedback is the result of a process that in turn influences the process itself.[31] The feedback might cause an effect of the process to be amplified (positive feedback) or dampened (negative feedback). A good example is the albedo feedback in the Arctic. As warming from the buildup of atmospheric CO_2 causes sea ice to melt, a higher proportion of solar energy gets absorbed by the ocean because there is less ice to reflect that energy, which enhances the warming, which causes the ice to melt even faster. The change in albedo amplifies the warming, so it is a "positive" feedback. The albedo feedback is one of the reasons why the Arctic is particularly sensitive to global warming.

Two additional and related positive feedbacks merit comment: atmospheric water vapor and clouds. Both are expected to increase as climate warms. The water vapor feedback comes about because of the combined facts that (1) the warmer it gets, the more water vapor can exist in the atmosphere, and (2) water is a greenhouse gas. In fact, water accounts for 50 to 60 percent of the greenhouse effect, in contrast to CO_2, which accounts for 15 to 20 percent. The water vapor feedback is most sensitive to changes in the water vapor content of the upper troposphere.[32] Climate models (chapter 9) generally indicate that the water vapor feedback amplifies warming about 50 percent above what it would otherwise be, making it by far the most important feedback.[33] The importance of the water vapor feedback in the real world remains uncertain, however.

If the water vapor feedback is most important, the cloud feedback is most uncertain because clouds exert two competing influences: they reflect solar radiation back to space, but by the greenhouse effect they trap IR radiation emitted from the surface below. Clouds are made of droplets of liquid water or particles of ice that form when the atmosphere becomes saturated in water. How clouds interact with solar and terrestrial radiation depends on the total mass of water (particularly cloud thickness), the size and shape of the droplets or particles, and the way they are distributed. In particular, albedo increases as the liquid-water content of clouds increases and as the droplet size decreases. Albedo also increases with increasing solar zenith angle. The dependencies are not linear, however. Low clouds typically reduce surface temperature better than high clouds because their albedos are usually higher, and they do not absorb long-wavelength radiation as effectively because they are warmer. In consequence, high clouds tend to be net warmers, whereas low clouds are net coolers.

TIPPING POINTS

The geological record of climate is replete with examples of abrupt and dramatic shifts in climate. A spectacular and well-documented example is the Younger Dryas (discussed more fully in chapter 6), which was an interval of extreme cold in the North Atlantic region (Greenland and northern and western Europe). It began abruptly about 12,900 years ago and ended just as abruptly 1,300 years later. In Greenland, the end of the Younger Dryas was marked by an increase in temperature of 7°C (13°F) and a doubling of the rate of snow accumulation in just a few decades and perhaps only a few years. The event seems to have been caused by a huge and sudden influx of freshwater into the North Atlantic, altering the ocean circulation pattern that would normally have brought heat to the region.

The abrupt beginning and end of the Younger Dryas did not come about because of any sudden changes in the magnitudes of the forcing factors, such as greenhouse gases or volcanic activity. Rather, the forcings provoked a sudden change in the internal dynamic of the climate system. Such a change is known as a *tipping point*: "the climate system is forced to cross some threshold, triggering a transition to a new state at a rate determined by the climate system itself and faster than the cause."[34] The abrupt changes may differ in their magnitudes and in their geographical reach, being global or even just regional. Although the North Atlantic region experienced intense cold during the Younger Dryas, for example, there is no record at all of this event in parts of the Southern Hemisphere.

The existence of tipping points and our inability to predict them represent significant elements in the risk associated with current climate change. We are imposing a change on the climate system, and we do not know how the system will react. At the same time, we do know that tipping points have been reached in the past and that they have caused dramatic climate shifts. To borrow an image conjured up by Wallace Broecker, a preeminent oceanographer at Columbia University, our modification of climate is like poking a fierce sleeping dog with a stick.

INERTIA

This chapter began with the notion of energy balance, so it shall end with a related characteristic of the climate system that has significant bearing on policy and our future. That characteristic is *inertia*, which means that it takes time for the climate system to reach a new balance in response to a factor that changes the radiative forcing.[35] Exactly how much time depends on how sensitive the climate system is to a particular forcing. A general estimate based on climate models (chapter 9) is that the global average temperature changes $0.75 \pm 0.25°C$ ($1.35 \pm 0.45°F$) for every 1 watt per square meter change in radiative forcing and that it takes 25 to 50 years to reach 60 percent of the equilibrium (final) temperature.[36]

Assuming that our knowledge of the magnitude of the forcing factors is approximately correct, in 2003 Earth trapped 0.85 ± 0.15 watt per square meter more than it radiated.[37] This amount is equivalent to a "committed warming" of $0.6°C$ ($1.1°F$). In other words, even if atmospheric CO_2 content had been held at 2003 levels and other forcing factors had remained constant, Earth would have unavoidably warmed by $0.6°C$ over the course of several decades. Of course, the CO_2 level is now higher and rising. With no practical hope of stabilizing the CO_2 level anytime soon, the committed warming is now greater, and as time passes, our ability to ameliorate the problem deteriorates. Our collective decisions on curbing the growth of carbon emissions need to take these sobering realities into account.

Perspective

This chapter set out the scientific framework for understanding how climate changes. The fundamental idea is that we can think of climate change in terms of radiation, or energy, balance. We account for the incoming and outgoing energy by determining and quantifying the factors external to the climate system that affect this balance—in other words, that exert a radiative forcing. The most important forcing factor, as we have seen, is atmospheric greenhouse gases, which tend to bring the amounts of incoming and outgoing energy back into balance by raising the temperature of the lower atmosphere.

To be sure, the magnitudes of the forcing factors, in particular the effects of atmospheric aerosols and changes in land use, are uncertain. Nevertheless, if climate change amounted simply to determining the forcing factors and their magnitudes, we would have a much more precise understanding of how climate will change as we continue to inject CO_2 into the atmosphere.

The climate system does not respond to forcing factors in any simple way, however. A variety of feedbacks determine the climate system's sensitivity and complicate its response to them. At the heart of these complexities are the dynamic, physical, and chemical interactions among the atmosphere, the ocean, and other parts of the climate system, as described in earlier chapters. How these interactions have played out in the past and how they are playing out in today's rapidly warming world concern us in the next chapter.

Bubbles in ice recovered by drilling from deep within the Greenland Ice Sheet

The bubbles are samples of past atmosphere. They—along with tree rings, ocean and lake sediments, cave deposits, corals, and other geologic objects—provide a record of past climate. Understanding how and why the climate has changed in the past provides important insights into how the present climate system works and how the climate will change in the future. (Photograph by D. Dahl-Jensen, University of Copenhagen, with permission)

6 LEARNING FROM CLIMATES PAST

HOW MIGHT WE SEEK to understand climate? One way, of course, is to observe the system in its current state: we determine radiative balance, investigate how forcing factors and feedbacks operate, study the various components of the carbon cycle, and so on. From these and other activities, we gain an understanding of the present climate system's character and dynamics. This approach, however, offers only a snapshot of a system that operates on a continuum of timescales that extends far beyond the decades or even centuries during which we have observed climate.

To understand today's climate, we must also look to other times. What were the causes of the large climate swings that occurred many thousands and even millions of years ago? Are those causes relevant to climate change today? How did the swings play out on a global scale, and do the long-term dynamics hold any lessons for present-day dynamics? Can we find events in the geological record of the distant past that in some ways mimic our present predicament and provide insight into what the future may hold? Through an investigation of such questions, it becomes apparent that the study of past climate, or *paleoclimatology*, is a fertile field for our growing knowledge of present climate.

A Lesson from the Distant Past: The Paleocene–Eocene Thermal Maximum

Let us begin with an extraordinary event that occurred during the Paleocene epoch, 55 million years ago, and that in some ways is analogous to present-day climate change. During the late Paleocene, climate had been slowly warming, but then a sudden, enormous mass of carbon flooded the ocean and atmosphere. In perhaps less than 1,000 years, an estimated 1,500 to 4,500 gigatons (billion metric tons) of carbon (equivalent to 5,500 to 16,500 gigatons of carbon dioxide [CO_2]) entered the climate system,[1] about the same amount of carbon projected to enter it as a consequence of human activities during the twenty-first century if fossil-fuel use continues to grow at the current pace. As atmospheric CO_2 content increased, the average global surface temperature rose 5 to 9°C (9 to 16°F),[2] and the ocean acidified.

These changes had a dramatic effect on terrestrial and marine ecosystems. For example, from the fossil record we know that the body sizes of mammals evolved through a transient decrease, and the ranges of midlatitude flora moved northward many hundreds of kilometers.[3] In the ocean, 35 to 50 percent of the benthic foraminifera species disappeared, but planktonic foraminifera diversified.[4] The Paleocene–Eocene Thermal Maximum (PETM), as it has become known, lasted around 120,000 years. It deserves examination because of its possible bearing on climate change in the coming centuries.

The geological evidence for PETM comes from deep-sea sediment cores and the terrestrial sediment record.[5] What one typically sees in the deep-sea cores is that Paleocene carbonate-rich mud ("carbonate ooze") is overlain at a sharp boundary by red clay from the Eocene, and then the red clay gradually grades back to carbonate ooze upward through the younger layers. The boundary between the carbonate ooze and the red clay is accompanied by a sudden decrease in the ratio of the isotopes carbon-13 to carbon-12 ($^{13}C/^{12}C$), which is indicative of the addition of methane (CH_4) or other forms of organic carbon rich in carbon-12 to the ocean–atmosphere system (figure 6.1).

The transition from carbonate ooze to red clay indicates that the ocean acidified, the process discussed in chapter 4: ocean carbonic acid (H_2CO_3) increased, and carbonate ion concentration (CO_3^{2-}) decreased, which is the expected consequence of rising atmospheric CO_2 content. The change in ocean composition caused a rise in the carbonate compensation depth (CCD) (chapter 4). Above this depth, calcareous sediments such as chalk and limestone form as the shells of planktonic foraminifera accumulate; below the CCD, the sediment consists of clay (for example, windblown dust) and the shells of silica plankton (for example, diatoms, radiolarians) because carbonate shells dissolve at that depth. The marine sediment records suggest that the CCD became shallower by more than 2,000 meters (6,600 feet) within a few thousand years.[6]

Where did all the carbon injected into the ocean and atmosphere come from? That remains a mystery. One idea is that it originated from the breakdown of methane hydrates that destabilized as the ocean warmed.[7] Methane hydrates are minerals ("ices"), the structures of which are characterized by cages of water molecules surrounding methane molecules. They commonly form in sediment where methane is produced by bacterial breakdown of organic material. The stability of methane hydrates is sensitive to temperature and pressure. For example, at 500 meters

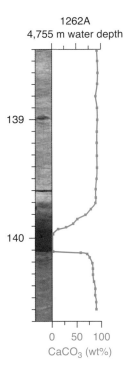

1262A
4,755 m water depth

139 —

140 —

0   50   100
CaCO$_3$ (wt%)

FIGURE 6.1

The record of the Paleocene–Eocene Thermal Maximum in a deep-sea sediment core

The sediment core displays an abrupt change in which Paleocene carbonate-rich mud is overlain by Eocene red clay, and then the red clay gradually grades back to carbonate mud upward through the stratigraphic section. The reason for the sudden change was that an increase in atmospheric CO$_2$ caused the oceans to become more acidic and the carbonate compensation depth to rise to shallower levels. The boundary is also accompanied by a sudden decrease in the ratio of carbon-13 to carbon-12, indicative of the addition of organic carbon to the ocean–atmosphere system. (After Zachos et al. 2005, with permission)

(1,640 feet) water depth, methane hydrates are stable to about 6°C (43°F), and at 2,000 meters (6,600 feet) they are stable to about 18°C (64°F).[8] In the present-day ocean, they can exist in the sediment at ocean depths greater than about 250 meters (820 feet).

Regardless of the source of the carbon, the sudden release seems to have been preceded by warming,[9] which suggests the worrying possibility that a warming climate may reach a threshold that triggers a much more dramatic shift. Was the PETM a harbinger for climate and ecosystem changes over the next several centuries? The answer to this question perhaps hinges on what caused the carbon release. There is currently no consensus on what that cause might have been.

Throughout much of the PETM, the climate remained warm.[10] The return to cooler conditions occurred over a period of about 40,000 years. During that time, ocean carbonate production and carbonate burial rate increased, and temperature decreased. Both changes were presumably brought about by an increase in the rate of silicate weathering, which draws down atmospheric CO_2 and increases the bicarbonate ion content of the ocean (chapter 4). In other words, the time it took the climate system to recover may indicate the timescale on which the global thermostat represented by the long-term carbon cycle takes over from short-term perturbations.

The Ice Age

Earth history has been characterized by long periods of stable climate punctuated by periods of fluctuating climate lasting millions to tens of millions of years. One such period appears to be the present "ice age," which began about 2.75 million years ago. The term *ice age* is somewhat misleading because it does not mean that the planet has been completely and perennially frozen. Rather, during this time the climate has fluctuated rapidly (in geological terms) between glacial intervals lasting tens of thousands of years and shorter interglacial intervals. During the glacial times, mid- and high-latitude land-surface temperatures were 5 to 15°C (40 to 60°F) lower than at present; temperatures during the interglacial times were in general not too different from those of today. The most recent glaciation reached its height about 21,000 years ago and then dissipated over the span of about 10,000 years into today's climate. The interglacial intervals amounted to about only 10 percent of the past million years, indicating that today's conditions have been relatively short-lived and are certainly not typical.

The world during the glacial periods was a much different place than it is today. When the most recent glaciation was at its most intense, sea level was about 120 meters (400 feet) lower than it is now because so much water, instead of being in the ocean, was tied up as ice on the continents.[11] Indeed, ice blanketed 30 percent of the land area. Ice covered Scandinavia, the North Sea, most of the British Isles, and parts of northern Europe. In North America, ice extended in an arc from Long Island in the east to central Illinois, across the Great Plains and the northern Rocky Mountains, and to Puget Sound in the west. These ice sheets shaped the landscape we see today in large parts of these northern regions.[12]

How did we come to recognize that our planet has experienced glacial periods? Somewhere in the valleys of the Swiss Alps nearly two centuries ago, naturalists and peasants alike were struck by the presence of large boulders in odd places, such as the flat, low expanses of river valleys.[13] They began to wonder how these boulders could have been transported to the middle of flat river plains, where there is nothing else but river sediment. Certainly they were not transported by the river itself, which if powerful enough to carry boulders would certainly have washed everything else away. It seemed possible that the boulders might have been carried to

FIGURE 6.2
Louis Agassiz
Louis Agassiz (1807–1873) was a pioneer in the study of ice ages. (Courtesy of South Caroliniana Library, University of South Carolina, Columbia)

their present locations by glaciers, remnants of which were at the time (and are even now) still present at the head of some valleys.

As the story goes, a mountaineer and hunter named Jean-Pierre Perraudin was one of the first advocates of this idea. The evidence caught the attention of Louis Agassiz (1807–1873), then a young professor at Neuchâtel (and already well known for his studies of fossil fishes) (figure 6.2). Agassiz knew that these "glacial erratics" also dotted the plains of northern Europe. In 1837, before the Swiss Society of Natural Sciences, of which he had just been elected president, he presented the seemingly outlandish notion that a great ice sheet had once covered large parts of Earth. Despite the skeptics, Agassiz then published his classic *Études sur les glaciers* (*Studies on Glaciers*, 1840), which gave birth to an entirely new line of research.

In North America, geologists were to recognize and map four great sets of glacial deposits, marking four major glaciations. In many parts of Scandinavia and around the Great Lakes in North America, odd lines of gravel were recognized as raised beaches, relicts from a time when the land had been depressed by the great weight of ice on it and was exposed when the ice melted and allowed it to rise. In fact, parts of Scandinavia are still rising. The maximum uplift in this part of the world, as measured in northern Sweden, is close to 300 meters (980 feet).

Fluctuations in Ancient Climate

How do we know what past climates were like? Climatic fluctuations on timescales of hundreds of thousands of years and longer are recorded primarily in the oxygen isotope record of deep-sea sediments. Because ocean sediments are such important climate records and oxygen isotopes represent one of the principal tools for their study, a brief digression is in order.

MARINE SEDIMENTS AND THEIR OXYGEN ISOTOPES

Every year, 6 billion to 11 billion metric tons of sediment accumulate on the ocean floor. As we learned in previous chapters, the sediment consists of the shells of dead planktonic (near-surface-dwelling) and benthic (deep-water-dwelling) organisms as well as of sand and mud washed off the continents. The marine sediment record is rather complete, although burrowing organisms and ocean currents commonly disturb it. The composition, abundance, and character of the fossils embedded in the sediment are useful for deducing climatic conditions. Eolian (wind-borne) sediment, iceberg-rafted debris, and river sediments also help to characterize climate. Thus an increase in eolian dust particles in sediment generally reflects a decrease in humidity and an increase in atmospheric circulation (strong winds), both indicative of cooling conditions. In fact, the evidence for Pleistocene cycling between glacial and interglacial conditions was first observed in deep-sea sediment cores as alternating carbonate-rich layers, which represent the warm parts of the cycle, and eolian sediment–rich layers.

One of the reasons deep-sea sediments are so important is that they help to relate what was happening in the atmosphere to what was happening in the oceans. For example, the records in sediment cores from the North Atlantic offer a clear comparison to atmospheric conditions recorded in ice cores from Greenland.[14] In sediments, the abundances of certain microorganisms are proxies for sea-surface temperature. The North Atlantic sediments also contain conspicuous layers of detrital (particles accumulated from the overlying water column) carbonate. These layers were deposited during cold periods and mark times of rapid and voluminous discharges of icebergs into the North Atlantic from the North American Ice Sheet.[15] They thus show how cooling can be accompanied by an influx of freshwater (from melting icebergs) into the North Atlantic.

Several key chemical characteristics of marine sediments serve as proxies of past climate. Particularly useful have been the oxygen isotopic compositions of carbonate shells of marine organisms such as foraminifera. Oxygen exists as three stable isotopes. Oxygen-16 is the most abundant isotope (99.8 percent of all oxygen); the other two are oxygen-17 and oxygen-18. The ratio of oxygen-18 to oxygen-16 ($^{18}O/^{16}O$) in shells depends on the $^{18}O/^{16}O$ ratio in the water in which the shells grew and the temperature of that water. The temperature of the deep ocean remains approximately constant at several degrees Celsius, so variations in the $^{18}O/^{16}O$ ratio of benthic (bottom-dwelling) foraminifera reflect the $^{18}O/^{16}O$ ratio of the water, not variations in the temperature at which the shells grew.

The $^{18}O/^{16}O$ ratio of ocean water depends on the proportion of water on the planet that exists as ice because glacial ice is enriched in oxygen-16 relative to seawater. Thus when ice caps are extensive, the ocean's $^{18}O/^{16}O$ ratio is higher than it is during times of relatively little ice, which shows up in the benthic foraminifera record. In other words, the $^{18}O/^{16}O$ ratio of benthic organisms is a measure of global ice volume.

The remains of planktonic organisms (those that live in the open ocean, usually near the surface) tell a different story. The $^{18}O/^{16}O$, magnesium/calcium, and strontium/calcium ratios in the shells of planktonic foraminifera may yield information on sea-surface temperature because these ratios are sensitive to water temperature. The records become available when the foraminifera die and accumulate on the ocean floor. Although there are a few localities where sea-surface temperature tends to be diagnostic of global climate, sea-surface temperature is typically sensitive to local climate, so the compositions of planktonic foraminifera shells provide information on local rather than global conditions.

FIVE MILLION YEARS OF CLIMATE

The record of climate during the Pliocene (5.3 million to 1.8 million years ago) and Pleistocene (1.8 million to 11,600 years ago) epochs is fairly complete. One such record comes from a core drilled through a continuous sequence of sediment depos-

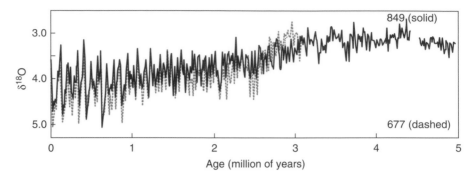

FIGURE 6.3

Variations in the oxygen isotopic ratio with depth of benthic foraminifera in two sediment cores from the equatorial eastern Pacific Ocean

The fluctuations in isotopic ratio reflect global changes in ice volume and thus provide information on climate. This core holds one of the most complete climate records of the past 5 million years. $\delta^{18}O$ is a measure of the ratio of oxygen-18 to oxygen-16 ($^{18}O/^{16}O$): the higher the $\delta^{18}O$ value, the higher the $^{18}O/^{16}O$ ratio. Specifically, $\delta^{18}O$ is the permil deviation of the $^{18}O/^{16}O$ ratio of the sample compared with a standard $= [(^{18}O/^{16}O_{sample} - {}^{18}O/^{16}O_{standard}) / {}^{18}O/^{16}O_{standard}] \times 1,000$. (After Mix et al. 1995)

ited from about 5 million years ago to the present in the equatorial eastern Pacific Ocean (figure 6.3).[16] The oxygen isotope profile of this core reveals numerous cycles consisting of warm (low $^{18}O/^{16}O$) and cold (high $^{18}O/^{16}O$) periods, and it mirrors the oxygen isotope record of sediment cores from other parts of the world. The time resolution, which is based on the spacing of analyzed samples, is about 4,000 years; that is, events 4,000 or more years apart can be distinguished from each other.

Some very interesting insights emerge from the oxygen isotope record. First, the early to mid-Pliocene, from about 5 million to 3 million years ago, was exceptionally warm, with the mean global temperature perhaps 2 to 3°C (3.6 to 6°F) warmer and sea level 15 to 25 meters (50 to 80 feet) higher than today.[17] At that time, ice covered eastern Antarctica,[18] but there was apparently no ice cap in the Northern Hemisphere. Beginning about 3 million years ago, the steady increase in the $^{18}O/^{16}O$ ratio upward through the core indicates a gradual expansion of the Southern Hemisphere ice cap and establishment of one in the Northern Hemisphere for the first time in nearly 200 million years.[19] At first, cooling occurred relatively rapidly, but then the expansion of the ice caps continued more gradually for at least another 2 million years.

During this long period of gradual cooling, a continuous cycling occurred between glacial and interglacial intervals as ice sheets periodically advanced and retreated. The cycles occurred with regularity at approximately 41,000-year intervals from about 1.6 million to 1 million years ago. After that, the cycles started to become much longer and less frequent, with long glacial periods that lasted on average about 80,000 to 90,000 years alternating with shorter interglacial periods that lasted about 10,000 years. The glacial periods ended relatively rapidly; the warm periods ended more gradually, with rapid cooling at first, but then slower cooling. This cycle happened again and again. The past 800,000 years have seen eight prolonged glaciations spaced about 100,000 years apart (plus two shorter glaciations, depending on how they are defined).

EARTH'S ORBITAL CHARACTERISTICS
AND MILANKOVITCH THEORY

This description begs the question: What might have caused the large and numerous cycles, not to mention their regularity? The answer, everyone agrees, is *orbital forcings*, which are embodied in what is known as Milankovitch theory. The notion that Earth's orbit around the Sun can affect climate was first advanced in 1842 by the French mathematician Joseph Adhémar (1797–1862). Nearly 80 years later, Serbian civil engineer and mathematician Milutin Milankovitch (1879–1958) quantified and formalized Earth's total orbital characteristics as a theory to explain glaciations.

Three characteristics of Earth's orbit change in cycles of different lengths (figure 6.4). First, the tilt of the axis of rotation relative to the solar plane (*obliquity*), currently 23.5 degrees, changes from 21.5 to 24.5 degrees and back again approximately every 41,000 years. The greater the tilt of spinning Earth, the more intense are the winters and summers of both hemispheres. The Northern Hemisphere is much more affected than the Southern Hemisphere, however, because the Southern Hemisphere has little in the way of high-latitude landmasses where ice sheets can grow.

Second, the *eccentricity*, or the degree that Earth's orbit around the Sun departs from a circle, changes with a period of about 96,000 years. As a result, Earth's distance from the Sun and thus the total amount of solar energy received by Earth vary throughout this cycle.

Third, Earth's axis of spin *precesses* relative to the orbital axis once about every 21,000 years. This means that the axis of spin itself rotates around another axis, just as the axis of a spinning top may slowly rotate around a second axis. Thus the Northern Hemisphere receives relatively large amounts of solar radiation when precession points the North Pole toward the Sun at the same time eccentricity brings Earth closest to the Sun.

The interplay of the three cycles creates, according to the Milankovitch theory, significant variations in *insolation* (the amount of solar energy received at any one location) in the summer at the northern latitudes and accounts for the waxing and waning of the great Pleistocene ice sheets.

Milankovitch theory gained currency with the discovery that the chronology of climatic variations recorded in deep-sea sediment cores extending back almost

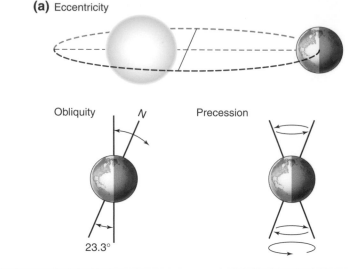

(a) Eccentricity

Obliquity

Precession

23.3°

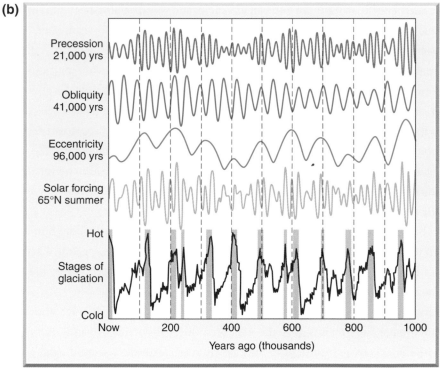

(b)

Precession
21,000 yrs

Obliquity
41,000 yrs

Eccentricity
96,000 yrs

Solar forcing
65°N summer

Hot

Stages of
glaciation

Cold

Now   200   400   600   800   1000

Years ago (thousands)

FIGURE 6.4

The Milankovitch cycles

Earth's orbital parameters of precession, obliquity, and eccentricity (*a*) and the way their individual, cyclical effects combine to influence summer insolation at 65°N latitude (*b*). The combined effects of the orbital parameters on insolation are known as Milankovitch cycles. The cycles of precession and obliquity are more important than that of eccentricity. These orbital variations were largely responsible for the cycles of Pleistocene climate. Alone, they are too small to account for the observed differences between glacial and interglacial periods, so other drivers in the climate system must have amplified the orbital-driven changes in insolation. (*a,* after Global Warming Art, http://www.global warmingart.com/wiki/Image:Milankovitch_Variations_png, original created by Robert A. Rohde)

450,000 years could be accounted for by periodic forcings occurring at intervals of 23,000, 42,000, and 100,000 years.[20] Because these intervals are nearly identical to the cycles in precession, obliquity, and eccentricity, the notion that orbital forcing has been largely responsible for the cycles of Pleistocene climate is now widely accepted.

Indeed, other observations support the theory. For example, the 41,000-year oscillations in early Pleistocene climate now appear to be well explained by the periodic change in obliquity,[21] and evidence of Milankovitch cycles has been found in older sedimentary sequences as well. Perhaps the most definitive test of Milankovitch theory comes from the timing of the glacial cycles in the different hemispheres. Thus the most recent glacial cycle ended about 15,000 years ago, corresponding to an increase in northern latitude summer insolation. The warming in Antarctica lagged behind the change in insolation, however, implying that Northern Hemisphere insolation change is the fundamental driver of the ice age cycles.[22]

Despite the acceptance of this theory, orbital parameters alone are not sufficient to explain the glacial and interglacial cycles because the orbital-induced variations in summer insolation at the northern latitudes are too small to account for the differences in the cycles. There must be some amplifier of orbital forcings internal to the climate system itself. One idea is that despite the small change in insolation, the warmth of the Northern Hemisphere summer is the critical factor because it determines the extent to which the snow melts and sea ice breaks up by the end of the summer. If each summer is the same as the previous summer, the climate continues as it has been. If each summer is slightly cooler than the previous summer, however, less snow melts and less sea ice breaks up, with the result that the Northern Hemisphere albedo, or the proportion of solar energy reflected back to space, increases. Increasing albedo represents a strong feedback that causes climate to cool as less energy is stored in the ocean and land surface and more is reflected back to space.

The importance of the internal dynamic of the climate system is further emphasized by considering the origin of the 100,000-year cycles. Eccentricity has a minuscule influence on the amount of solar radiation reaching Earth, much less than either obliquity or precession. Why, then, is the 100,000-year cycle so prominent in ice cores and other climate records, and how is it that the glacial terminations can be so abrupt if the orbital forcing is so small?[23] It appears that the abrupt glacial terminations of the past 700,000 years can be statistically associated with every second or third obliquity maximum, but not with precession or eccentricity.[24] The glacial terminations occur *on average* every 100,000 years, but there is some variation in their frequency. The climate system appears to "skip one or two obliquity beats before deglaciation."[25] In other words, the terminations occur closer to alternating 80,000-

and 120,000-year intervals than to 100,000-year intervals. The theory is that this variation may be due to the interplay between high polar insolation at times of high obliquity and how the ice sheets melt.[26]

THE PLEISTOCENE ATMOSPHERE

Several ice cores drilled into the East Antarctic Ice Sheet—notably the Dome Concordia (or Dome C), Vostok, and Taylor Dome cores—together provide an astonishing, continuous record of the atmospheric concentrations of CO_2 and other greenhouse gases for the past 800,000 years (figure 6.5).[27] The time interval covers nine glacial–interglacial cycles, and the mean temporal resolution (the minimum time that two features in the record can be distinguished) is about 1,000 years. The record has been obtained by measuring the air in bubbles trapped in the ice. Ice forms as the snow laid down each year compacts and recrystallizes during burial by younger layers of snow. Newly fallen snow is porous and mixed with air. As the snow recrystallizes, some of the air is trapped to form bubbles.

For the past 800,000 years, atmospheric CO_2 varied between about 170 and 300 parts per million (ppm) in approximately 100,000-year cycles. In other words,

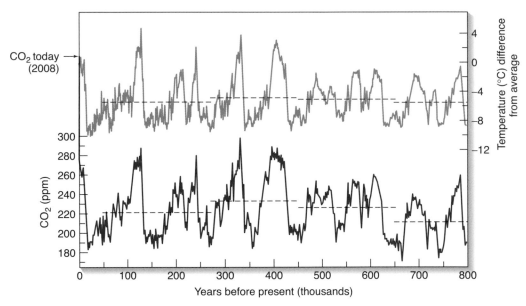

FIGURE 6.5

The 800,000-year record of atmospheric CO_2

Based on analyses of air trapped in Antarctic ice cores compared with the temperature record derived from the composition of ice, it was found that during the entire 800,000 years, the atmosphere's CO_2 content changed in concert with temperature and remained between 170 and 300 ppm. In 2008, the atmosphere contained 385 ppm CO_2. (Adapted by permission from Macmillan Publishers Ltd: Nature, Lüthi et al. 2008, copyright 2008)

the present atmospheric concentration of CO_2 (385 ppm) is substantially higher than at any time during these 800,000 years. Furthermore, the fluctuations in CO_2 content display a very close correspondence to variations in local temperature.[28] The correspondence implies a strong, stable coupling between CO_2 and temperature. Close inspection of the data suggests that the CO_2 changes lag temperature changes by centuries, although this connection cannot be seen in figure 6.5. Whether the lag is real is uncertain because the lag is near the limit of temporal resolution, and the local temperature variations may not be precisely in phase with global ones. If it is indeed real, it would suggest that a buildup of greenhouse gases in the atmosphere did not initiate global warming. More likely, warming was initiated by another mechanism, and then rising CO_2 promoted and amplified the warming. The exact mechanism that ties atmospheric CO_2 content to temperature is not known. Almost certainly, however, this mechanism has something to do with the exchange of CO_2 between the atmosphere and the ocean.

Another remarkable feature of the record is that the CO_2 content of the atmosphere never fell below 170 ppm during glacial periods or rose above 300 ppm during interglacial periods. This suggests that a set of governing processes have determined the amount of CO_2 in the atmosphere by transferring carbon back and forth between the atmosphere, on the one hand, and the ocean and land reservoirs, on the other. How well these processes are operating in today's world in which atmospheric CO_2 content is so much higher than in the past remains an unanswered but important question.

The methane record is similar. Over the past 800,000 years until the beginning of the industrial era, the methane content in the atmosphere ranged from 0.320 to 0.790 ppm. Again, the present methane content (1.77 ppm) is higher than at any other time in the past 800,000 years. Atmospheric methane is believed to mirror the extent of natural wetlands, the main source of methane, and is high during warm periods and low during cold periods. Indeed, as for CO_2, the methane variations closely match those in local temperature.

Finally, the ice core samples yielded a partial record of atmospheric nitrous oxide (N_2O) content, which also follows the trend of the other gases and is now higher than at any other time than in the past.[29] The nitrous oxide generated in soil accounts for about two-thirds of that in the atmosphere, with the ocean accounting for the rest.

The Climate of the Past 100,000 Years

Although deep-sea sediments and other records show that the Pleistocene has been characterized by relatively long, large, and global climate cycles, records of the past

100,000 years reveal more local fluctuations on much shorter timescales. These variations have less to do with orbital forcings than with the climate system's internal dynamics. The outstanding record comes from ice cores taken from Greenland.

ICE CORES FROM GREENLAND

In 1989, two teams, one European and the other American, began drilling into Greenland ice (figures 6.6 and 6.7). By the summer of 1993, they had recovered two complete drill cores through the entire ice sheet. The American Greenland Ice Sheet Project Two (GISP2) core penetrated 3,053 meters (10,016 feet), and the European Greenland Ice Core Project (GRIP) drilled 3,029 meters (9,940 feet) of ice before hitting bedrock. It quickly became apparent that each of the cores held a continuous, 108,000-year record of climate.[30] The record amazed climatologists and other Earth scientists alike because it revealed climate oscillations of far greater magnitude and rapidity than anyone had imagined could have occurred during that period or that have occurred since the beginning of human agricultural activity, let alone recorded history.

The Greenland ice cores can be dated with relatively good accuracy.[31] Annual layers are visible and can be counted in ice as old as about 40,000 years. The layers are defined mainly by variations in the cloudiness of the ice, reflecting differences in bubble and ice-grain size. Events that appear in the ice and the independent knowledge of dates in which these events occurred provide a time calibration; for example, sulfuric acid–bearing layers can be identified as having originated from known volcanic eruptions. By combining these and other methods, scientists can know the age of 10,000-year-old ice within about 200 years, although the errors expand somewhat the farther one goes back in the record.[32]

GISP2, located far above the Arctic Circle (at 72.6°N, 38.5°W), and GRIP, located 28 kilometers (17 miles) to the east of that, are on the very top of the Greenland Ice Sheet at an elevation of 3,200 meters (10,500 feet). The average annual temperature on this lofty plateau is a frigid −31°C (−24°F), and the ice is well below the freezing point all the way down to the bedrock, which eliminates any possibility that the climate record has been disturbed by melting—at least recently, anyway. The sites were chosen for an additional reason: they are very close to the *ice drainage divide* (the location where ice flows in opposite directions to the seas on either side of Greenland) and to where the bedrock is nearly flat. This choice was important. Ice may undergo slow flow, folding, and fracturing that can disrupt the delicate annual layers and destroy or repeat parts of the record. Fortunately, the disruption is limited to the lowest 200 meters (700 feet) of the ice cap, the base of which is about 250,000 years old, leaving undisturbed the upper 2,800 meters (9,200 feet).

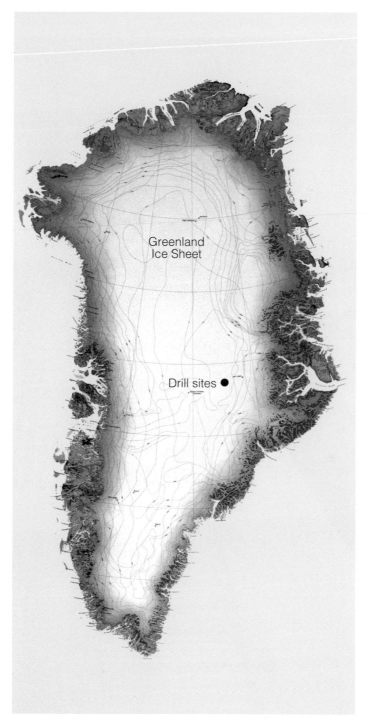

FIGURE 6.6

The location of Greenland Ice Sheet drill sites

The drill sites are on the very top of the Greenland Ice Sheet at an elevation of 3,200 meters (10,500 feet) above sea level. Each drill core penetrated nearly 3,100 meters of ice before reaching bedrock, providing a detailed record of climate during the past 110,000 years. (Map by the Geological Survey of Greenland)

FIGURE 6.7

A Greenland ice core

A section of Greenland ice is removed from the core barrel. (National Oceanic and Atmospheric Administration, Climate and Global Change Program, and National Geophysical Data Center)

We infer the point of this limitation because down to 2,800 meters the GISP2 and GRIP records match perfectly, but are inconsistent below that. This serves to illustrate why two cores were necessary: they served to verify the record by showing where the ice is and is not disturbed.

READING THE GREENLAND RECORD

Greenland ice bears witness to past climate in a variety of ways. The $^{18}O/^{16}O$ ratio of the ice is directly related to air temperature. When liquid water evaporates to water vapor or when water vapor condenses to liquid water, the heavy oxygen-18 isotope becomes slightly enriched in the liquid, and the light oxygen-16 isotope becomes slightly enriched in the vapor. As water condenses from the atmosphere, the remaining water vapor thus becomes progressively more enriched in oxygen-16 and depleted in oxygen-18. More water condenses from the air that provides the snow to Greenland during cold periods than during warm periods. Therefore, water vapor and ice formed during cold periods contain relatively more oxygen-16 and have a lower $^{18}O/^{16}O$ ratio than do water vapor and ice formed during warm periods.

The concentrations of calcium, sodium, and chlorine in the ice are also good climate indicators. Calcium is present mainly in the form of carbonate and represents atmospheric dust. As the circulation system that transports air to Greenland widens (which it does during cold times), it scours dust from a wider landmass. Periods of cold climate are also drier, leading to more extensive arid regions and more dust. Sodium and chlorine come from sea salt. Both dust and sea salt reach Greenland primarily in the late winter to early spring because at that time of year atmospheric circulation over North America is most intense and storms are more frequent.

The temperature, dust, and sea salt records show that the climate oscillated between short warm periods lasting hundreds to thousands of years and longer, more stable cold periods during which temperatures on the Greenland ice cap were as much as 20°C (36°F) colder than today (figure 6.8). In all, there were an astonishing 23 warm periods in the span of nearly 100,000 years.

As an aside, it is worth mentioning some other phenomena recorded in Greenland ice. Ice layers of high electrical conductivity record volcanic eruptions (figure 6.9).[33] Electrical conductivity is mainly a measure of acidity, the most important component of which is sulfuric acid (reported as sulfate in figure 6.9). Explosive volcanic eruptions can inject large amounts of sulfur dioxide into the atmosphere (chapter 5). Most of the major eruptions of the past 2,000 years show up in the ice, from the small 1963 eruption of Surtsey, Iceland, to the gigantic 1815 eruption of Mount Tambora

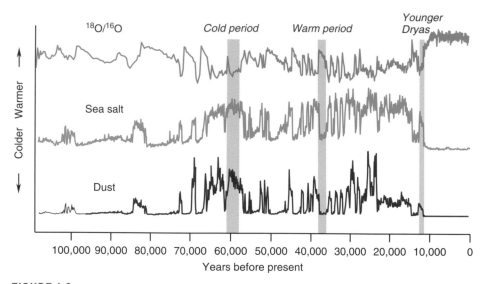

FIGURE 6.8

The record of temperature in the Greenland ice cores

The temperature record for Greenland can be determined from the ratio of oxygen-16 to oxygen-18 and the amount of sea salt and dust in the Greenland ice cores. Cold periods correspond to low $^{18}O/^{16}O$ ratio and high sea salt and dust contents. The climate through most of the 108,000-year record was characterized by abrupt swings between warm and cold periods, during which temperatures were as much as 20°C (36°F) colder than they are today. The abruptness of the swings implicates the ocean as having a major influence on the climate of the North Atlantic region. (Data from Mayewski et al. 1997)

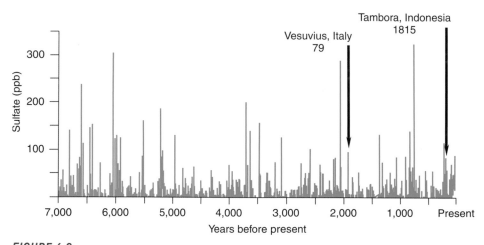

FIGURE 6.9

The record of major volcanic eruptions in the Greenland ice cores

The size of the spikes in sulfate content in the Greenland ice cores does not correlate with the size of the eruptions because the eruptions occurred in different locations, and the amounts of sulfuric acid deposited on the ice depended on the atmospheric circulation and weather patterns of the time. (Data from Zielinski et al. 1994)

in Indonesia. More ancient eruptions are also in evidence, such as the Mazama eruption, which occurred sometime around 5677 B.C.E. and created Crater Lake in Oregon.[34]

The ice cores also record human activities. In addition to recording volcanic eruptions, the sulfate and nitrate contents of the ice document atmospheric pollution transported from the middle latitudes.[35] Their concentrations begin to increase in ice formed in the late nineteenth century as the Industrial Revolution began to pick up steam. The passage of the Clean Air Act (1974) in the United States is also in evidence. Somewhat more arcane but no less interesting is the copper content of the ice, which documents pollution from copper smelting that began 2,500 years ago, about the time that coin usage began in Greece, and continued to medieval times.[36]

ABRUPT CLIMATE SHIFTS

Perhaps more astonishing than the number and extremes of climate shifts in the 100,000-year Greenland record is the abruptness of those shifts (see figure 6.8). The warm periods, which are known as Dansgaard-Oeschger (D-O) events after the investigators who first recognized them in the Greenland ice cores, typically began with rapid warming, but subsequent cooling was much slower. In fact, the warming typically occurred in only a few decades, with air temperature rising 9 to 16°C (16 to 29°F) and snow accumulation rates doubling.[37] The D-O events appear to involve rapid changes in ocean heat transport.

The best documented of the abrupt shifts is the most recent one—the Younger Dryas (chapter 5), glacial conditions that began about 12,900 years ago and ended 11,600 years ago, a span of approximately 1,300 years (figure 6.10).[38] In Greenland, the end of the shift was marked by an increase in temperature of 7°C (13°F) and a doubling of the rate of snow accumulation. This change occurred in a few decades or less and possibly in just a few years. It was far greater and occurred more rapidly than any climate change experienced since the rise of organized human society.

One idea to account for the Younger Dryas starts with events in North America about 15,000 years ago.[39] At the time, the ice sheet covering the continent began to retreat ever more rapidly, and meltwater began to collect in Lake Agassiz, an enormous body that covered part of southern Manitoba and drained down the Mississippi River into the Gulf of Mexico. But as the ice continued to melt, other drainages, originally blocked by ice, began to open. When the ice dam finally broke, the water level of Lake Agassiz fell some 40 meters (130 feet) in just a few years, and a vast amount of freshwater spilled into the North Atlantic, possibly through Hudson Bay. The flood of freshwater covered the warmer but more saline (and thus more

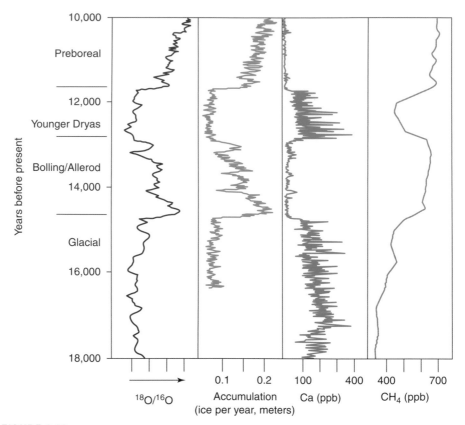

FIGURE 6.10

The record of the Younger Dryas in the Greenland ice cores

The end of the Younger Dryas was marked by an increase in temperature of 7°C (13°F) and a doubling of the rate of snow accumulation in a few decades and possibly in only a few years. Ca = calcium; CH_4 = methane. (Data from Mayewski and Bender 1995)

dense) ocean water that had been warming the North Atlantic. This event initiated glacial conditions by shutting down the global ocean conveyor system (chapter 3). The Younger Dryas ended when the amount of freshwater flowing into the North Atlantic dwindled, and warm ocean water reclaimed the surface of the North Atlantic, warming the atmosphere again.

An interesting event occurred several thousand years later that supports this general idea. The Greenland ice shows a period of unusually cold climate beginning about 8,500 years ago. It lasted for several hundred years and represents the coldest period in the entire Holocene (the past 11,600 years). Fossil and other evidence from North Atlantic sediment cores indicate that two massive outpourings of freshwater flooded the North Atlantic, one 8,490 years ago and the other 8,290 years ago.[40] The timing is coincident with the final drainage of Lakes Agassiz and Ojibway, the two lakes in central Canada remaining from the earlier glacial floods.

LINKS BETWEEN THE NORTH ATLANTIC AND THE ANTARCTIC

Climate shifts in Antarctica are clearly coupled in time to shifts in Greenland. The coupling is demonstrated by the fact that eight of the nine warm periods lasting for 2,000 years or more according to Greenland ice also show up in Antarctic ice. But the records are not identical. Although the onsets of North Atlantic D-O events were usually marked by dramatic warming in just decades, the corresponding temperature changes in Antarctica were much less dramatic and took centuries.[41] More curious, however, is that the timings of the warming events in the North Atlantic and those in Antarctica do not exactly correlate.

The difference in timing is best illustrated by an ice core recently recovered from Dronning Maud Land (DML) in eastern Antarctica (known as the European DML, or EDML, core).[42] The Greenland warm periods correlate well with Antarctic warm periods (figure 6.11). However, each Antarctic warming started well before the respec-

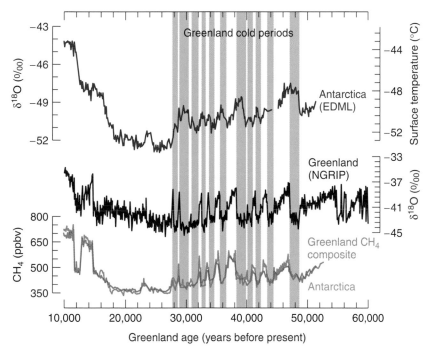

FIGURE 6.11

The correlation of Antarctic EDML and Greenland NGRIP ice cores

Temperature increases upward. The Greenland warm periods (*unshaded*) correlate well with Antarctic warm periods, but the latter generally started well before the warming in Greenland and during the Greenland cold periods (*blue shading*). In addition, the level of warming in Antarctica was linearly proportional to the duration of the cold periods in Greenland. The synchronization error is 400 to 800 years. $\delta^{18}O$ is a measure of the oxygen-18 to oxygen-16 ($^{18}O/^{16}O$) ratio and correlates with increasing temperature (for the definition of $\delta^{18}O$, see figure 6.3). (Adapted by permission from Macmillan Publishers Ltd: *Nature*, EPICA Community Members 2006, copyright 2006)

tive D-O event and during a Greenland cold period. In addition, the level of warming in Antarctic was linearly proportional to the duration of Greenland cold periods.

How are the climates of Greenland and Antarctica linked? One idea, known as the *bipolar seesaw*, is that slowing of the North Atlantic meridional overturning circulation (the northward flow of warm surface water in the North Atlantic and the southward return of cold North Atlantic Deep Water [NADW], described in chapter 3) causes heat to build up in the Southern Ocean because the Atlantic currents are no longer transporting as much heat northward.[43] To appreciate this connection, one must remember that the northward flow in the Atlantic Ocean is responsible for one-quarter of the global meridional heat transport. The longer and slower the circulation, the longer the Greenland cold period, the more heat that builds up in the South Atlantic, and the warmer it gets in Antarctica, as observed.

Although the bipolar seesaw model offers a simple explanation for the coupling of the polar Northern and Southern Hemisphere climates, it does not explain why warming begins in Antarctica before it begins in Greenland. The difference presumably has something to do with how heat diffuses in the ocean, which is not fully understood.[44]

Marine sediment cores provide some insight into the ocean's role. The compositions of benthic foraminifera and certain characteristics of the organic chemistry of sediments off the coast of Spain indicate that during previous North Atlantic warm periods, the deep water there originated from the north.[45] The intensity of these northerly bottom currents gradually decreased as the North Atlantic climate gradually cooled (see figure 6.11). The subsequent, more rapid coolings that terminated the warm periods were accompanied by abrupt inflows of water from the south. Finally, new cycles of warming are indicated by abrupt returns to warm conditions in the North Atlantic and shifts in the source of bottom water back to the north. These observations are consistent with the bipolar seesaw model.

Changes elsewhere on the globe appear to have been associated with changes in the polar regions. For example, the chemical composition of corals indicates that surface waters in parts of the equatorial Pacific Ocean were relatively cool during the Younger Dryas;[46] and fluctuations in alpine glaciers, mountain snow lines, and vegetation type in the Andes of South America appear to correspond to temperature changes in Greenland.[47] These and other changes have been interpreted to indicate a southward shift of the Intertropical Convergence Zone (ITCZ) and thus the belts of tropical rainfall during the Younger Dryas rather than to demonstrate general global cooling.[48] A shift in the ITCZ is consistent with the fact that the Younger Dryas is not recorded in glacial advances in New Zealand.[49] In short, the millennial-scale changes recorded in Greenland and Antarctica ice generally appear to reflect swings in climate patterns rather than uniform global-warming and -cooling events.

The Climate of the Holocene: The Past 12,000 Years

The termination of the Younger Dryas 11,600 years ago is considered the start of the Holocene epoch, which is by far the longest stable and warm period in the entire 108,000 years of the Greenland ice core record. To be sure, climate has fluctuated during the Holocene. The previously mentioned cold period beginning 8,500 years ago was followed by a warm interval from 7,000 to 5,000 years ago, during which time glaciers in the European Alps were even less extensive than they are now. That particular glacial retreat can be attributed to orbital forcing. Other warm intervals, however, appear to be mainly regional and are not synchronous, suggesting that they reflect ephemeral reorganizations of the climate system rather than external forcings. But all these changes are small compared with the ones that occurred before this point. In the absence of forcings related to human activity, Milankovitch theory would predict that the current temperate conditions will continue for at least another 30,000 years and probably much longer.[50]

TREE RINGS, LAKE SEDIMENTS, CORALS, AND CAVES AS PROXIES FOR CLIMATE

Compared with climates of more distant times, the Holocene climate is naturally better known. It is worth exploring some of the techniques used to infer past climate because they provide the basis for both understanding Holocene climate and relating it to human history. The techniques apply to tree rings, lake sediments, corals, and cave deposits, which, in conjunction with radiocarbon dating, provide both dates and information on past environmental conditions.

In many tree species, the annual seasonal changes influence the growth rate of the cambium, which lies just underneath the bark, and thus the width the cambium attains during any one growing season (figure 6.12). Such tree-ring chronology has been extended back 11,000 years by correlating rings in trees with different but overlapping ages.[51] It has served not only to date events, but also to correct the radiocarbon clock.

Tree rings provide a chronology of changes in moisture and other environmental characteristics in addition to temperature, but extracting such information is not a straightforward process. First, trees display a complex response to changing climatic conditions. Second, one can readily imagine that at any given site some trees occupy more desirable real estate than do others. As a result, there are significant differences in growth rate because of such local factors as soil drainage and availability of light, so that an individual tree records only the most local of conditions. The derivation of a detailed climate record from tree rings thus requires careful examination of a large number of trees.

FIGURE 6.12

Cross section of a tree trunk being prepared for exhibition

The width of a growth ring is the result of temperature, moisture, and other characteristics of the local environment. The rings form annually, so they provide a detailed chronology of changing conditions. Tree-ring chronology has been extended back 11,000 years by correlating rings in trees with different but overlapping ages. The inset shows a detail of the trunk section in the foreground. (Photograph by R. Mickens, American Museum of Natural History)

Lake sediments tell their own story of the recent past climate. In many large lakes in the midlatitudes, a high carbonate content in sediment correlates with extensive global ice volume, reflecting a relatively low quantity of water entering the lake and more brackish water, leading to a high carbonate precipitation rate. Conversely, low-sediment carbonate content correlates with warm periods, when lake water level was high, the water was relatively fresh, and the carbonate-precipitation rate was low.

When it comes to small lakes at high latitudes, especially those in valleys with glaciers, sediment sequences commonly consist of alternating dark and light layers, known as *varves* (figure 6.13).[52] In the winter when the lakes are sealed under ice, no clastic sediment enters the water. At the same time, microscopic organisms such as diatoms die (because of the lack of light for photosynthesis), and fine clay suspended in the water accumulates on the lake bottom, producing a dark layer. When spring arrives, sand and silt from river runoff enter the lake along with nutrients from streams fed by melting snow, which promote summer blooms of organisms. The sediment settles throughout the summer to form the light layer, which is then covered the next winter by another layer of dark clay and organic debris. The utility of varves in climate studies is that they represent an absolute chronology because they are annual. In addition, the thickness of dark layers reflects summer biological productivity, and thickness of the light layers indicates the amount of meltwater entering the lake.

Short-term changes in climate are also documented by the skeletal remains of coral (figure 6.14). Coral polyps build "skeletons" by extracting calcium carbonate ions from seawater.[53] Coral formed in winter and coral formed in summer have different densities, which result in annual growth rings and thus create a reference chronology. Among the indicators of environmental conditions are the strontium/calcium ratio, the $^{18}O/^{16}O$ ratio, and the concentrations of certain trace elements, such as cadmium and barium, in the carbonate skeletal material.

Strontium/calcium and $^{18}O/^{16}O$ ratios typically reflect the temperature of the seawater in which the corals grew,[54] and changes in seawater trace-element concentrations reflect processes such as upwelling and changes in the amount of windblown sediment or river runoff entering the ocean. The concentration of cadmium, for

FIGURE 6.13
Varves exposed on the campus of the University of Massachusetts, Amherst
Varves form in many high-latitude lakes, indicating that this area in Amherst was once a lake bed. The dark layers accumulate in the winter as diatoms and other microscopic organisms in the water die and as fine clay settles out of the water. The light layers represent sand and silt deposited in the spring and summer from river runoff fed by melting snow. (Photograph by J. Beckett, American Museum of Natural History)

FIGURE 6.14
Coral

Compositional differences among annual growth rings record differences in temperature and other environmental conditions. This sample is on display in the Gottesman Hall of Planet Earth at the American Museum of Natural History, New York. (Photograph by D. Finnin, American Museum of Natural History)

example, is low in shallow waters from which it is removed by biological processes, but it is high in deeper waters where dying organisms have released it. Thus a sudden increase in cadmium in the coral's growth rings may indicate that the coral became bathed in water upwelling from the depths. Barium behaves like cadmium, but it is much more sensitive to the same processes and exhibits much wider variations over the seasons. For example, barium concentrations in Galápagos corals record El Niño events because of changes in the pattern of ocean upwelling around the islands.[55]

As for caves, some are literally vaults holding detailed records of past climate. They exist where groundwater percolates through limestone, dissolves some of the calcium carbonate through which it passes, drips into caves, and reprecipitates the carbonate to form deposits known as *speleotherms* (the dramatic examples of which are known more popularly as the hanging *stalactites* and upright *stalagmites*). Under the right circumstance, the carbonate faithfully records the $^{18}O/^{16}O$ ratio of the rain from which the groundwater originated and thus, as described earlier, the regional climatic conditions. Caves in China, for example, have provided detailed records of changes in the Asian monsoon.[56]

DATING MATERIALS WITH CARBON-14

Most carbon (98.89 percent) consists of carbon-12; practically all the remainder is carbon-13. Both isotopes are stable. However, about one of every trillion carbon atoms is the radioactive isotope carbon-14, which forms in the upper atmosphere when cosmic ray neutrons emanating from the Sun interact with the isotope nitrogen-14. Carbon-14 decays at a constant rate back to nitrogen-14, with a half-life of $5,730 \pm 40$ years (that is, after 5,730 years, half of the carbon-14 originally present has decayed to nitrogen-14; in another 5,730 years, one-quarter [half of a half] remains; in yet another 5,730 years, one-eighth [half of a half of a half] remains; and so on). In the atmosphere, carbon-14 mixes with the rest of the carbon and is taken up by living organisms. Consequently, while organisms are alive, they possess the same ratio of carbon-14 to carbon-12 ($^{14}C/^{12}C$) as the atmosphere. When organisms die, however, they effectively become isolated because their carbon exchange with the atmosphere ceases. Thus the $^{14}C/^{12}C$ ratio in a dead organism decreases as carbon-14 decays and thus becomes an indicator of when the organism lived.

Radiocarbon dating is most practical for materials less than about 30,000 years old, although in some circumstances it can be extended to materials as old as 60,000 years. For older samples, too little carbon-14 generally remains for accurate analysis. Trees, charcoal, shells, bones, soil, pollen, and coral are among the obvious materials amenable to radiocarbon dating. A less obvious but important application is the dating of ice, groundwater, and ocean water (all of which contain carbon in solution, mainly in the form of bicarbonate). Some of these waters are very old, and dating them provides information on how and where they circulate.

But this description is too simple. For one thing, the production of atmospheric carbon-14 through time has not been absolutely constant, but has varied by as much as 5 percent over the past 1,500 years because of fluctuations in the solar cosmic ray flux. Fortunately, the tree-ring chronology gives absolute and accurate dates because annual tree rings can simply be counted. Carbon-14 analysis of the rings used to establish the chronology gives us the record of fluctuations in carbon-14 production and provides a means for calibrating carbon-14 dates.

Added to this problem are the very recent changes in atmospheric carbon-14 content. Nuclear weapons produce neutrons, and the testing programs carried out in the 1950s doubled the amount of carbon-14 in the atmosphere. (Atmospheric carbon-14 reached a maximum in 1963 in the Northern Hemisphere and has since been decreasing.) However, the increased use of fossil fuels since 1900 has decreased the relative amount of carbon-14 in the atmosphere by about 2 percent because these fuels originate almost exclusively from old coal, oil, and gas containing no carbon-14. Yet another complication arises because some organisms actually use more than just atmospheric carbon to grow. This is particularly the case for marine organisms, which commonly obtain part of their carbon from deep, upwelling water. The bicarbonate dissolved in this water tends to be old and thus depleted in carbon-14. These organisms' shells, in consequence, typically possess carbon-14 ages that are several hundred years older than real ages and that must therefore be adjusted.

CLIMATE FLUCTUATIONS OF THE PAST MILLENNIUM

The past 1,000 years of climate are best known, and the proxy records are sufficiently detailed that we can discern spatial variations in climate. This is also a time during which human history has been influenced by climate change in well-documented ways. Even though the millennial changes have been small compared with those in the more distant past, they suggest some of the changes that may occur in the future.

Researchers have attempted to synthesize the proxy data to come up with a precise millennial record of global (more particularly, Northern Hemisphere) temperature. These efforts are of interest because they provide information on how sensitive climate is to both natural forcings (mainly volcanic and solar) and anthropogenic forcings. One well-known effort, published in two papers in 1998 and 1999, was made by Michael Mann, Raymond Bradley, and Malcolm Hughes.[57] They derived their reconstruction from tree-ring, coral, and other temperature proxies, which they calibrated by comparing them with the instrumental records of the nineteenth and twentieth centuries. Consistent with orbital forcing, the reconstruction showed gradual cooling (but interrupted by several multidecadal intervals of warmth and cold) from about the year 1000 to the beginning of the twentieth century. This long-term cooling is followed by a sharp increase in temperature in the twentieth century. The shape of the trend came to be described as a hockey stick.

This reconstruction and subsequent reconstructions have demonstrated that the twentieth century was warmer than any other time during the previous 400 years. The data also suggest that the late twentieth century was warmer than any other time during the previous 1,000 years. Because the uncertainties in reconstructed temperature increase the farther we go back in time, however, the comparison beyond 400 years ago is not definitive (figure 6.15).

Mann, Bradley, and Hughes's reconstruction quickly drew attention for another reason: although it placed the twentieth-century warming in context, it did not clearly distinguish the Medieval Warm Period from the Little Ice Age.[58] In Europe, the Medieval Warm Period lasted from about 850 to 1200, and the Little Ice Age

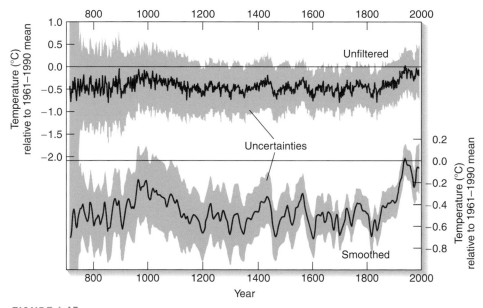

FIGURE 6.15

A 1,200-year reconstruction of Northern Hemisphere temperatures relative to the 1961–1990 mean

The top panel shows annual variations; the bottom panel displays the annual data smoothed. Note the uncertainties (*light blue*). The late twentieth century was warmer than any time during the past 400 years and probably warmer than any time during the past 1,000 years. This reconstruction, based on tree rings, clearly distinguishes between the Medieval Warm Period, which lasted from about 850 to 1200, and the Little Ice Age, from about 1350 to 1850. (After D'Arrigo, Wilson, and Jacoby 2006)

from about 1350 to the mid-nineteenth century.[59] Although both are well documented in Europe and elsewhere around the North Atlantic, the end of the Medieval Warm Period and particularly the beginning of the Little Ice Age are ill defined because they appear to have occurred at different times in different places. The warmth around 1,000 years ago may have been caused by an increase in solar irradiance, whereas the Little Ice Age appears to have coincided with an increase in the intensity of atmospheric circulation and may be related to decreasing solar output supported by relatively frequent volcanic eruptions.[60]

Other Northern Hemisphere temperature reconstructions now extend back before the year 800. A recent reconstruction based on high-latitude tree-ring data, for example, suggests that the late twentieth century was a substantial 0.7°C (1.3°F) warmer than the years 950 to 1100 when comparing decadal averages (see figure 6.15).[61] It documented an extended interval of cooling from about 1100 to 1400 and cool conditions extending to 1850. The relatively clear distinction between the Medieval Warm Period and the Little Ice Age in Mann, Bradley, and Hughes's reconstruction may be due to the fact that the tree-ring data come from polar to near-polar regions mostly above 60°N, a region more sensitive to climate change than are other parts of the globe.

Lessons Learned

Several general lessons may be gleaned from this discussion of paleoclimate. First, certain past events offer an analogy to the changes occurring today. The PETM is one such event. It teaches us that the injection of large quantities of CO_2 into the atmosphere can significantly heat the planet and alter both terrestrial and ocean ecosystems, and that it takes tens of thousands of years for the climate system to recover to its former state.

Second, climate change has occurred on a number of different scales. The past million years has been dominated by 100,000-year cycles of change, and within these long cycles there are shorter, millennial-scale ones, which are defined mainly in ice core records of the past 100,000 years.

Third, climate may change due to a variety of reasons. Although the PETM was caused by the buildup of CO_2, the 100,000-year and 41,000-year cycles evident in geological records back to about 1.6 million years are clearly related to orbital forcings that caused global warming and cooling by changing the radiative balance. In contrast, the millennial fluctuations appear to reflect large-scale reorganizations of the climate system rather than wholesale global change. Nonetheless, the detailed reconstructions of climate over the past millennium, made possible mainly by analysis of tree rings, show relatively subtle changes in climate that appear to be related to natural forcings.

We also saw that climate change can be both gradual and abrupt, which leads to the notion that the climate system contains tipping points, as discussed in greater length in chapter 5. Tipping points are inherently difficult to predict. They represent a significant uncertainty in how climate will change in the future and thus add to the risk we bear.

The Sahel of Africa

Global warming may bring long and severe droughts to many parts of the world, such as south-western North America and the Sahel of Africa, the southern borderland of the Sahara Desert. Indeed, as suggested by the dying trees in this photograph, the Sahel has experienced a series of long and severe droughts in the past 30 years, bringing famine and other hardships to the region. (Photograph by David Haberlah, http://www.flickr.com/, with permission)

7 A CENTURY OF WARMING AND SOME CONSEQUENCES

IN DISCUSSIONS OF GLOBAL WARMING, two questions always arise: Is the warming real? If so, why is it occurring? These questions now have definitive answers based in *observation*. The fact of the matter is that warming has been measured on land, in the atmosphere, in the ocean, and even in the ground, as described in this chapter. In addition, ice is melting, which is the subject of the next chapter. In other words, all of the components of Earth's surface (geosphere, atmosphere, hydrosphere, and cryosphere) have been warming.[1]

As to why, *all* the observations of warming are consistent with the buildup of greenhouse gases as the primary culprit. First of all, let us recall that the atmospheric concentrations of carbon dioxide (CO_2) and other greenhouse gases are indeed rising and changing Earth's radiation balance (chapter 5). These changes and the basic fact that greenhouse gases absorb infrared (IR) radiation leads to the expectation that they *will* cause warming. Furthermore, we know of no other obvious external forcing that can account for the observed changes. Solar irradiance, for example, has remained essentially unchanged (except for fluctuations associated with the 11-year sunspot cycle) since it has been continuously measured by satellite.

And how do we know that the excess CO_2 is coming from human activity? First, in 2008 anthropogenic CO_2 emissions amounted to about 36 billion metric tons (or 36 gigatons) per year. Of that, 29 gigatons originated from the burning of fossil fuels, an amount we know with some accuracy because we know how much fuel we are consuming. (The remaining 7 gigatons, which is generated by cement production, deforestation, and agricultural activities, we know with less accuracy.) Second, the atmosphere's carbon isotopic composition is changing. Fossil fuels possess much lower ratios of carbon-13 to carbon-12 ($^{13}C/^{12}C$) than do the atmosphere and the ocean. Consistent with the buildup of atmospheric CO_2 as a result mainly of fossil-fuel burning is the observation that the atmosphere's $^{13}C/^{12}C$ ratio has been in steady decline since systematic measurement began in 1977.[2]

Warming will surely have consequences, some of which are described in this chapter and the next. Among them are drought, decreased water supplies, extreme and unusual weather events, rising sea level, decrease in biodiversity, spread of disease, and other ecosystem changes. Some of these consequences may significantly harm the well-being of human society. Although warming is the fact, the consequences are the fears because they are just beginning to be felt, and how they will play out remains uncertain. Bearing in mind this distinction between facts and fears, let us investigate what is happening and what may happen.

A Century of Warming and Its Causes

The warming that has been occurring since the beginning of the twentieth century is unusual. For one, it has been very rapid—in fact, about 10 times more rapid than the warming at the end of the most recent glacial period 15,000 years ago.[3] Another is that the warming has occurred more or less everywhere—it is a global, not a regional, phenomenon.

Despite fluctuations through time, global mean surface air temperature remained approximately constant through the latter part of the nineteenth century to about 1910, after which rapid warming occurred in two distinct intervals (figure 7.1). In the first interval, which commenced in 1910 and lasted until about 1940, temperature increased by more than 0.1°C (0.2°F) per decade. The second interval began in the late 1970s and continues. During this period, from 1979 to 2006, land-surface air temperature increased at an even more blistering pace of 0.27°C (0.49°F) per decade.[4] The intervening interval, from 1940 to 1979, was characterized by a stable temperature or even slight cooling, possibly due to atmospheric aerosols from human activities (chapter 5). In a little more than a century, the global mean land-surface air temperature has increased about 1°C (1.8°F), a change well beyond any errors in measurement or uncertainties resulting from geographical variability.[5]

An important point to be drawn from the global record of warming is that generalizations about long-term trends cannot be based on how conditions change from one year to the next or even on changes over two- or three-year intervals. The short-lived effects of volcanic eruptions and natural internal variations of the climate system, such as those represented by El Niño–Southern Oscillation (ENSO) events, disturb the long-term warming trend, but they do not change it.

Warming has also affected nearly the entire globe, with land areas warming somewhat more than ocean areas (figure 7.2). It should be noted that because of the considerable variability from place to place, global change cannot be characterized by measuring the changing conditions at any one location. The global record is just that—a record integrated over the entire planet. Nonetheless, the fact that warming

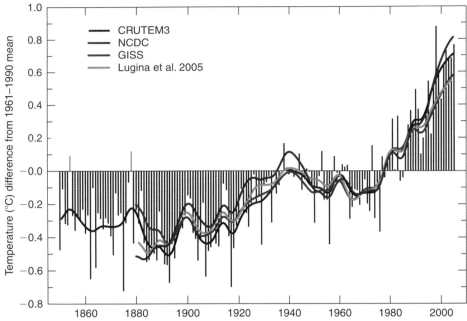

FIGURE 7.1

The change in the average global surface air temperature relative to the 1961–1990 mean

Annual averages are represented by the vertical bars; the curves are the running 10-year means from four analyses of global temperature data. The recent warming is notable because it is global and rapid. (After Trenberth et al. 2007, with permission)

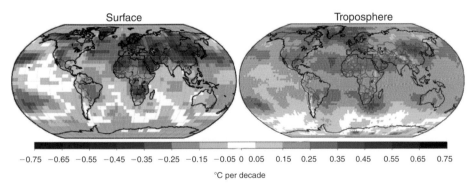

FIGURE 7.2

Geographic trends in temperature, 1979–2005

The trends are expressed as the amount of warming or cooling in °C per decade. The trends for Earth's surface (*left*) are based on measurements, and those for the troposphere (*right*) are based on satellite data. The maps show that the warming has been nearly global, but that it is not the same everywhere. (From Trenberth et al. 2007, with permission)

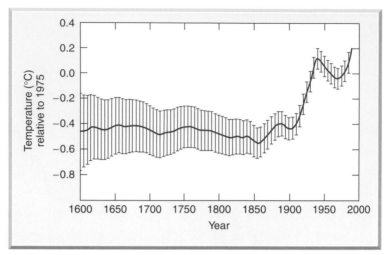

FIGURE 7.3

The change in global temperature deduced from the glacier advances and retreats, 1600–2000

Temperature is relative to the 1975 global average. The importance of this record is that it is independent of the instrumental one. The bars are estimates of the errors. (After Oerlemans 2005, with permission)

is occurring almost everywhere suggests that it is not due to internal reorganization of the climate system, which would not cause the warming to be so pervasive.

If there is any doubt that the instrumental records truly reflect global conditions, we can also deduce global temperature change from the rates at which glaciers have retreated and from the warming of the near surface of Earth itself. Despite the fact that glaciers are in different parts of the world at different elevations and subject to different meteorological conditions, their rates of retreat indicate a mean global temperature increase since the latter part of the nineteenth century of nearly 0.8°C (1.4°F), almost identical to the instrumental record (figure 7.3).[6] The importance of the glacier record is in its wide coverage and independence from instrumental data.

Evidence for global warming also comes from observed increases in temperature of the subsurface as measured in boreholes.[7] The warmth of the atmosphere penetrates only slowly into soil and rock, but it also dissipates only slowly. For example, the cold of the most recent glacial maximum 21,000 years ago commonly shows up in thermal profiles at several kilometers depth.[8] Although thermal profiles are not particularly well resolved in time, they do have the advantage of multiplicity— measurements exist for more than 600 boreholes, so the geographical coverage is extensive. The borehole data indicate that global climate has undergone a long-term warming of about 1°C (1.8°F) over the past 500 years, with about half of the warming occurring within the past century.

Corroborating the air temperature increase at the land surface are changes in other parts of the climate system. For example, radiosonde (weather balloon) measurements, which have been made since the late 1950s, and satellite observations, which exist from 1979 onward, demonstrate that the lower and middle troposphere have also been warming (figure 7.4).[9] At the same time, the lower stratosphere is cooling, for two reasons.[10] First, there is less stratospheric ozone, a greenhouse gas, due to breakdown of ozone by manufactured chemicals (chapter 2). More important, however, is that as the CO_2 content of the stratosphere increases, IR emissions from CO_2 also increase. In other words, cooling occurs because energy is increasingly radiated out of the stratosphere, both downward and upward. The changing thermal structures of the troposphere and stratosphere are also causing the boundary between them (the tropopause) to rise. These three phenomena—warming of the troposphere, cooling of the stratosphere, and rising of the tropopause—are expectations within the basic theory that warming is being driven by buildup of CO_2 in the atmosphere.[11]

The atmospheric temperature trends are also sensitive to natural phenomena, however. Two distinct warming events in the stratosphere were the result of large

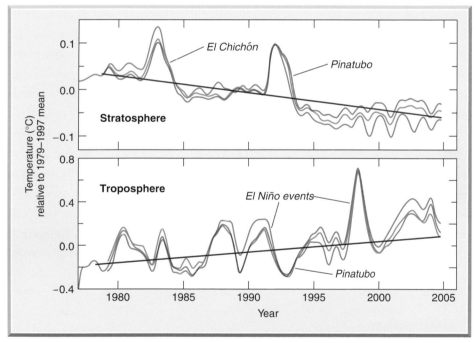

FIGURE 7.4

The globally averaged temperatures of the lower stratosphere and the middle troposphere from various data sets

The curves represent seven-month running averages expressed relative to the 1979–1997 mean. It can be seen that the stratosphere (*upper panel*) is cooling and the troposphere (*lower panel*) is warming. The ephemeral warming events in the stratosphere were caused by injection of aerosols from volcanic eruptions (El Chichón, Pinatubo), whereas the warming events in the troposphere were due to El Niño events. Three sets of measurements are shown, corresponding to different color curves. (After Lanzante et al. 2006)

volcanic eruptions, which caused corresponding cooling in the troposphere below (see figure 7.4). In the troposphere, the several-year temperature fluctuations were caused by El Niño events. For example, the significant warming in 1998 is related to the unusually intense El Niño of that year.

The observed cooling of the stratosphere is worth noting for another reason: it is inconsistent with warming being driven by an increase in solar irradiance, a common misconception among global-warming skeptics. Were global warming due to increased solar output, the stratosphere would warm, not cool. In fact, the stratosphere does display fluctuations in temperature coinciding with the 11-year sunspot cycle (chapter 5), but these fluctuations are minuscule compared with the other changes and do not show up in figure 7.4.

An interesting characteristic of the warming since 1975 is that both the annual average maximum and the annual average minimum temperatures have increased, but that minimum temperature has increased more than maximum temperature.[12] This distinction can be clearly seen from records showing a trend to fewer cold days and nights and another trend to more warm days and nights (figure 7.5).[13] But it is also evident that there have been even fewer cold nights than cold days, and that there have been more warm nights than warm days. During the day, the Sun and greenhouse gases together keep the surface and troposphere warm; at night, however, practically the only warmth comes from the insulation offered by the greenhouse gases. Again, these observations are consistent with warming due to greenhouse-gas buildup.

In step with the changes in air temperature over land during the past century, global sea-surface temperature has increased about 0.7°C (1.3°F).[14] The heat has also "penetrated" into the deep ocean. Thus all the oceans have warmed by a combined estimated average of 0.037°C (0.066°F) per decade from 1955 to 1998, down to a depth of 3,000 meters (9,800 feet).[15] This warming is substantially more than can be accounted for by natural, internal variations in the climate system. Yet again it is in accord with the predictions of climate models in which the warming is due to greenhouse-gas buildup.[16]

Precipitation and Drought

Temperature change is not the only game in town. Rather, changes caused by the buildup of greenhouse gases seep throughout the entire climate system, and some of them are more important than the warming itself. So it is with changes to the hydrologic cycle, in particular with changes in patterns of precipitation and drought. Although these changes are just beginning to show themselves, they should engender particular concern because of their direct and potentially severe negative impacts on agriculture, water supplies, and ecosystems.

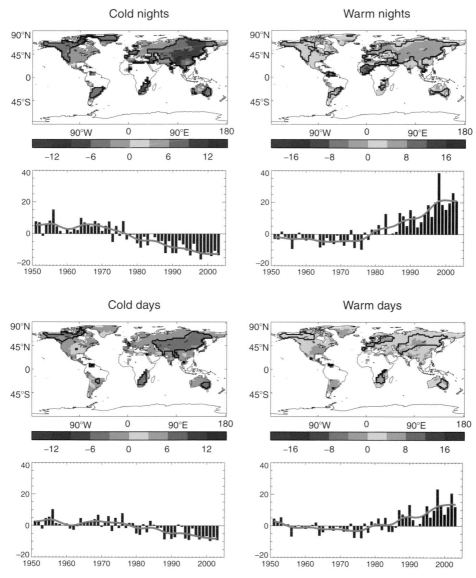

FIGURE 7.5

Changes in the number of cold and warm days and nights per year relative to the mean values, 1951–2002

The data are in days per decade, and the red curve represents the smoothed trend. (After Alexander et al. 2006)

Warming influences precipitation and drought in several ways. First, warming affects large-scale climate patterns. In general, the expectation is that as Earth warms, the downward limbs of the Hadley cells will migrate poleward, extending the regions of desert farther poleward as well (chapter 2).[17] In addition, warming increases sea-surface temperature, which may cause changes in large-scale circulation patterns, bringing more or less precipitation to different parts of the globe.

Warming also increases evaporation. Although most of the evaporation is from the ocean, evaporation over land is important because of its drying effect on soils. Warming also enhances the atmosphere's ability to hold water (the amount of water that air can hold increases 7 percent for every 1°C [1.8°F] rise in temperature). Thus warming should result in more atmospheric water vapor, an expectation now confirmed regionally by surface observation[18] and more globally by satellite data.[19]

One might expect further that more moisture in the atmosphere will lead to more precipitation. Indeed, consistent with this expectation are satellite observations that suggest that between 1987 and 2006 global mean precipitation increased 7.4 ± 2.6 percent per degree Centrigrade.[20] This correlation hardly tells the whole story, however. Weather systems are in part driven by water vapor. A more moist atmosphere results in more intense precipitation and floods, but the events are shorter and more infrequent, and thus the incidence of drought increases.[21] The commonly sited example is the western European summers of 2002 and 2003. The summer of 2002 saw heavy rains and widespread floods, but in 2003 the region was hit by a record heat wave and drought.

The Palmer Drought Severity Index illustrates these points (figure 7.6).[22] This index is a measure of soil moisture calculated from a combination of precipitation records and estimates of evaporation based on temperature. Although there were no twentieth-century trends in overall global precipitation, the index shows a decrease in soil moisture worldwide in that century. Perhaps it is premature to draw a firm conclusion from the latter trend given the short amount of time in which measurements have been taken; nonetheless, the trend is consistent with analyses of extreme events noted earlier.

Atmosphere temperature and precipitation are related in other ways, at least regionally. Over land, dry summers are warm, and wet summers are cool.[23] When conditions are wet, heat goes into evaporation rather than into warming, but when they are dry, heat warms the land surface and air. The situation is somewhat different in the winter. At high latitudes, cold air cannot hold much moisture; hence, when it is cold, it is also dry, and when it is warm, there is more precipitation.

KANSAS, 1935

In Kansas and neighboring states, the year 1935 was the driest of dry years. Severe drought had persisted off and on, but mostly on for four years. If the spring wheat sprouted at all, it died in the parched summer months. The plowed fields, bare now, were easy pickings for the dry winds that sucked up the soil, deposited it as great dust drifts that buried everything, and eventually dispersed it afar. (In fact, the dust during those years periodically darkened the skies of New York, Washington, and

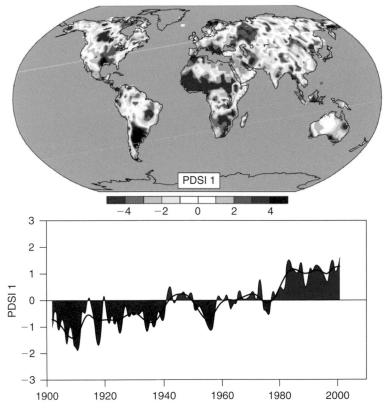

FIGURE 7.6

The change in the spatial and globally averaged Palmer Drought Severity Index, 1900–2002

The PDSI is a measure of soil moisture based on precipitation and temperature records and is expressed relative to local mean conditions: (*map*) reds and oranges are drier, and blues and greens are wetter, than the average; (*graph*) red is drier and blue is wetter than the average. The black line in the graph is the 10-year running average. (After Trenbreth et al. 2007:FAQ fig. 3.2)

other cities in the east.) The date April 14, 1935, became known as "Black Sunday," a particularly dismal day that spawned the name Dust Bowl thanks to a news reporter's imaginative flair (figure 7.7).

The Dust Bowl was the region most severely affected by the drought—western Kansas, southeastern Colorado, neighboring New Mexico, and the Oklahoma and Texas panhandles—but the drought affected a much larger region across the Great Plains. As the farms collapsed, so did the rural economy. There was nothing to do but leave, so 400,000 people fled the region in the decade to 1940. The Dust Bowl, arriving as it did in the Great Depression, has been etched in American memory by John Steinbeck in *The Grapes of Wrath* and by the photographs taken by Dorothea Lange, Arthur Rothstein, Walker Evans, and others.[24] What was the cause of the Dust Bowl, and does it hold lessons for the future?

FIGURE 7.7
Dust storm, April 18, 1935
This dust storm is about to engulf Stratford, Texas. (National Oceanic and Atmospheric Administration, http://www.photolib.noaa.gov/, photograph theb1365)

The prolonged drought, of course, was one culprit, but first we should recognize the other factors at play. Westward migration during the late nineteenth century had brought increasing numbers of farmers and cattle ranchers to the Great Plains. It was not until the beginning of World War I that agriculture really began to expand, however, driven by skyrocketing demand for wheat. After the war, farmers found themselves in territory all too familiar to history. The copious production could no longer be absorbed, demand decreased, prices began to fall, farmers planted more and more wheat to stay afloat, and the increased production forced prices even lower. Despite the dwindling prices, most farms endured through the mostly wet years of the 1920s, but then drought arrived in 1931.

Not that the 1930s marked the first drought that the West had ever seen. A previous drought had, after all, hit as recently as 1910, although it was much shorter.[25] The indigenous grasses had also naturally evolved to survive the droughts, and when droughts did arrive, the grass stubble protected the soil from erosion. By the time 1931 rolled around, however, the grass was gone, and deep plowing, a technique more appropriate for the wetter East, from where almost everyone had come, had left soil particularly vulnerable to wind erosion. Further, the use of eastern strains of wheat not resistant to drought added to the problem. As the drought deepened, the farmers kept plowing and planting, which only exacerbated the situation. They had no choice.

THE OCEAN–DROUGHT LINK

For thousands of years, the Great Plains and more generally the American Southwest and parts of Mexico and Central America have experienced periodic droughts of various lengths, from annual events to multidecadal "megadroughts" (figure 7.8).[26] Numerous studies have now made clear that these droughts are linked to conditions in the equatorial Pacific Ocean and to a lesser extent in the equatorial Atlantic Ocean.[27] The tropical Pacific Ocean holds an enormous amount of heat, and the warm sea-surface temperature drives atmospheric convection that transports both heat and moisture to extratropical regions. The tendency is that when tropical Pacific sea-surface temperature is warm, the American Southwest experiences relatively wet conditions, but when sea-surface temperature is cool, the Great Plains are dry.

Exactly why these connections exist is not clear. One possibility is that sea-surface temperature, by its subtle effect on ocean and atmosphere circulation, determines the location of springtime storm tracks. Regardless, the decreasing moisture in the Great Plains during times when the tropical Pacific Ocean is cool appears to be magnified by variations in soil moisture.[28] As noted, soil moisture depends on

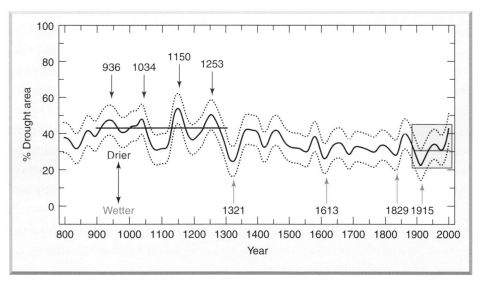

FIGURE 7.8

Records of drought in the American West, 800 to present, in terms of drought area relative to the long-term mean

The horizontal black line, the long-term mean, is a smoothed reconstruction, and the dashed lines represent the error at the 95 percent confidence interval. The black curve is based on tree-ring data, with drought defined in terms of the Palmer Drought Severity Index. The arrows indicate the central dates of the significant dry or wet periods; note the long periods of drought between 900 and 1300. Coastal regions and the Pacific Northwest are not included in this record. (After Cook et al. 2004, with permission)

evaporation rate, which in turn depends on the balance between precipitation and temperature. The Great Plains appear to be particularly sensitive to changes in soil moisture. In the specific case of the Dust Bowl, the effects of unusually cool tropical Pacific sea-surface temperature and dry soil were augmented by unusually high sea-surface temperature in the tropical Atlantic.[29] The warm Atlantic waters appear to have led to a pattern of atmospheric circulation that blocked the normally northward flow of moisture into the Great Plains from the Gulf of Mexico in the summer and autumn.

If the tropical oceans play a fundamental role in how global climate reacts to climate forcings, then the obvious question is how the buildup of greenhouse gases will affect those ocean regions. The American Southwest experienced severe droughts lasting several decades centered on the years 936, 1034, 1150, and 1253, a time interval broadly coincident with the Medieval Warm Period (chapter 6).[30] One idea is that the warming during this time led surprisingly to a cooler Pacific sea-surface temperature by increasing the vigor of easterly blowing winds. These winds caused increased upwelling of cold water in the eastern equatorial Pacific and then dragged this cold water westward, resulting in relatively long and possibly intense La Niña–like conditions (chapter 3).

The Medieval Warm Period was not forced by atmospheric CO_2 buildup, but the current CO_2-induced warming may have a similar influence on the equatorial Pacific. Climate models generally predict that subtropical regions will become drier and drier during the twenty-first century as the downward limbs of Hadley cells migrate poleward, as noted. This zonal shift, along with persistent La Niña (cold eastern equatorial Pacific) conditions, may contrive to produce drought conditions in the American Southwest not seen since the Medieval Warm Period.[31]

What, then, are the lessons of the Dust Bowl and other droughts in western North America? First, they emphasize the importance of conditions in the equatorial oceans as well as the connections between the ocean and the extratropical land regions. Second, warming may increase the probability of the occurrence of long and intense Dust Bowl–like droughts in the American Southwest by decreasing tropical Pacific sea-surface temperature and by magnifying the effects of large-scale zonal shifts in precipitation patterns. Third, the American Southwest has experienced severe megadroughts lasting several decades. Megadroughts are unknown in recent history, and when one does occur, it will likely have an enormous negative impact on economic life and society.

As a final note, the American Southwest is just one of a number of regions in which droughts are clearly linked to conditions in the ocean. For example, unusually warm North Atlantic sea-surface temperatures in the summer of 2005 have been implicated in the severe drought that occurred at that time in the Amazon,[32] and the periodic and sometimes severe droughts that have hit Southeast Asia and the Sahel of Africa (the southern borderland of the Sahara Desert) also have ocean links.[33]

WHY SOME WATER SUPPLIES ARE IN JEOPARDY

Warming affects water in other ways, particularly in regions where snow accumulates in the winter and melts in the spring. In such regions, winter precipitation affects the amount of spring runoff. More important, temperature affects the timing of that runoff—the warmer it is, the sooner the snowpack melts. If existing reservoirs are not large enough to capture this "early water," the water will simply flow into the ocean and be unavailable when it is needed most—in late summer and autumn. It turns out that about one-sixth of the world's population lives in regions where the water supply is dominated by snow or glacier melt and where summer storage capacity is limited, so the problem of early runoff is potentially serious.[34] The regions at various levels of risk include nearly all parts of the globe north of about 40°N latitude and parts of southern South America.

Areas of the American West are thus at risk. One climate model suggests that by 2050 the western United States will be 0.8 to 1.7°C (1.4 to 3.1°F) warmer than at present, although the region as a whole will experience little change in precipitation (as opposed to what will happen in the drought-prone region discussed earlier, which includes the Great Plains, the American Southwest, and northern Mexico, but excludes the Northwest and coastal regions). The past several decades especially have seen declining mountain snowpacks and other changes in the hydrology in the West.[35] More warming will result in more precipitation falling as rain and less as snow, and if the temperature projections given here hold true, maximum stream flow will come about a month earlier in 2050 than it does now, but because there is not enough storage capacity to handle the early water, it will be lost. This situation will affect the Columbia River, for example, whose summer water supply will no longer be sufficient to provide water release for both salmon runs and hydroelectric power. We will be forced to make a choice. The example emphasizes that it does not take a decrease in precipitation to put water supplies at risk; warming alone is enough to diminish the availability of water for all its uses.

Another part of the globe faces the same risk—northern India, western China, and neighboring lands of the so-called Himalaya–Hindu Kush region, basically the region that depends on water from melting glaciers of the Himalayas and the Tibetan Plateau (figure 7.9). To understand how important that meltwater is, consider that it constitutes as much as 70 percent of the summer flow of the Ganges River. The few observations that exist indicate that maximum runoff is now occurring earlier

FIGURE 7.9

Near the headwaters of the Indus River, Karakoram Range, Pakistan

Ice on the distant peaks of the Himalayas is the source of much of the water in the Indus River.
(Photograph by E. Wesker, http://www.euronet.nl/users/e_wesker/, with permission)

in the season and that total runoff has substantially increased because of increased melting.[36]

The snow and ice on the Himalayas and the Tibetan Plateau hold a great deal of water, so the region is unlikely to experience water shortages for the next several decades. Rather, the problem will emerge several decades from now as the glaciers start to disappear. Then the water supply will decrease rapidly "from plenty to want in perhaps a few decades or less."[37] The seriousness of the situation arises from the fact that we are talking about going from "plenty to want" for hundreds of millions of people, so although the danger is not imminent, avoiding significant economic, social, and human costs later on will require advanced planning. Furthermore, that planning will have to take into account the fact that the water resources will not recover—the glaciers have accumulated over centuries, so they are effectively irreplaceable (at least for a long, long time). When the water is gone, it is gone.

Severe Storms and Other Extreme Events

In the wake of Hurricane Katrina, which destroyed much of New Orleans in August 2005, scientific and public attention alike focused on the question of the effect of warming on hurricane activity. Although the answer to this question is a matter of current debate,[38] the more general question of global warming's impact on the frequency and severity of storms is less controversial because indices representing worldwide occurrences of extreme weather events, such as days of heavy precipitation and consecutive dry days, have displayed consistent patterns of increase in these occurrences over several decades.[39] In fact, the past three decades have seen a fourfold increase in the incidence of such events compared with the earlier part of the twentieth century. Katrina and other recent and unusual events, such as the heat wave that struck Europe in the summer of 2003, have prompted a related question: Can any specific event be blamed on global warming?

Although it may be tempting to do so, attributing a specific cause to an extreme weather event is simply not possible for two reasons. First, extreme events by their very nature result from a combination of causes that together have low probability of occurring. Second, extreme events are expected even in a perfectly stable climate.[40] A warming climate, however, can increase the *probability* of extreme events.

Consider that the summer of 2003 in Europe was the hottest in more than 600 years, if the long record of the timing of the grape harvest in Burgundy is any indication of temperature trends. Summer temperature exceeded the average by more than 5°C (9°F); rainfall was 50 percent below normal; and the unanticipated heat wave is said to have killed 22,000 to 45,000 people in a two-week period.[41] The event's intensity was extremely unlikely,[42] which can be seen by comparing the 2003

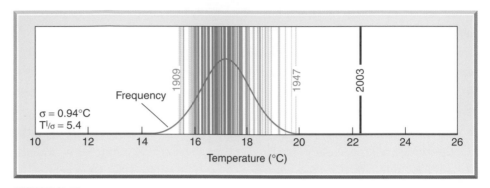

FIGURE 7.10

The extreme summer temperature in 2003, compared with summer temperatures from 1864 to 2002, Switzerland

The green curve is a fitted normal distribution of temperatures around the mean value of 17.2°C (63°F) and yields a standard deviation of 0.94°C (1.7°F) (two-thirds of the time, the average summer temperature is within 0.94°C of the mean). The 2003 temperature (*red line at 22.3°C*) is 5.4 standard deviations above the mean, a statistically extremely unlikely event. (Adapted by permission from Macmillan Publishers Ltd: Nature, Schär et al. 2004, copyright 2004)

summer temperature with summer temperature records from several cities in Switzerland that go back to 1864 (figure 7.10). The 2003 average summer temperature of 22.3°C (72.1°F) was 5.1°C (9.2°F) higher than the mean summer temperature of 17.2°C (63°F) for this 137-year period. Based on the summer temperature record, the likelihood of a summer as hot as that in 2003 is 0.00006 percent.[43]

At least two factors appear to have conspired to account for the heat wave. A shift in weather patterns had brought a prolonged high-pressure system over Europe preceding and during the heat wave, and the clear skies and lack of rain had dried out the soil. Consequently, less heat than normal was taken up in evaporation and transpiration, making more heat available to raise the temperature.

Extreme events are by definition unusual and, of course, may occur in the absence of greenhouse-gas warming. Global warming, however, makes extremely improbable events less improbable. The example of the European heat wave in 2003 perhaps illustrates this point. In any case, this particular event as well as Hurricane Katrina illustrate the devastation that extreme events can bring.

More Lessons from the American West: Ecosystem Responses

Warming can affect ecosystems in a variety of ways. It can have a direct impact on plant photosynthesis and respiration, increase susceptibility to pests and disease, force a redistribution of species upslope or toward higher latitudes, and modify phenology (the timing of organisms' life-cycle events, such as egg laying, flower blooming, and spring migration). The subject is enormous, but a few examples serve to illustrate both its scope and its significance.

TREES VERSUS PESTS

The first example concerns piñon pines. From 2000 to 2003, a severe drought in the American Southwest resulted in an infestation of bark beetle in 2002 and 2003 that caused the die-off of piñon pines over 12,000 square kilometers (4,700 square miles) in Colorado, New Mexico, and Arizona. Northern New Mexico was hit particularly hard. In one intensely studied site there, more than 90 percent of the trees died; regionally, mortality rates ranged from 40 to 80 percent.[44] Other coexisting plant species were also affected, although not to the same extent.

Many of the trees that died in 2002 and 2003 had survived the 1950s drought and others before it. This fact might come as a surprise given that the most recent drought was not too different from earlier ones in terms of length and amount of precipitation. The drought of 2000 to 2003 was warmer, however, which magnified the effect of low precipitation by causing soils to dry out even more and by stressing trees more than they would have been otherwise. Thus it was the warmer conditions in conjunction with the drought, not the drought alone, that led to the high mortality. The effects of such die-off typically ripple through the ecosystem. It may lead, for example, to increases in storm runoff and erosion and to changes in the food web, in species structure, and in land-surface properties, all of which may affect the local climate through feedbacks.

The pine forests of British Columbia are also facing a warming-related plight. The problem there is not too little water, but too much.[45] The culprit in this case is the fungus *Dothistroma septosporum*, which causes a disease known as needle blight. The disease is common at low-incidence levels in temperate forests worldwide, but it has recently become a scourge in plantations of nonindigenous pine in Brazil, southern Africa, and New Zealand, where the devastations have had significant negative economic impacts.

In British Columbia, needle blight attacks lodgepole pines. The disease was first reported in the early 1960s, but it was not until 1997 that the first signs of epidemic appeared. In surveys of managed and natural forests conducted from 2002 to 2004, trees in 92 percent of the 40,000 hectares (99,000 acres) surveyed were infected, and 9 percent of that area required replanting because of the high mortality rate.

The fungus spores appear in spring from dead leaves of the previous year. The fungus then infects the pine needles, causing them to die and reducing photosynthesis. This infection happens only in moist conditions; because the fungus's life cycle is one to two years, an epidemic requires several years of moist summers. The appearance of the current epidemic coincided with more summer rains beginning in the mid- to late 1990s, but with no significant changes in temperature.

In these examples from New Mexico and British Columbia, the climate changes and associated ecosystem responses occurred in years, not decades and longer.

Short-term fluctuations in climate are common, of course, but what these examples nonetheless illustrate is the kind of permanent changes that will occur as the globe warms and long-term changes set in.

FOREST FIRES

This brings us to a completely different, if indirect, way that climate change can affect ecosystems: warming and drought also increase the incidence of wildfires (figure 7.11). A recent study of 34 years of wildfire and climate history in the western United States has uncovered some remarkable statistics.[46] The frequency of large fires was nearly four times greater in the 14-year period from 1987 to 2003 than in the preceding 14 years. In addition, the total area burned was more than six times greater; the burn time (time from discovery to control) increased from 7.5 to 37.1 days; and the length of the fire season was longer by 78 days per year. During the years of increased fire activity, spring and summer temperature was on average 0.87°C (1.6°F) higher than in the earlier period, and throughout the latter period the incidence of fires from year to year was strongly correlated with spring and summer temperature. In addition, the increased frequency of fires was particularly marked in areas with elevations higher than 1,700 meters (5,600 feet).

FIGURE 7.11

Fire in Yellowstone National Park, 1988

A crown fire approaches the Old Faithful Photo Shop and Snow Lodge. (National Park Service, http://www.nps.gov/archive/yell/slidefile/fire/wildfire88/index.htm, photograph 13744)

The observations indicate that mountain snowpack plays an important role. The time that the snowpack melts is determined primarily by spring and summer temperature. When the snowpack melts early, soils dry out, and dry conditions set in earlier. Furthermore, the disappearance of snow exposes a larger region at relatively high elevation to fire earlier in the season.

It should be added that on a global scale, fires represent a substantial addition of carbon to the atmosphere. In fact, a recent estimate suggests that fires in the continental United States contribute 60 megatons of carbon to the atmosphere per year, which amounts to about 4 percent of U.S. fossil-fuel carbon emissions.[47] So wildfires are not trivial in this respect.

Perspective

It is worth reemphasizing a point made at the beginning of this chapter—that global warming is a phenomenon we can *observe*. Further, we are quite certain that the warming is due to buildup of greenhouse gases because (1) it is an expected physical consequence of the *observed* increase in CO_2 content of the atmosphere; (2) all the *observations* of warming and other changes on the planet are consistent with greenhouse-gas warming; and (3) no other drivers of climate that we know of can account for the changes.

I emphasize the observational basis for our understanding because all too often I have heard perfectly thoughtful people voice the suspicion that natural causes may account for the recent warming—after all, climate has changed naturally in the past. Again, this suspicion has no basis given what we observe. Furthermore, climate skeptics are prone to grasp at local or short-term trends—for example, the relatively cool winter of 2007/2008 in some parts of the Northern Hemisphere—to proclaim the end to global warming or even that warming is a myth. This chapter should make clear why such proclamations reflect misunderstanding of the climate system.

The consequences of warming are only beginning to be felt, and how they will play out remains uncertain. Some of these consequences are potentially dire and may have significant negative impacts on our well-being: extensive and severe drought that may affect agriculture; significant decreases in water supplies for the large proportion of world population that depends on ice and snowmelt for its water; increased incidence of severe weather events, and displacement of huge numbers of people as sea level rises; spread of plant and animal disease; collapse of whole ecosystems; and significant decreases in biodiversity. These consequences are the risks involved in global warming, and we need to recognize them as such in formulating rational responses to an uncertain future.

Melting ice at the edge of the Greenland Ice Sheet

Global warming is causing the great ice sheets on Greenland and Antarctica to melt, which will likely cause significant rise in sea level. During the most recent interglacial period, 125,000 years ago, when temperatures were not too different from what they are today, sea level stood 4 to 6 meters (13 to 20 feet) above its present position. Sea level may ultimately rise the same amount in the next several centuries, with potentially serious consequences. The water in the photograph is charged with "rock flower," produced by the grinding of the rock by the flowing ice. (Photograph by Minik Rosing, University of Copenhagen, with permission)

8 MORE CONSEQUENCES: THE SENSITIVE ARCTIC AND SEA-LEVEL RISE

THE ARCTIC IS UNIQUE. The vast, deep, and mysterious Arctic Ocean is shrouded in sea ice (figure 8.1) and surrounded by barren tundra and scrub forest tenaciously sprouting from a layer of soil that lies on permanently frozen ground. This frozen world is "framed in extremes."[1] Long, cold winters with little light transform abruptly into light-filled summers when snow and ice melt and the tundra comes alive.

But the Arctic is warming at an alarming rate. For those of us who do not live in the far north, we see it most dramatically in images of the steady disappearance of sea ice and of polar bears clinging to what ice remains. In concert with the decline in sea ice, the permafrost is also melting, and these changes are having a cascading, in some cases profound, effect on the physical environment and ecosystem. What is happening in the Arctic, and why?

The Arctic is not the only place on the planet where ice is a significant part of the landscape or seascape. This chapter also addresses what is happening to ice in Greenland and Antarctica. At both ends of the planet, loss of ice from ice sheets perched on land is contributing to sea-level rise, which has the potential to be catastrophic for the rest of the world.

The Shrinking Arctic Ice Cap

An understanding of the Arctic Ocean might begin with an examination of how the atmosphere and the ocean circulate in this part of the world. This process is actually quite important because the natural changes in circulation patterns result in considerable variability in the Arctic climate and point to mechanisms of likely future changes.

FIGURE 8.1
Sea ice

The sea ice near Tigvariak Island, Beaufort Sea, Alaska, 1950. (Photograph by Rear Admiral H. D. Nygren, National Oceanic and Atmospheric Administration, http://www.photolib.noaa.gov/, photograph corp1079)

THE ARCTIC OSCILLATION

The Arctic displays two dominant and distinctive "regimes" of atmospheric and ocean circulation. The two regimes oscillate back and forth every five to seven years, with the oscillations described by an index known as the Arctic Oscillation.[2] In the negative phase of the Arctic Oscillation, the atmosphere and ocean surface circulate in the clockwise (*anticyclonic*) direction (figure 8.2a).[3] The circulation draws freshwater from Siberian rivers and sea ice into the Beaufort Sea, where the ocean circulates in a gyre. The Beaufort Sea gyre thus acts as a reservoir for relatively fresh water. Also, because sea ice is trapped in the gyre, it thickens with each succeeding year as more ocean water freezes. During the negative phase of the Arctic Oscillation, atmospheric pressure over the Arctic Ocean is high, the winters tend to be cold, and the sea ice extends over a relatively large area of the ocean.

When the Arctic Oscillation is positive, the atmosphere and ocean currents shift to the counterclockwise (*cyclonic*) direction (figure 8.2b), effectively limiting the clockwise-circulating Beaufort Sea gyre to a smaller region against North America and isolating it from the rest of the Arctic Ocean. The effect is to enhance the so-called Transpolar Drift Stream, which is a rapid flow of ice through the central Arctic Ocean, across the pole, and out through the Fram Strait. The important consequence of these motions is that they flush both sea ice and freshwater entering the

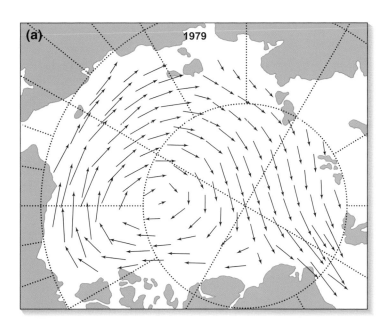

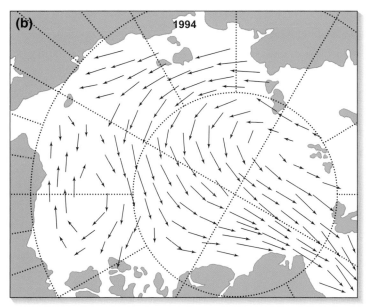

FIGURE 8.2

The contrasting patterns in ocean surface currents and ice flow during the 1979 anti-cyclonic and 1994 cyclonic climate regimes in the Arctic Ocean

In the anticyclonic regime (*a*), the atmosphere and ocean surface circulate in the clockwise direction; freshwater accumulates in the Beaufort Sea gyre; relatively high atmospheric pressures and low winter temperatures prevail; and sea ice tends to age and thicken. In the cyclonic regime (*b*), atmosphere and ocean currents result in a strong Transpolar Drift Stream, which flushes sea ice out through the Fram Strait into the North Atlantic, and the atmospheric circulation draws warm air masses into the Arctic from Russia. The two regimes alternate every five to seven years in what is known as the Arctic Oscillation. (After Rigor, Wallace, and Colony 2002)

Arctic Ocean from Siberian rivers out through the Fram Strait and into the North Atlantic. At the same time, the atmospheric circulation draws warm air masses into the Arctic Ocean from Russia, resulting in more rapid summer melting and less extensive and thinner sea ice. The flow of water out of the Arctic Ocean through the Fram Strait is replaced by saltier water entering through the Bearing Strait from the Pacific Ocean.

Because of the dominant negative and positive phases of the Arctic Oscillation, the Arctic Ocean's temperature and ice cover have swung back and forth between two conditions in a regular, natural beat that has no doubt persisted much longer than the 100 or so years over which Arctic climate has been observed.

THE LOSS OF SEA ICE

Like the rest of the planet, the Arctic has been warming; in fact, it has been warming at an even faster clip. From 1975 to 2005, Arctic surface temperature increased about 0.7°C (1.3°F) per decade,[4] which is more than double the 0.3°C (0.5°F) per decade rise in global surface air temperature. Arctic surface air temperature is closely linked to sea-ice cover. Obviously, the warmer it is, the more the ice melts, but there is more to the situation than that. The characteristic ocean circulation of the warm, positive phase of the Arctic Oscillation drags ice out of the Arctic Ocean, another mechanism of ice loss. There is also the important albedo (the proportion of solar energy reflected back to space [chapter 5]) feedback between melting and temperature. Ice is more reflective than water, so as the proportion of open water increases and that of ice decreases, more energy is absorbed by the ocean. Heat thus builds up both in the ocean and in the atmosphere and accelerates the melting. In short, the melting of sea ice amplifies greenhouse-gas warming.[5]

In the twentieth century, ice cover in the Arctic Ocean has typically reached a maximum of about 16 million square kilometers (6.2 million square miles) in March and then has shrunk to about 6 million square kilometers (2.3 million square miles) by September. (The large swing results because in the winter the Arctic receives no sunlight, but in the summer sea ice is particularly sensitive to the factors mentioned earlier.) Since 1979, the areas of both March and September sea ice have been shrinking each year at an increasingly rapid rate; by the summer of 2005, a new low of 5.6 million square kilometers (2.2 million square miles) of ice cover had been reached. But that was nothing compared with the completely unexpected drop in the summer of 2007, when ice cover shriveled to a mere 4.1 million square kilometers (1.6 million square miles).

Why the unusually rapid decline? Two factors may have contributed. First, 2007 appears to have brought an unusual combination of a strong Transpolar Drift Stream (see figure 8.2) and an influx of warm water flowing into the Arctic Ocean through the Bering Strait.[6] More fundamental, however, was the paucity of old, thick ice.[7] In 1988, most of the ice floating on the Arctic Ocean was more than 10 years old and several meters thick. In 1989, the Arctic experienced one of its periodic transitions from the negative to positive phase of the Arctic Oscillation. The result was that within a year most of the old ice had disappeared through the Fram Strait; in fact, by the summer of 1990 only about 30 percent of the sea ice was old (figure 8.3).

Such an abrupt depletion is probably not unprecedented, given the nature of natural climate oscillations. In the absence of warming, however, sea ice would likely build in age and thickness again with the return to the negative phase of the Arctic Oscillation. With warming, the proportion of old and thick ice has continued to

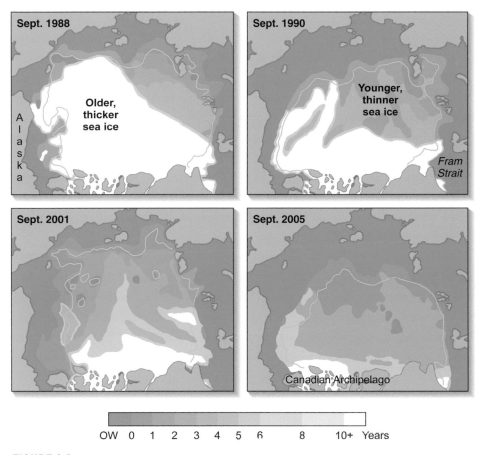

OW 0 1 2 3 4 5 6   8   10+ Years

FIGURE 8.3
The shrinking extent of old, thick sea ice in the Arctic, September 1988–September 2005
The area of old, thick ice has been rapidly shrinking as the result of prolonged melting and flow of ice out of the Arctic through the Fram Strait. OW = open water. The pale green line encloses the region of 90 percent ice concentration. (After Richter-Menge et al. 2006, as modified from Rigor and Wallace 2004)

dwindle, so that now there is hardly any remaining.[8] Although the events of 2007 may have been unusual, the steady decline of old and thick sea ice suggests that within several decades the Arctic Ocean may actually become ice free in summer.

IMPACTS

What are the likely impacts of declining sea ice?[9] The introduction of more freshwater into the North Atlantic may slow thermohaline circulation and thus the ability of that part of the ocean to remove carbon dioxide (CO_2) from the atmosphere (chapter 3). However, more open water means that the Arctic itself may become a sink for CO_2, although how important this sink might be probably depends on whether cold Arctic Ocean surface water eventually ends up somewhere in the deep ocean. How the decline in sea ice will affect midlatitude weather patterns is not known at this point. It may modify patterns such as the North Atlantic Oscillation (chapter 2) and thereby shift storm tracks over Europe, bringing more moisture to some regions and less to others.

Apart from the physical changes the shrinking sea ice will impart to ocean circulation, it will also have numerous ecological and other impacts, and the poster child for such impacts is the polar bear. Sea ice is an important hunting ground for polar bears, which feast mainly on seals, so as sea ice disappears, especially along coastal areas, so will the polar bear habitat. As a consequence of its loss of habitat, the polar bear was officially recognized in the United States as a threatened species in 2008. Peoples indigenous to the north also hunt on sea ice, so traditional cultures are affected by the shrinking of sea ice as well.

Less obvious but nonetheless serious from an environmental point of view is that the reduction of sea ice will cause the ocean to become significantly more wavy, which in turn will result in more coastal erosion.[10] Particularly susceptible to erosion are long stretches of the coastlines of Alaska and Siberia. However, a more open Arctic Ocean means that fisheries may become more extensive and that marine transportation will become much more efficient by the opening of northern sea routes.

Melting Permafrost and Changing Tundra

Interwoven with warming of the Arctic are a host of related environmental and ecological issues that bear on climate through their effect on carbon emissions to the atmosphere.[11] Most notably, warming has a significant effect on *permafrost*.

Permafrost underlies about 25 percent of the land area in the Northern Hemisphere. A map of the distribution of permafrost in Siberia, where most of the permafrost exists, shows its astonishing extent (figure 8.4). The term *permafrost* refers

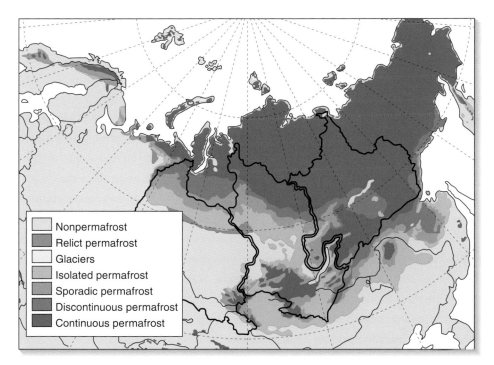

FIGURE 8.4

The distribution of permafrost in the Russian Arctic

Permafrost underlies about 25 percent of land area in the Northern Hemisphere and holds some 900 gigatons of organic carbon, which is more than the entire living biomass. Much of this carbon might be released to the atmosphere if the permafrost melts. (After Zhang et al. 2005)

to soil or rock that has been below 0°C (32°F) and frozen for at least two years. In most areas, it is overlain by a layer of soil—known appropriately enough as the *active layer*—that freezes in winter and thaws in summer. The thickness of the active layer and thus depth to permafrost depend on a variety of factors. Chief among them is summer air temperature, but winter snow cover, vegetation cover, and the amount of water in the soil also affect the depth to permafrost. To give a sense of what is typical, in Siberia's enormous Lena River basin the average depth to permafrost is 1.7 meters (5.6 feet).[12] The permafrost layer itself is usually tens to hundreds of meters thick, but in some locales it may be much thicker—for example, up to 1,500 meters (4,900 feet) thick in parts of Siberia that have remained mostly ice free (and thus without an insulating layer) throughout the Pleistocene.[13]

Permafrost is now melting. The Lena River basin study, for example, found that on average the active layer extended 0.32 meter (1 foot) deeper in 1990 than in 1956. In the same period, the average air temperature of eastern Siberia increased by 0.9°C (1.6°F).[14] Similar observations have been made in Alaska and northwestern Canada. Because permafrost depth more or less follows summer temperature when averaged over a decade or more, it is possible to use climate models to project how permafrost will change. By one such (admittedly uncertain) projection, the present 10.5 million square kilometers (4.1 million square miles) of permafrost will shrink to about 1 million square kilometers (386,000 square miles) by the beginning of the twenty-second century.[15]

THE EFFECTS OF MELTING PERMAFROST ON CLIMATE

The melting of permafrost may cause a significant increase in carbon emissions, which in turn will aggravate greenhouse-gas warming. About half the carbon in permafrost is contained in frozen loess, of which there are particularly vast deposits in Siberia. *Loess* is windblown dust, and frozen loess is known as *yedoma*. Yedoma represents "relict soils" of the glacial steppe-tundra ecosystem.[16] It typically consists of 50 to 90 percent ice and 2 to 5 percent carbon, which is much more carbon than exists in typical tundra soil. By one estimate, the worldwide yedoma reservoir contains 500 billion metric tons (gigatons) of carbon mainly in the form of organic debris, with an additional 400 gigatons contained in nonyedoma permafrost.[17] This is more carbon than contained in either the atmosphere (760 gigatons) or the living biomass (600 gigatons) (chapter 4). The melting of permafrost will release some of this carbon to the atmosphere as the organic material oxidizes to form CO_2 or methane.[18] If this release happens over the next century, as seems possible, it will exacerbate greenhouse-gas warming.

Another way that warming influences climate is through its effect on snow cover. The disappearance of snow earlier in the year is particularly important because it decreases albedo when insolation is relatively high. For example, at the North Slope of Alaska, the date of snow melt now arrives about eight days earlier than it did 40 years ago, resulting in an increase in energy input to the ground of two watts per square meter per year.[19]

THE EFFECTS OF MELTING PERMAFROST ON LOCAL HYDROLOGY

Warming and melting of permafrost have an important effect on the local hydrology. Permafrost acts as an impermeable barrier to downward infiltration of water, so soils above the permafrost layer are usually wet in the summer. As permafrost thaws, more water is available and the area of standing water and lakes increases with the development of *thermokarsts*, which are pits and depressions formed by subsidence as the underlying permafrost melts. The thermokarsts fill with water to form ponds, bringing about ecosystem changes. With further thawing, the permafrost layer drops farther below the surface, and the surface water table drops.

When regions in the permafrost layer eventually melt completely through to the base, water may drain into the ground below the permafrost. This stage results in the eventual drying out of soils and disappearance of lakes.[20] The problem of drying is further aggravated by thaw-enhanced river runoff. Again, this effect has been documented in Siberia, where runoff increased by 2 ± 0.7 cubic kilometers (0.5 cubic mile) per year from 1936 to 1999.[21]

Warming also causes drying by decreasing soil moisture, which depends on the balance between precipitation and evapotranspiration (the combined loss of water from evaporation and transpiration by plants). One study from Alaska documented a 41-year decrease in this "surface-water balance" brought on not by decreased precipitation, but by increased evapotranspiration.[22] The latter was itself caused by higher summer air temperature, which had been rising about 0.04°C (0.07°F) per year.

The drying out of Arctic soils by these various mechanisms increases the rates of plant and soil respiration more than the rate of vegetative growth. The result is a net loss of CO_2 from tundra ecosystems. One study of the Alaskan tundra puts the loss there at 40 grams of carbon per square meter per year.[23] Fortunately, there are signs that the rate of CO_2 loss is decreasing as the ecosystem slowly adjusts to new moisture levels.

THE GREENING OF THE ARCTIC

Earlier snowmelt also means a longer growing season, which together with warming has resulted in a "greening" of the Arctic. This effect has been documented by satellite observations of tundra regions. For example, in northern Alaska vegetation greenness increased 17 percent from 1981 to 2001, corresponding to an increase of an estimated 171 ± 81 grams per square meter (15 ± 7 grams per square foot) in summer plant biomass (even though there is a net annual loss of biomass carbon for the reasons stated earlier).[24] The warming also changes the ranges of shrubs and trees, which are reflected by the encroachment of the boreal forest on the tundra. The encroachment represents an increase in vegetation productivity and thus in uptake of atmospheric CO_2. However, more southerly forested areas of the Arctic are displaying just the opposite response—they are becoming browner rather than greener, apparently because of increasing "moisture stress."[25]

Ice Loss Beyond the Arctic and Sea-Level Rise

Ice is not restricted to the Arctic Ocean. The two large ice sheets of the world, meaning the broad expanses of ice resting on solid ground, are in Greenland and Antarc-

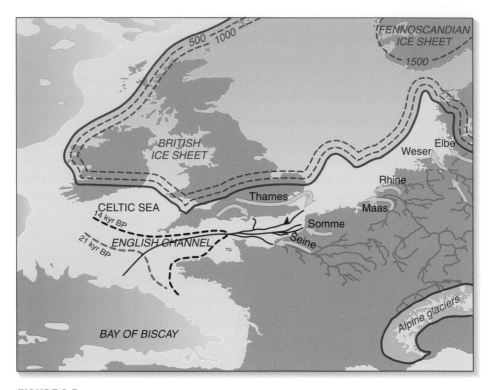

FIGURE 8.5

*The English Channel and surrounding region and the maximum extent of ice about 21,000
years ago at the most recent glacial maximum*

Around 21,000 years ago, sea level was 120 meters (394 feet) below its present-day level, no water
separated the British Isles and continental Europe, and what is now the English Channel was a major river valley into which the Seine, Rhine, Thames, and other rivers flowed. The 400-kilometer-
long (250-mile) Channel River flowed into the Atlantic Ocean to the southwest. In the north, an
enormous, continuous sheet of ice covered Scandinavia, most of the North Sea, and the British Isles.
The shore lines at 21,000 and 14,000 years ago, when sea level was 75 meters (246 feet) below present sea level, are shown. The dashed curves within the ice indicate its estimated thickness in meters.
Kyr B.P. = thousands of years before the present. (After Ménot et al. 2006, with permission)

tica. Together, these two ice sheets contain 30 million cubic kilometers (7.5 million
cubic miles) of ice, representing nearly 80 percent of the freshwater on the planet.
Just as the Arctic Ocean is sensitive to warming, so are the ice sheets. Their fate in
a changing climate merits serious attention because as the ice sheets disappear, sea
level will rise. This problem is different than the disappearance of Arctic Ocean sea
ice, which when it melts has no effect on sea level because the ice is floating on the sea.

Sea level has exhibited substantial fluctuations throughout the Pleistocene. For
example, about 21,000 years ago, at the most recent glacial maximum, sea level
stood approximately 120 meters (400 feet) below its present level (figure 8.5).[26] With
the subsequent melting of the Northern Hemisphere ice cap, sea level rose quickly,
but then, according to geological evidence, the rising slowed so that over the past

3,000 years sea level rose about 1 to 2 centimeters (0.4 to 0.8 inch) per century before the twentieth century. Tide gauge data, however, put the twentieth-century rise
at a more rapid and accelerating 15 to 20 centimeters (6 to 8 inches), apparently
from increased glacial discharge.[27] Global warming is expected to cause sea level to
rise more rapidly in the future. It is no exaggeration to state that the potential for
warming-induced, catastrophic sea-level rise has engendered considerable alarm in
the scientific community.

If we neglect for the moment the Greenland and Antarctic ice sheets, sea level is
expected to rise mostly from thermal expansion of the ocean (water expands when
heated) and to a lesser extent from the melting of mountain glaciers and permafrost.[28] Based on a variety of climate models and assumptions about the growth of atmospheric CO_2 content over the twenty-first century (chapter 9), the 2007 report of
the Intergovernmental Panel on Climate Change (IPCC) projected that sea level will
increase between 20 and 60 centimeters (8 and 24 inches) by 2100.[29] Other estimates
suggest the rise may be as high as 1.4 meters (4.6 feet).[30] Such increases stretched
over a century may not seem like much, but in fact they would create havoc in many
regions by steady erosion of coastlines and by increasing the chances of catastrophic
floods associated with unusually high tides and storms. For example, as noted in
chapter 1, along the coast of China a sea level rise of just 0.5 meter (20 inches) will
likely affect all regions that are lower than 10 meters (33 feet) above sea level.[31] That
region is home to 144 million people, or 11 percent of the current population of the
country. But these consequences would still be minuscule compared with what
might happen should the Greenland and Antarctic ice sheets shrink significantly.

THE GREENLAND ICE SHEET

The Greenland Ice Sheet covers some 1.7 million square kilometers (656,000 square
miles). In the unlikely event that it were to melt completely (which would take centuries), sea level would rise 7 meters (23 feet). There are two reasons for the growing
alarm over the fate of the Greenland Ice Sheet. First, Greenland's climate is particularly sensitive to global warming due to the albedo feedback. In addition, much of
the ice sheet is presently at high elevation, which keeps its surface cold, but as the
ice sheet shrinks and elevation decreases, the surface will become warmer, enhancing melting further. It has been estimated that if Greenland warms by 2.7°C (4.9°F)
above 1990 levels, ice loss by melting will exceed snowfall.[32] According to nearly all
projections (described in chapter 9), this 2.7°C threshold will be exceeded by the
end of the twenty-first century. Melting, however, is not the entire story.

The second reason for alarm is that the Greenland Ice Sheet also appears to be
losing mass by the flow of its marginal glaciers into the ocean. This process is poorly
understood, so the rate of ice disappearance cannot now be accurately determined.

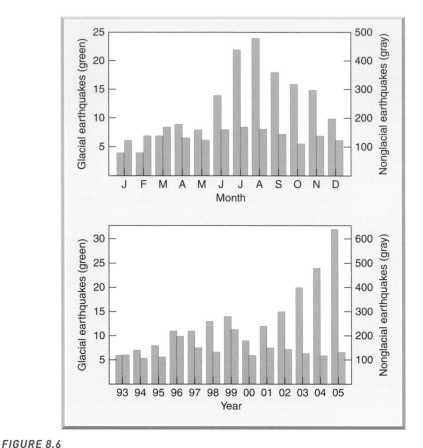

FIGURE 8.6

The increase in the number of glacial earthquakes in Greenland, 1993–2005

The earthquakes are most frequent in the summer. The increased activity indicates that the glaciers are sliding more rapidly. (After Ekström, Nettles, and Tsai 2006, with permission)

Furthermore, the 2007 IPCC projections of rising sea level did not take this process into account. Recent observations of glacial flow include "glacial earthquakes," which are small tremors typically lasting for several tens of seconds and produced by the periodic sliding of the glaciers as they move down valleys. Most occur in July and August and originate from the three largest Greenland glaciers: Kangerdlugssuaq, Jakobshavn, and Helheim. For the years 1993 to 2001, there were 6 to 15 glacial earthquakes per year; since then, the number has been steadily rising so that in 2005 there were 32 such events (figure 8.6).[33] The rise presumably indicates increased glacier flow from increased summer meltwater, which reduces friction at the glacier base.

A similar picture emerges from measurements of the height of the Greenland Ice Sheet surface obtained from satellite radar data. Ice loss calculated from these observations more than doubled, from 91 ± 31 cubic kilometers (22 cubic miles) per year in 1996 to 224 ± 41 cubic kilometers (54 cubic miles) per year in 2005.[34] Along the coast, meanwhile, the glaciers are flowing and losing ice much more rapidly. Flow rate at the nose of Jakobshavn glacier (also known by its Greenlandic name, Ilulissat), for example, increased 95 percent from 1996 to 2005, corresponding to an increase in ice loss from 24 to 46 cubic kilometers (6 to 11 cubic miles) per year. At the same time, the glacier's floating ice tongue has been breaking up and retreating (figure 8.7).

Satellite-based measurements of gravity, which provide an estimate of total mass of ice, corroborate the radar data.[35] One such study estimated that Greenland lost 248 ± 36 cubic kilometers (59 cubic miles) of ice per year from April 2002 to April 2006; furthermore, the mass loss has been accelerating such that between May 2004 and April 2006 the loss was more than double the loss between April 2002 and April 2004.[36] Nearly all the ice was lost from southern Greenland. Another estimate puts ice loss at 239 ± 23 cubic kilometers (57 cubic miles) and yet another at 101 ± 16 cubic kilometers (24 cubic miles) per year, reflecting the uncertainty in the estimates based on measurement of the gravity field.[37]

FIGURE 8.7

The steady retreat of the Jakobshavn (Ilulissat) glacier, Greenland

The lines show the calving front at different years. This glacier is one of the largest in the world, draining 7 percent of the Greenland Ice Sheet area. (NASA Goddard Space Flight Center, Scientific Visualization Studio, http://svs.gsfc.nasa.gov/vis/a000000/a003300/a003374/)

THE WEST ANTARCTIC ICE SHEET

If the Greenland Ice Sheet is the 800-pound gorilla of sea-level rise, the West Antarctic Ice Sheet is the wildcard (figure 8.8). Unlike the East Antarctic Ice Sheet, the base of which is mostly above sea level, the entire base of the West Antarctic Ice Sheet is below sea level, making it sensitive to ocean temperature. (Except for its margins, however, the West Antarctic Ice Sheet is grounded; that is, it is resting on solid earth.) The Antarctic Peninsula of West Antarctica also extends much farther north than any part of the East Antarctic Ice Sheet. These two factors combine to make the West Antarctic Ice Sheet much more sensitive to fluctuations in climate than the East Antarctic Ice Sheet, which appears to have persisted intact for millions of years.[38]

The Antarctic Peninsula is a rather cold part of the world where the temperature remains well below freezing for most of the year, so one might not expect that the approximately 2°C (3.6°F) warming that has occurred there since the 1950s would have much influence on the extent of the ice sheet. Nevertheless, a number of

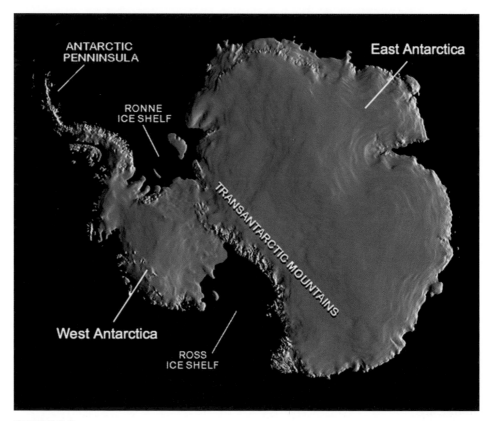

FIGURE 8.8

Antarctica and its major ice shelves

The map is based on satellite data. (United States Geological Survey, http://terraweb.wr.usgs.gov/projects//Antarctica/AVHRR.html#titlestereo, map I-2560)

studies have documented the retreat of ice shelves (that is, the floating ice around the ice sheet). The ice shelves act as buttresses to the glaciers that feed into them, so as the ice shelves retreat, the glaciers have been flowing more rapidly.

A dramatic demonstration of the importance of the ice shelves' buttressing effect occurred in February and March 2002, when 3,250 square kilometers (1,255 square miles) of the Larsen B ice shelf shattered spectacularly (figure 8.9). To emphasize the magnitude of this event, the shattered ice shelf was nearly as large as the state of Rhode Island (which is about 4,000 square kilometers [1,560 square miles]). The breakup appears to have been unprecedented in the Holocene.[39]

The immediate consequence of Larsen B's breakup was that because it could no longer act as a buttress to nearby land-based glaciers, they began to flow much more rapidly.[40] With increased flow, the glaciers' noses break up more rapidly, and the glacier front retreats inland. Glacier retreat appears to be widespread. One study, for example, found that of 244 mapped glaciers on the Antarctic Peninsula, the vast majority (87 percent) were retreating and that the boundary between retreating and advancing glaciers has been moving steadily south (toward the pole).[41]

The same process seems to be happening to the West Antarctic Ice Sheet where it flows into the Amundsen Sea. There, the grounded West Antarctic Ice Sheet has been disappearing at the rate of 51 ± 9 cubic kilometers (12 cubic miles) per year for the past decade, and at the same time the glaciers have been discharging ice into the ocean at an accelerating rate. Satellite measurements of the change in surface elevation of the ice shelves indicate that they thinned at rates up to 5.5 ± 0.7 meters (18 feet) per year from 1992 to 2001, which amounts to a loss of 1 to 7 percent of their thickness in the nine-year period, or 45 ± 5 billion metric tons of ice per year.[42] The thinning, due to ocean warming, has caused the ice sheets' *grounding line* (the line between floating and grounded ice) to retreat, which in turn has reduced both lateral and basal traction, explaining why the glaciers have been flowing more rapidly.

As in the case of Greenland, satellite observations measuring the gravity field indicate that the ice mass of the West Antarctic Ice Sheet has diminished by 148 ± 21 cubic kilometers (35 cubic miles) per year.[43] This diminishment in ice mass corresponds to a sea-level rise of 0.4 ± 0.2 millimeter (0.016 inch) per year from 2002 to 2005.

Lessons for Sea-Level Rise from the Most Recent Interglacial Period

Insight into how rapidly and how much sea level may rise comes from the study of the most recent interglacial period, which began about 136,000 years ago and lasted for about 20,000 years.[44] During that time, sea level was 4 to 6 meters (13 to 20 feet) higher than it is now, as evidenced by the existence of 125,000-year-old corals now exposed several meters above present sea level in the Florida Keys (figure 8.10).

FIGURE 8.9
The breakup of the Larsen B ice shelf, Antarctica, February and March 2002
The collapse of the Larsen B ice shelf, nearly the size of the state of Rhode Island, was followed by more rapid flow of the glaciers feeding into the area. This satellite image was taken on March 7, 2002, and is of an area approximately 149 by 186 kilometers (93 by 116 miles). It is a false-color composite designed to enhance the visibility of the region of breakup. (NASA/GSFC/LaRC/JPL, MISR Team, http://visibleearth.nasa.gov/view_rec.php?id=2573)

FIGURE 8.10
Fossil brain coral from Windley Key, Florida

This coral grew 125,000 years ago and is now several meters above sea level. At the time of its growth, sea level was about 1 meter (3.3 feet) above the coral and 4 to 6 meters (13 to 20 feet) above its present-day level. (Photograph by Dan Muhs, United States Geological Survey, with permission)

Average global surface temperature then was probably not much greater than it is today. This interglacial corresponded to and was apparently a result of the substantial increase in high-latitude insolation brought about by an increase in Earth's tilt (chapter 6). This increased insolation would have affected mainly the Northern Hemisphere, so probably the Greenland Ice Sheet was much diminished at the time.

Model simulation of warming caused by higher summer insolation suggests that the Greenland Ice Sheet did not disappear completely, but that Northern Hemisphere ice fields (mainly Greenland, but also Iceland) contributed 2.2 to 3.4 meters (7.2 to 11.2 feet) to sea-level rise, with the remainder contributed by the melting of Antarctic ice.[45] The simulation also indicates that by 2100, if CO_2 emissions grow by 1 percent per year, Greenland will be as warm as or warmer than it was 125,000 years ago and that by 2130 (when atmospheric CO_2 content will be three times higher than it is today) it will be much warmer, and the ice will disappear at an even faster rate.[46] If Earth reacts the same way it did during this past interglacial period, we shall eventually see the same 4- to 6-meter (13- to 20-foot) sea-level rise.

From a practical point of view, it is just as important to know the rate of sea-level rise as the magnitude because the rate will determine how well society will be able to adapt to that rising level. Sea level is currently rising at 2.6 ± 0.04 millimeters (0.1 inch) per year.[47] Careful dating of corals suggests that at the beginning of the most recent interglacial period, sea level rose a "catastrophic" 20 millimeters (0.8 inch) per year, or 2 meters (6.6 feet) per century.[48] If this were to happen over the next century, it will seriously tax human society's ability to adapt, flooding vast stretches of what are now coastal areas, displacing hundreds of millions of people, and engendering strife.

Perspective

This chapter illustrates how sensitive the high-latitude regions are to global warming and how they may influence conditions on the rest of the plant. Warming in the Arctic, for example, amplifies global warming by means of two important feedbacks. Warming decreases albedo as areas of ice- and snow-cover shrink, which allows the exposed ocean and land surfaces to absorb more solar energy, causing further warming and melting. Warming also melts permafrost and changes surface hydrology, enhancing CO_2 and methane emission rates to the atmosphere and again magnifying greenhouse-gas warming.

At the same time, warming will bring widespread and significant changes to the Arctic's physical environment and its ecosystems. The effects of warming ripple beyond the few examples mentioned here; nearly all plant and animal communities will be affected, as will human communities—for example, buildings and other structures will slowly collapse when the permafrost beneath them melts. Further, the specific responses will likely differ considerably from one Arctic region to another, so how changes in the Arctic will affect the global carbon budget remain unclear.

Ice loss from the Greenland and West Antarctic ice sheets will also reverberate around the globe by causing sea level to rise more than it would due to just warming of the oceans and with potentially serious consequences. No one is suggesting that sea level will rise many meters within the next century, but the problem is that we do not know either how much sea level will rise or, more important, the rate of rise. During the most recent interglacial period 125,000 years ago, sea level stood 4 to 6 meters (13 to 20 feet) higher than its present position. The possible rate of rise in our future, however, is really anyone's guess. Particularly troubling is that we may reach a point of no return in this century, when the disappearance of much of the Greenland Ice Sheet will become irreversible no matter what we do.

We see the effects of global warming around us. What the future will bring depends on the growth of greenhouse-gas emissions; projecting that future hinges on our ability to determine how the climate system will respond to this growth.

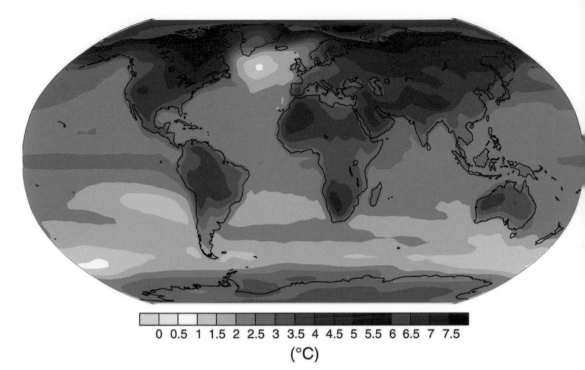

0 0.5 1 1.5 2 2.5 3 3.5 4 4.5 5 5.5 6 6.5 7 7.5

(°C)

A multiclimate model average projection of global surface temperature change in the decade 2090–2100 compared with the years 1980–1999

Climate models provide projections of future climate. Although the models are only approximate representations of nature, uncertainty about how the future will play out depends more on the uncertainty of how the rate of greenhouse-gas emissions will change than on imprecision of climate models or on uncertainties in the science. The projection assumes that greenhouse-gas emissions will grow at a moderate rate in the twenty-first century. (After Solomon et al. 2007:fig. SMP 6)

9 CLIMATE MODELS AND THE FUTURE

UP TO NOW in this book, the focus has been on past and present climate and how the climate system works. The workings of the climate system are at the heart of our ability to make realistic projections of how climate will change in the future. To make such projections, we need models, so this chapter describes the general nature of climate models. Models, of course, are not reality, so inevitably the question of how well climate models represent the real world arises. The models possess both intrinsic and practical limitations, which, along with the uncertainties about how the climate system works, translate into uncertainty in projections. Nonetheless, experience suggests that climate models provide some useful and insightful information about our future.

The big question of what the future climate will be hinges less on uncertainties associated with science-based model projections than on uncertainty about future levels of greenhouse-gas emissions. Accordingly, we must be able to relate greenhouse-gas emissions to climate change, which is one of the important utilities of climate models. This question also brings us to a final question of how we go about developing sensible policies to respond to climate change. The response has to be based in risk assessment, which we touch on as a way of determining the climate-change targets needed to guide policy.

What Are Climate Models?

Models of the climate system evolved from early computer models that sought to forecast weather. The basic processes, which are represented by mathematical equations in the models, are the motions of air and water around the planet in response to uneven heating by the Sun, the Coriolis forces, and other factors, with the motions limited by conservation of mass, energy, and momentum.[1] The most complex models, known as *atmosphere–ocean general circulation models*, treat the coupled circulation of the atmosphere and the ocean, but simpler models of just atmospheric circulation are also of great use. The models produce a three-dimensional change in temperature and other fundamental characteristics of the climate system over time.

To solve the equations that describe their motions, models represent the atmosphere and the ocean as three-dimensional grids of points, such that each grid point represents a certain volume and has a specified set of properties. In the case of the atmosphere, for example, the properties include barometric pressure, wind velocity, humidity, and temperature. As the properties of one point change, so do those of all the neighboring points. The models calculate how the properties of each point change with time in response to external influences, such as the amount of solar energy received, and to changes in the neighboring points' properties.

The distances between grid points are important. The closer the grid points, the better the model's spatial resolution—in other words, the size of features that can be distinguished. Good spatial resolution is necessary for depicting certain features, such as patterns of tropical rainfall, that can influence projections of the future. To illustrate, in the National Center for Atmospheric Research CCSM3 model, the atmosphere volume is represented by a 2.8×2.8 degree (about 300×300 kilometers [180×180 miles] at the equator) horizontal grid of points. The vertical dimension is accommodated by a stack of 26 more closely spaced grids. The ocean volume in this model is represented by a 1×1 degree horizontal grid of points (locally the spacing is less) stacked 40 high. These spacings determine the resolutions of atmosphere and ocean phenomena that can be modeled. Smaller features, such as individual storms or ocean eddies, are not seen in the model simulations. The time steps (a typical step is a half-hour) are also important in producing realistic solutions. However, available computer power sets practical limits to spatial and temporal resolutions.

One challenge in climate models is to represent the feedbacks realistically. To understand why, imagine that one of our grid points represents a volume of air over the North Atlantic Ocean: How are we going to determine the temperature of that volume? Suppose the air is colder than the ocean, so heat moves directly from the ocean to the atmosphere. The properties of the water and the atmosphere, on the one hand, and the equation describing heat transfer, on the other, are well known, so the amount of heat that will flow between the two can readily be calculated. However, heat is also transferred by evaporation of the water, which is dependent on temperature. The water vapor forms clouds, which shade and thus cool the ocean surface, so we have to know the proportion of clouds. Here we must estimate. Perhaps we have an equation that relates the fraction of clouds to relative humidity, which is known as a *parameterization*. It may be based on theory or on observations; in any case, the parameterization is only an approximation of reality and perhaps not a very good one. Different climate models have different parameterizations relating the system's different characteristics, and for this reason they yield somewhat different results.

Specifying a set of properties for our volume of atmosphere for given values of ocean temperature, ocean albedo, sea-surface roughness, wind velocity, cloud cover, atmospheric aerosol load, atmospheric greenhouse-gas content, solar radiation, and so on leaves plenty of room for uncertainty. Indeed, much of the work in climate modeling revolves around developing realistic parameterizations relating the various feedbacks. The interaction of the small-scale physics from which local characteristics are computed and the extension of these features in space result in extremely complex climate models, which is to be expected because, after all, the models are constructed to mimic an extremely complex real world.

Are Climate Models Credible?

Can climate models produce sensible results? Two important characteristics of the models suggest they can and do. First, the models mimic the "emergent" character of the real climate system. In other words, large-scale features develop as a consequence of the system's complexity, not because the models embody any mathematical description of the phenomena. For example, the Intertropical Convergence Zone (ITCZ) of tropical rainfall is a feature of the real climate system and results from a combination of processes, such as the Coriolis force, the seasonal cycle of insolation, and the convective motions in the atmosphere (chapter 2). The ITCZ also appears in the models and results only from the forces acting on each grid point, not because ITCZ-like features have been built into them.[2]

One may wonder how climate models can be meaningful in light of the fact that weather is chaotic. In chaotic systems, small differences in initial conditions lead to large differences in how the systems evolve. The evolution of chaotic systems is therefore predictable only in the short term, and the predictions rapidly lose accuracy and thus meaningfulness the farther out into the future we go. For that reason, the weather forecasts for today or tomorrow are now fairly accurate, but forecasts four or five days out are not. Climate models, in contrast, do not display chaotic behavior.[3] Instead, they produce stable climates—that is, ones that persist through time. This is their second most important characteristic. Moreover, as time passes, the models also display familiar natural phenomena, such as seasonal cycles, trade winds, modulations in the jet stream, common weather patterns, cyclonic storm patterns, and even events like El Niño–Southern Oscillation (ENSO), but without changes in long-term average conditions (unless external forcings are folded into the model).[4]

That climate models display the same emergent behavior and reach stable states as the real climate system does mean that the models allow researchers to explore how the real system works. Thus one can query a model regarding what happens to the climate system in response to a specific forcing, such as (no surprise) an increase in the greenhouse-gas content of the atmosphere—which is in fact what modelers do. In this way, the model becomes the basis for projections.

Climate models cannot be evaluated precisely. They can, however, be at least generally evaluated by being tested against the climate of the recent past, particularly of the past century or even of the past 25 years, which is well documented from satellite observations. Thus when fed data on greenhouse-gas emissions and other known quantities, the models closely simulate climate of the past century. For example, they capture changes due to external forcings, such as the global cooling due to the eruption in 1991 of Mount Pinatubo.[5]

In addition, certain periods of large climate shifts in the more distant past, such as the end of the most recent glacial maximum, serve as "benchmarks" for climate models. In general, the models are adept at simulating these shifts, although the climates of more remote periods are obviously far less precisely known than that of the previous century.

An interesting characteristic of climate models is that the average of a number of models better represents the basic features of the observed climate system than does any individual model. The implication is thus that the simulations have myriad small biases, but that in aggregate many of the biases cancel each other out. For this reason, the 2007 report of the Intergovernmental Panel on Climate Change (IPCC) simulated climate with an "ensemble" of 23 well-tested models.[6]

Modelers can also glean valuable information by producing multiple simulations with the same model. A climate model's output is a combination of natural climate variability and external forcings. The former is random and thus unpredictable, representing model "noise." Running a number of simulations under identical forcings is a way to obtain a measure of the noise and to distinguish it from the effects of external forcings.

A particular model's ability to simulate the present climate is not necessarily a good criterion with which to judge its ability to predict future climate correctly. The climate system is now operating and will continue to operate beyond the bounds of conditions with which we have experience. In fact, for the future there is no single "best model." One model may be more accurate under one set of conditions, but another model may be more accurate under another set. As a corollary, the range of model output is not a real measure of uncertainty, which might in reality be greater.

Despite the growing confidence in models and their "credibility," they are not portrayals of reality because they contain many uncertainties and do not include all natural phenomena. For example, water vapor and cloud feedbacks (chapter 5) are significant sources of uncertainty. Also, most models do not now include the poorly understood feedbacks involving the terrestrial biosphere. Model simulations are more realistic for some parameters—notably changes in global mean temperature—than for others. Among the latter, for example, is change in sea level, which is unlikely to be accurately predicted because the models cannot consider all the important mechanisms of ice sheet loss (chapter 8).

Simulating the Twentieth Century: Anthropogenic Versus Natural Causes of Climate Change

As noted in chapter 7, the observed changes in climate, particularly over the past three or four decades, are consistent with the assertion that warming is the result of human activities—it is being driven mainly by anthropogenic forcings rather than by natural forcings. Model simulations support this assertion. But attributing the warming to human activities requires modelers to distinguish between the effects of external forcings, be they natural or anthropogenic, and of natural internal climate variability. The distinction thus must be made for a specific timescale because internal climate variability operates across a wide range of timescales, as described in chapter 1. Even so, attribution can seldom, if ever, be made with absolute certainty, so in practice it means being able to state that an observed change is consistent with the attributed cause.

The most general measure of warming is global mean surface temperature. In an effort to identify the cause of warming, the 2007 IPCC report compared 58 simulations from 14 models of twentieth-century warming.[7] When the models were run with both anthropogenic and natural forcings, the simulations closely replicated the observed warming. But when they were run with only natural forcings—that is, volcanic eruptions and variations in solar irradiance—the simulations displayed a significant and increasing diversion from the observed trend as of around 1960 (figure 9.1). In other words, to reproduce the actual late-twentieth-century warming, the models *required* the input of anthropogenic forcings, specifically those due to greenhouse gases and aerosols.

If the warming is due primarily to anthropogenic forcings rather than to natural internal climate variability, then other signatures of the warming should also be apparent in the model simulations. The spatial pattern of warming is one such diagnostic. The entire globe has essentially warmed, with the Arctic in particular and to a lesser extent the Antarctic warming much more rapidly than the middle and tropical latitudes. This pattern is not expected from natural modes of climate variability such as modulations of the North Atlantic Oscillation (chapter 2) and ENSO (chapter 3). Simulations that included anthropogenic as well as natural forcings rep-

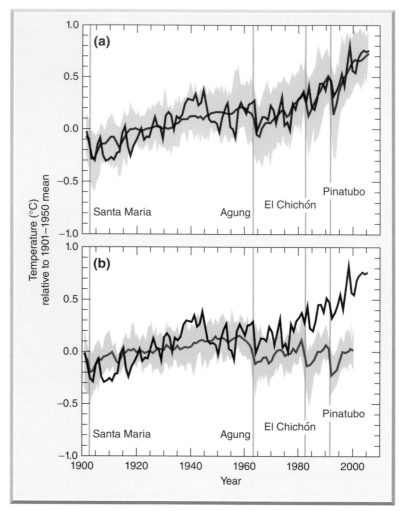

FIGURE 9.1

Modeling global mean surface temperature

The observed change in global mean surface temperature (*black line*) during the twentieth century was compared with (*a*) 58 simulations (*range shaded in light orange*) from 14 climate models that included both anthropogenic and natural forcing factors, the ensemble mean of which is shown by the heavy red line; and (*b*) 19 simulations (*range shaded in light blue*) from 5 climate models that included only natural forcing factors, with the ensemble mean indicated by the heavy blue line. The vertical lines indicate times of major explosive volcanic eruptions. Only the simulations that include anthropogenic forcings (*a*) reproduced actual climate observations. (After Hegerl et al. 2007:fig. 9.5)

licated the observed spatial pattern of warming; those with no anthropogenic component did not (figure 9.2). This result has also been found to be the case in analyses of regional patterns of warming.[8]

Several other diagnostic, temperature-based indexes have also been developed. They include the land–ocean temperature contrast, the Northern Hemisphere's north–south temperature gradient, temperature difference between hemispheres, and temperature variation with the seasons.[9] These indexes' utility is that they should display a coherent response to greenhouse-gas warming, but not to natural variability. The indexes change coherently through the twentieth century, and model simulations indicate that anthropogenic forcing can account for all the changes.

The other observed changes described in chapter 7—fewer cold days and nights, more warm nights, warming of the troposphere but cooling of the stratosphere, rise of the tropopause, increasing heat content of the ocean, and even heat waves—arise only in simulations that include anthropogenic forcings.[10] Although no one claims that climate models are precise representations of nature, they do provide strong, credible evidence that the warming of the late twentieth century is driven by anthropogenic forcings.

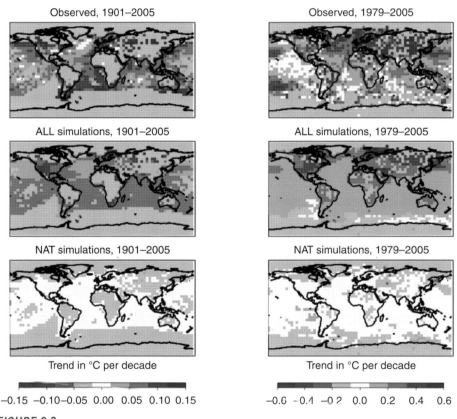

FIGURE 9.2

Spatial distribution of observed warming compared with the distribution obtained from model simulations, 1901–2005 and 1979–2005

The top maps ("Observed") show the observed gridded temperature change. The middle maps ("ALL simulations") show the average gridded temperature change of 58 simulations from 14 models that included both anthropogenic and natural forcings and most closely resemble the observed temperature change (*reds*). The bottom maps ("NAT simulations") show the average gridded temperature change of 19 simulations from 5 climate models that included only natural forcings. (From Hegerl et al. 2007:fig. 9.6)

Chapter 7 also noted that the probability of the occurrences of highly unusual events, such as the European heat wave in 2003, has increased because of anthropogenic forcings. Observations support this assertion; in addition, model simulations show that distinct geographical patterns of extreme heat emerge in a world driven by greenhouse gases, but not in one driven only by natural forcings.[11]

Peering into the Future

Credible predictions of future climate obviously depend on the availability of climate models that are adequate representations of nature. They also obviously depend on the rate at which greenhouse gases continue to be emitted. Patently not obvious, however, is what emission rates will actually be because they will depend on a complex group of interacting factors, such as the increase in population, the growth of the economy, the distribution of income, the policies of governments, the pace of technological change, the adoption of new technologies, and so on.

EMISSION SCENARIOS AND TEMPERATURE PROJECTIONS

It is, of course, impossible to know how the various factors affecting emission rates will play out to shape our future world. In order to provide a means of making model projections of future climate, the IPCC and other agencies have developed a set of "emissions scenarios": "Scenarios are [alternative] images of the future. . . . They are neither predictions nor forecasts. Rather, each scenario is one alternative image of how the future might unfold."[12]

In its 2007 report, the IPCC chose three previously developed (in 2000) emission scenarios as a basis for making model projections.[13] The three scenarios (described in detail in note 13) were chosen to cover a widely varying future world, ranging from one in which emissions are high (A2) to one in which they are medium (A1B) and then to one in which they are low (B).[14] To emphasize again, the future is hardly certain, so wishful thinking aside, no one scenario is more or less probable than any other. In the IPCC's modeling effort,[15] the projected increase in global mean surface temperature by the year 2100 for the high-emissions (A2) scenario is 3.6°C (6.5°F) higher than the 1980 to 1999 average; for the medium-emissions (A1B) and low-emissions (B1) scenarios, the projected mean temperature increases are, obviously, less: 2.8°C (5°F) and 1.8°C (3.2°F), respectively (figure 9.3).

There are several things to note about these projections. First, these temperature increases refer to the mean of an ensemble of the 23 models. To reiterate a point made earlier, the ensemble mean has been found to be a more accurate representation of simulations of past climate, so that may be true of future projections as well.

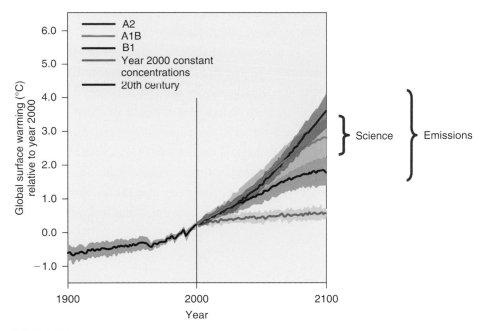

FIGURE 9.3

Projected changes in mean global surface temperature under three emissions scenarios

The heavy lines are the ensemble means of 23 model simulations run as part of the report of the Intergovernmental Panel on Climate Change (Meehl et al. 2007), and the shading refers to the ±1 standard deviation range of individual models. Also shown is the change in temperature if the greenhouse-gas content of the atmosphere had been held constant at year 2000 concentrations. The brackets to the right of the graph show the approximate uncertainties in future temperature changes resulting from lack of scientific knowledge and from growth of emissions, assuming that the three emission scenarios represent the possible range in emissions growth. (After Intergovernmental Panel on Climate Change 2007:fig. SPM 5)

Second, the temperatures change with time, as shown in figure 9.3. Thus the temperatures projected for the high- and medium-emissions (A2 and A1B) scenarios do not significantly diverge from each other until the late twenty-first century. This outcome reflects the mitigating effect of aerosols on warming (chapter 5), which is greatest in the high-emissions (A2) scenario (that is, the atmosphere in the A2 world contains more surface-cooling aerosols than it does in the A1B world). Another feature is that the projected temperatures under all scenarios are about the same until about 2030, after which the full benefit of the low-emissions (B1) scenario starts to become apparent. One reason for this outcome is that at least for the next decade we will be experiencing warming that is already "committed" or "in the pipeline" from recent greenhouse-gas emissions (chapter 5), as is illustrated by the curve in figure 9.3 showing how warming would have been dampened had the greenhouse-gas content of the atmosphere been contained at year 2000 concentrations. According to this projection, there would have been about 0.4°C (0.7°F) of committed warming in the system had greenhouse gases been kept at 2000 levels. But the levels are now higher, and the committed warming is thus greater.

Third, figure 9.3 shows for each scenario the shaded range in model results,[16] which by 2100 are approximately ±0.5°C (0.9°F) for each scenario. Again, recall that this uncertainty is not the true one, which may be somewhat larger; it simply reflects the difference among models. Nonetheless, this spread of 1°C (that is, ±0.5°C) is probably a reasonable minimum estimate of the uncertainty, given that we do not understand the climate system completely. Contrast this 1°C uncertainty with the spread in projected global temperature for the three emission scenarios. In the latter case, the spread is nearly 2°C (3.6°F). In other words, uncertainty about how the future plays out depends more on the growth of emissions than it does on lack of scientific knowledge.

OTHER PROJECTED CHANGES

Changes in global mean surface temperature do not tell the whole story, of course. There are also distinct geographical patterns of temperature change, with the greatest warming occurring over land, especially over the Arctic, and the least warming occurring over the Southern Ocean and the North Atlantic (figure 9.4). The warming is also accompanied by other projected changes, which, not surprisingly, are the ones that are already beginning to occur.

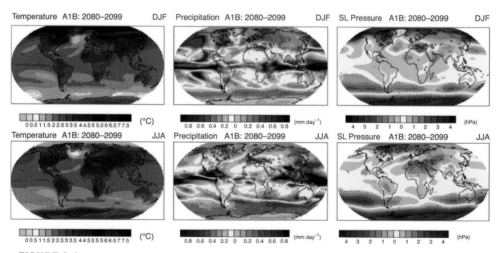

FIGURE 9.4

Projected changes in temperature, precipitation, and sea-level air pressure for winter (December–February) and summer (June–August), 2080–2099

The patterns of these changes, under the ICCP medium-emissions scenario, in part reflect large-scale changes in atmospheric circulation. The stippled regions denote those changes for which there is a relatively high degree of agreement among the different model simulations. hPa, hPascal = millibar. (From Meehl et al. 2007:fig. 10.9)

Most ominously, precipitation patterns are projected to change, with a large increase in rainfall in equatorial regions, less precipitation in the midlatitudes, and somewhat greater precipitation at high latitudes (see figure 9.4). The changing patterns of precipitation are reflected in changes in air pressure, which themselves are an indication of how atmospheric circulation patterns may be changing (see figure 9.4). In particular, the sea-level pressure changes indicate a poleward expansion of the Hadley cells (chapter 2), which in turn forces a poleward shift of midlatitude storm tracks, in part accounting for the higher precipitation at high latitudes and the drop in precipitation at midlatitudes. The latter will occur especially in the summer, with an attendant decrease in soil moisture because of the combined effects of less rain and higher temperature. The precipitation will be concentrated in more intense but less frequent events, with longer periods of no precipitation. These factors imply that the midlatitudes will experience a much greater risk of drought. The models also project a future with even more frequent and more intense heat waves, fewer and less severe cold spells, a steady decrease in the daily temperature range as nights become warmer, and a decrease in the number of frost days throughout the high and middle latitudes.

The model simulations also indicate that the oceans will continue to warm. The warming will initially occur slowly and be restricted mainly to the ocean mixed layer, typically the upper 100 meters (330 feet) or so, but later in the twenty-first century it will begin to extend into the deep ocean. Recall that the slow warming of the ocean represents committed warming, which cannot be stopped. Because the ocean is slow to warm, the surface water remains cooler than its equilibrium temperature with the atmosphere would dictate, which keeps the atmosphere cooler than it would otherwise be.

The ocean and living biota are projected to become progressively less efficient at removing carbon dioxide (CO_2) from the atmosphere, a potentially important feedback that will force CO_2 levels higher than they would otherwise be. As the atmospheric CO_2 content rises, the ocean will become more acidic, upsetting the biological cycling of carbon in ways that are not understood (chapter 4). Warming will bring widespread thawing of permafrost as well as emissions of CO_2 and methane from the carbon now held in permafrost, another potentially important feedback (chapter 8).

The extreme warming of the Arctic results in loss of sea ice, as the people and wildlife that live there are already experiencing (chapter 8). Under the high-emissions (A2) scenario, models indicate that the Arctic becomes ice free in the summer by the latter part of the twenty-first century. But the extent of sea ice has been shrinking much faster over the past several years than projected (chapter 8), suggesting that an ice-free Arctic may be upon us much sooner than that.

The models also indicate that sea level will rise. For example, under the medium-emissions (A1B) scenario, models project that by 2100 sea level will be 0.21 to

0.48 meter (8.3 to 18.9 inches) higher than the 1980 to 1999 average sea level. This projection considers only sea-level change from thermal expansion of the ocean, melting of glaciers and ice caps, and changes in rates of snowfall. It does not consider disintegration of parts of the Greenland and West Antarctic ice sheets. The increased flow of ice and other changes to those ice sheets, as we have seen (chapter 8), may begin to have an important influence on sea level in this century, but will likely have greater influence in subsequent centuries.

Perspective

How do we go about developing a rational policy response to climate change in view of the uncertainties about what the future will bring? First, we should recall the argument that future climate depends on future greenhouse-gas emissions and realize that these emissions are closely tied to our energy future. In other words, rational response to climate change essentially involves developing rational energy policy.

One way of figuring out what to do in the face of uncertainty is through risk assessment. At the most basic level, this assessment involves (1) identifying a consequence of climate change and then calculating its cost, (2) determining the probability that the consequence will come about as function of temperature increase,[17] and (3) establishing the cost of preventing the consequence (or, alternatively, the cost of adaptation) as a function of the probability.[18] Needless to say, numerous complexities accompany such an analysis. Some consequences, such as destruction of sensitive ecosystems, are far more likely to occur with a small degree of warming than are other consequences, such as widespread drought with a substantial negative impact on agriculture. Also, we can only consider the *probabilities* that events will occur, and these probabilities increase with warming.

Then there is the matter of determining costs. Those associated with the consequences of warming are difficult to assess because they depend, in part, on the degree of warming; the costs of mitigation and adaptation efforts are even more difficult to determine for the same reason. The point, however, is that assessing risks in an uncertain and probabilistic world is a rational basis for determining target temperatures above which mitigation and/or adaptation costs to society are less than the costs of consequences.

Risk assessment is a formidable subject far beyond this book's scope. An essential part of assessing risk, however, is being able to relate greenhouse-gas emissions to climate change, which is what climate models allow us to do. The IPCC's effort to relate emissions and global temperature change is illustrated in table 9.1, which shows the changes in year 2050 emissions as a percentage of year 2000 emissions that will bring about various target temperature increases.

TABLE 9.1 **CHANGES IN THE GROWTH OF CARBON DIOXIDE (CO_2) EMISSIONS REQUIRED BY 2050 TO BRING ABOUT SPECIFIC WARMING TARGETS**

Additional Radiative Forcing (W/m^2)[a]	Atmospheric CO_2 Contents (ppm)[b]	CO_2-eq. Contents (ppm)[c]	Target Temperature Increase[d] (°C)	Peak Year for Emissions	Change in Emissions by 2050 (% of 2000 Emissions)
2.5–3.0	350–400	445–490	2.0–2.4	2000–2015	−85 to −50
3.0–3.5	400–440	490–535	2.4–2.8	2000–2020	−60 to −30
3.5–4.0	440–485	535–590	2.8–3.2	2010–2030	−30 to +5
4.0–5.0	485–570	590–710	3.2–4.0	2020–2060	+10 to +60
5.0–6.0	570–660	710–855	4.0–4.9	2050–2080	+25 to +85
6.0–7.5	660–790	855–1130	4.9–6.1	2060–2090	+90 to +140

[a] Calculated radiative forcing from greenhouse-gas buildup. W/m^2 = watts per square meter.

[b] ppm = parts per million by volume.

[c] CO_2-eq. is the concentration of CO_2 that would have the same radiative forcing as the forcing due to all of the greenhouse gases (CO_2, methane, nitrogen oxide, ozone, halocarbons).

[d] The equilibrium global mean temperature (the temperature at the time the climate finally stops changing—that is, the time that all the committed warming has occurred) above the preindustrial temperature, based on a best estimate of climate sensitivity to radiative forcing.

Source: The data were derived by the 2007 IPCC report from numerous studies. See T. Barker, I. Bashmakov, L. Bernstein, J. E. Bogner, P. R. Bosch, R. Dave, O. R. Davidson, B. S. Fisher, S. Gupta, K. Halsnæs, G. J. Heij, S. Kahn Ribeiro, S. Kobayashi, M. D. Levine, D. L. Martino, O. Masera, B. Metz, L. A. Meyer, G.-J. Nabuurs, A. Najam, N. Nakićenović, H.-H. Rogner, J. Roy, J. Sathaye, R. Schock, P. Shukla, R. E. H. Sims, P. Smith, D. A. Tirpak, D. Urge-Vorsatz, and D. Zhou, "Technical Summary," in *Climate Change 2007: Mitigation of Climate Change. Contribution of Working Group III to the Fourth Assessment Report of the Intergovernmental Panel on Climate Change*, edited by B. Metz, O. R. Davidson, P. R. Bosch, R. Dave, and L. A. Meyer (Cambridge: Cambridge University Press, 2007), 25–94.

Two points are evident. First, limiting the global mean temperature increase to, say, 3°C (5.4°F), which is a commonly cited, prudent target temperature increase, will require limiting atmospheric CO_2 content to less than about 480 parts per million. To achieve this goal, global CO_2 emissions by 2050 will have to be *reduced* below the year 2000 emission level or at best to that level—not just slowed and certainly not allowed to grow. Second, due to the inertia in the climate system, the reduction will have to begin soon—probably within the next decade—if the target is to be achieved. Perhaps the particular goal of limiting temperature increase to less than 3°C is impractical. With this question in mind, in any case, we can now time to turn our attention to our energy future.

Trains loaded with coal departing from the Rawhide coal mine near Gillette, Wyoming

Climate is inextricably linked to energy production because how the climate changes will depend primarily on how the rate of CO_2 emissions from the burning of fossil fuels changes. The main challenge for the twenty-first century is how to meet the world's insatiable appetite for electricity without bringing on catastrophic climate change. Keys to the generation of clean electricity are efficient burning of coal and natural gas in combination with CO_2 capture and sequestration, nuclear power, and solar and wind power. The Rawhide and other coal mines in the Powder River Basin of Wyoming provide about 20 percent of the coal demand in the United States. (Photograph by Jeremy Foster, with permission)

10 ENERGY AND THE FUTURE

AS A WORLD SOCIETY, we have some control over our future climate by how we produce and consume energy simply because energy production accounts for about 80 percent of carbon dioxide (CO_2) emissions. Most of the rest comes from deforestation and land-use change related to agriculture, so it is in tackling energy use that we will make the biggest dent in carbon emissions.

A look at world energy consumption in 2005 by fuel type reveals what probably most of us already suspected. Fossil fuels—oil (37 percent), coal (27 percent), and natural gas (23 percent)—together accounted for 87 percent of the world's energy (figure 10.1). The total consumption globally that year was 488 billion billion joules (or exajoules; for an explanation of the units and measures applied to energy, see table 10.1).

Carbon dioxide emissions from fossil-fuel burning for 2005 were about 29 billion metric tons (gigatons, see table 10.1). (Another 7 gigatons are thought to be generated by deforestation and agricultural activities, bringing total anthropogenic emissions to about 36 gigatons of CO_2 per year.) Recall from chapter 9 (and the estimates in table 9.1) what it would take to limit the increase in average global surface temperature to less than 3°C (5.4°F) above preindustrial temperature: cutting CO_2 emissions to 2000 levels or lower by 2050, corresponding to emissions of about 25 gigatons or less of CO_2 per year.[1] Furthermore, if energy consumption continues to grow at its present pace and energy production patterns do not significantly change, the emissions rate from burning fossil fuels will *grow* more than 50 percent to well more than 40 gigatons per year by 2030 (figure 10.2).[2] According to table 9.1, even if emissions were to stabilize but not decrease by then, global mean temperature would increase by more than 5°C (9°F) sometime later in the twenty-first century.

What are the solutions to this dilemma? An examination of CO_2 emissions by sector reveals that electricity generation accounts for about one-third of present-

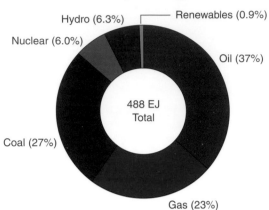

Total global energy

Hydro (6.3%) — Renewables (0.9%)

Nuclear (6.0%)

Oil (37%)

488 EJ
Total

Coal (27%)

Gas (23%)

FIGURE 10.1

The fuels used to produce all energy worldwide, 2005

Total global consumption of fuels in 2005 amounted to 488 exajoules (10^{18} joules), 87 percent of which was generated by fossil fuels. (Data from Energy Information Agency, http://www.eia.doe .gov/)

TABLE 10.1 **SOME COMMON UNITS OF ENERGY AND POWER**

Prefixes

Kilo (k)	10^3
Mega (M)	10^6
Giga (G)	10^9
Tera (T)	10^{12}
Peta (P)	10^{15}
Exa (E)	10^{18}

Units and Some Common Amounts

Joule (J) = basic unit of energy
   exajoule = 10^{18} joules
British thermal unit (Btu) = energy needed to heat 1 pound of water 1°F = 1,055 joules
   1.055 exajoule = 10^{15} (1 quadrillion [quad]) Btu
Toe = tons of oil equivalent = 41.868×10^9 joules
   1 million toe = 41.868 petajoule
Watt (W) = unit of power (work) = energy per unit time = 1 joule/sec
   kilowatt = 1,000 watts, megawatt = 10^6 watts, gigawatt = 10^9 watts
Watt hours (Wh) = energy = 1 W delivered over 1 hour = 1 joules/sec $\times$ 3,600 sec/hr = 3,600 joules
   1 kilowatt hour = 3.6×10^6 joules, 1 megawatt hour = 3.6×10^9 joules, 1 gigawatt hour = 3.6×10^{12} joules; 1 kilowatt hour = 3,413 Btu
Metric ton (t) = 1,000 kilograms

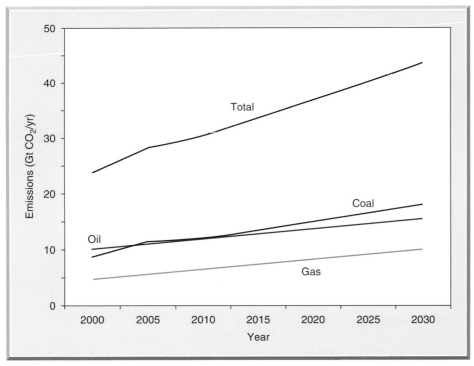

FIGURE 10.2

Emissions of CO$_2$ from fossil-fuel burning according to fuel type, 2000–2005 and projected to 2030

Emissions are expressed as gigatons (10^9 metric tons) of CO$_2$ per year. The business-as-usual projection is not a portrayal of the future. It simply suggests what might happen under the assumptions that no major conflicts or other events will disturb the world economy and that no changes in government policies will occur. (Data from Energy Information Agency, http://www.eia .doe.gov/)

day CO$_2$ emissions. More important, emissions from this sector are growing far more rapidly than emissions from other sectors (figure 10.3). (The emissions rate from transportation is also growing rapidly, but future transportation will likely derive more of its power from electricity and alternative fuels.) Therefore, although avoiding significant climate change will require mitigation efforts on other fronts as well, the crux of the problem is how we are going to meet the world's electricity needs without bringing on catastrophic climate change.

Most electricity, as it turns out, comes from coal (40 percent), natural gas (20 percent), nuclear power (15 percent), and hydropower (16 percent), with oil (6.9 percent) and renewable sources other than hydropower (2.2 percent) playing only minor roles.[3] Hydropower resources are limited and thus cannot accommodate the growth in demand. Renewable sources include geothermal, biomass (wood, landfill gas, agriculture by-products, and the like), wave and tidal, solar, and wind power. According to a recent report, even with major investment geothermal power is

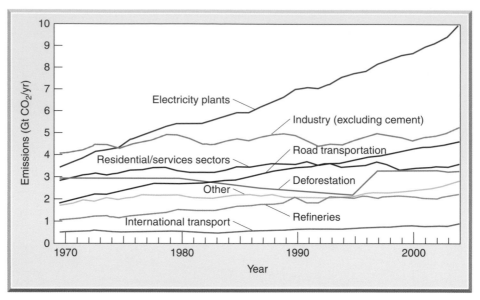

FIGURE 10.3

Annual emissions of CO₂ from various sectors

Emissions are expressed as gigatons of CO_2 per year. The category *deforestation* includes burning of wood for fuel; *other* includes domestic surface transportation, nonenergy use of fuels, cement production, and venting/flaring from oil production; *international transport* includes aviation and marine transport. (After Rogner et al. 2007:fig. 1.2)

unlikely to account for more than 5 percent of electricity needs in the United States by 2050,[4] and presumably that is true for the rest of the world as well; biomass, which currently accounts for about 20 percent of electricity generated in the United States from renewable sources (excluding hydropower), cannot substantially expand; and wave and tidal power are insignificant as a proportion of total world electricity capacity. This leaves wind and solar power, both of which are on the verge of major expansion. Central to the production of "clean" electricity, then, are (1) more efficient burning of coal and natural gas in combination with CO_2 capture and sequestration, (2) nuclear power, and (3) solar and wind power, so they are the focus of this final chapter.

Coal: A Vast Enterprise

When it comes to coal, there are certain inescapable realities. Coal is cheap and abundant, especially in the countries that are and will continue in the foreseeable future to be the largest energy consumers (the United States, China, and to a lesser extent India). Furthermore, the production of energy from coal is supported by a large, extant infrastructure. Coal will therefore continue to provide a large proportion of the world's energy in the decades to come. The consumption of coal has dramatically increased in the past several years. Indeed, in 2006 consumption jumped by an astronomical 4.5 percent, compared with the 10-year average growth rate of 2.8 percent before this year.[5]

The burning of coal presents significant problems. First, coal produces more CO_2 per unit energy than do other fossil fuels because it has a much higher ratio of carbon to hydrogen. In comparison with natural gas, for example, coal burning generates about 170 percent more CO_2 for an equivalent amount of energy, depending on coal type. Second, coal burning produces sulfur dioxide, nitrogen oxides (collectively, NO_x gases), particulates, mercury, cadmium, uranium, and other toxic pollutants. Third, in addition to having negative environmental, health, and safety consequences, mining and transport of coal are themselves energy intensive.

COAL SUPPLY AND DEMAND

"The scale of the enterprise is vast." So states a recent study on energy from coal.[6] To emphasize the point, the study begins with some startling statistics. Among them are that the equivalent of more than five hundred 500-megawatt (see table 10.1) coal-fired power plants in the United States are producing in aggregate about 1.5 gigatons of CO_2 per year, and, more striking, China has been constructing the equivalent of two 500-megawatt coal-fired power plants *per week*, a capacity comparable to the entire annual power grid of the United Kingdom.

World coal reserves (that is, the amount for which recovery is reasonably certain) are estimated to be 905 gigatons, an amount sufficient to power the world for the next 200 years.[7] Just four countries—the United States (27 percent), Russia (17 percent), China (13 percent), and India (10 percent)—hold 67 percent of the recoverable reserves, with most of the remaining reserves concentrated in just a few other countries (table 10.2). This situation makes coal even more localized than oil in terms of geographical distribution, although it is concentrated in more politically stable parts of the world. Coal has another important advantage: it is inexpensive. In 2007, electricity generated from coal cost anywhere from about 10 to 30 percent of that generated by oil and natural gas.

Projections have coal consumption increasing rapidly over the next two decades, due almost entirely to growth in China and the United States (figure 10.4). In China, annual consumption is projected to rise from 43 to 100 exajoules worth of coal from 2004 to 2030, reflecting a combination of the country's rapid economic growth and its lack of indigenous oil and natural-gas reserves. China's coal-fired electricity-generating capacity in 2004 was about 271 gigawatts; by 2030, that capacity will have nearly tripled to 770 gigawatts, consuming 59 exajoules worth of coal annually. Much of the rest of China's coal is used by the industrial sector—for example, in the production of steel and pig iron (China is the world's leading producer of both). By

TABLE 10.2 *WORLD'S RECOVERABLE COAL RESERVES IN GIGATONS AS OF JANUARY 2003*

Region/Country	Bituminous and Anthracite[a]	Sub-bituminous	Lignite	Total
United States	112.2	100.1	30.4	250.9
Russia	49.1	97.4	10.4	157.0
China	62.2	33.7	18.6	114.5
India	90.1	0.0	2.4	92.4
Non-OECD[b] Europe, Eurasia	45.4	17.0	28.4	90.8
Australia, New Zealand	38.6	2.4	38.0	79.1
South Africa	47.2	0.2	0.0	47.3
OECD Europe	17.7	4.5	17.1	39.3
Brazil	0.0	10.1	0.0	11.1
World total	479.7	270.4	155.0	905.1

[a] Anthracite, bituminous, and lignite are different coal types with decreasing carbon and heat contents.

[b] OECD = Organization for Economic Cooperation and Development.

Source: Energy Information Agency, *International Energy Outlook 2007* (Washington, D.C.: Department of Energy, 2007), available at http://www.eia.doe.gov/oiaf/ieo/index.html.

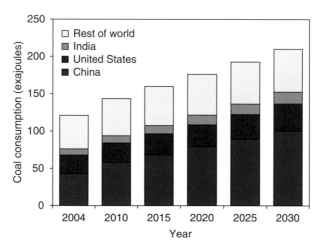

FIGURE 10.4

Current and projected coal consumption in India, the United States, China, and the rest of the world

Coal consumption is expressed in units of exajoules. (After Energy Information Agency, http://www.eia.doe.gov/)

2030, China's industrial coal use is expected to reach 39 exajoules, nearly 40 percent of the country's total consumption.

With the assumption that existing laws and policies do not change, coal consumption in the United States is projected to rise at an annual rate of 1.5 percent, from 24 exajoules in 2004 to 36 exajoules in 2030, when it will provide 57 percent of the electricity produced. Meanwhile, India's coal consumption is projected to increase from 8.5 to 15.8 exajoules in the same period, again mostly reflecting growth in electricity-generating capacity.

Are these projections meaningful? Probably not. Despite the enormous supply of coal and the growing demand for electricity, other factors will shape future coal consumption. Concerns about climate change and other environmental issues, on the one hand, and consequent changes in government policy, on the other, will surely affect consumption. One day we may see, for example, some sort of cap-and-trade or tax system for carbon that will significantly change the economic equation. These strategies may be important for coal for two reasons. First, there is a great need to increase the efficiencies of coal-fired power plants, particularly in the United States. Second, the climate and environmental concerns will spur development of CO_2 capture and long-term storage.

EFFICIENCIES OF COAL-FIRED PLANTS AND CO_2 CAPTURE

In the United States, most coal-fired power plants are of the subcritical pulverized coal (PC) type.[8] In this process, powdered coal and air are injected into the combustion furnace; the heat is used to produce superheated, high-pressure steam; the steam drives a turbine, which produces electricity; and the steam exiting the turbine is condensed to liquid water and returned to the furnace. Along with pollutants (dominantly sulfur and nitrogen oxide gases), the flue gas consists of nitrogen (N_2) from the combustion of air and 10 to 15 percent CO_2. Subcritical PC units have maximum generating efficiencies (the amount of electricity produced per unit of thermal energy in the fuel) of about 34 percent. More efficient are units that use supercritical to ultrasupercritical PC technologies (table 10.3), in which the steam is brought to higher temperatures and pressures and then returned to the furnace as steam rather than as liquid water.[9] The most advanced designs promise generating efficiencies up to 46 percent, which would represent a decrease in CO_2 emissions of about 20 percent compared with the emissions from subcritical plants.

As an alternative to PC combustion, coal may be burned with limestone in a circulating fluid bed. Although generation efficiency is not much better in this process than for subcritical PC units, circulating-fluid-bed units have the advantage of being suitable for low-grade coals and even peat and waste fuels. More important,

TABLE 10.3 **COMPARISON OF PERFORMANCE AND COST OF SOME COAL-FIRED, ELECTRICITY-GENERATING TECHNOLOGIES**

	Subcritical Pulverized Coal (PC)		Super-critical PC		Ultra-super-critical PC		Subcritical Circulating Fluid Bed		Sub-critical PC-oxy	Integrated Gas Combined Cycle (IGCC)	
CO_2 capture?	No	Yes	No	Yes	No	Yes	No	Yes	Yes	No	Yes
Efficiency (%)	34.3	25.1	38.5	29.3	43.3	34.1	34.8	25.5	30.6	38.4	31.2
CO_2 emitted[a]	931	127	830	109	738	94	1,030	141	104	832	102
Cost[b]	4.84	8.16	4.78	7.69	4.69	7.34	4.68	7.79	6.98	5.13	6.52

[a] In units of grams per kilowatt hour.

[b] Cost refers to cost of electricity (COE) in cents per kilowatt hour. The COE is the constant dollar electricity price required over the life of the plant to provide for all expenses and debt and bring in an acceptable rate of return to investors.

Source: Massachusetts Institute of Technology, *The Future of Coal: Options for a Carbon-Constrained World* (Cambridge, Mass.: MIT, 2007), available at http://web.mit.edu/coal/.

with this method the sulfur gases are captured by the limestone,[10] and little or no nitrogen oxide is generated, thereby eliminating most pollutants.

Capture of CO_2 is technically viable, but it comes with a cost of a significant decrease in generating efficiency (in addition to the capital cost). The common process involves chemical absorption of flue gas CO_2 into a solution containing amine, a nitrogen-bearing organic compound. The solution must be heated to release the CO_2, which then must be compressed for storage. For subcritical PC units built to incorporate CO_2 capture, generating efficiency is expected to be about 25 percent, but units retrofitted but not originally designed for CO_2 capture will have a much lower efficiency of about 14 percent.

The essential difficulty for efficiency is that the low CO_2 content of the flue gas requires that the CO_2 be concentrated. Two technologies exist to get around this problem. One is to burn the coal in an oxygen-rich gas rather than air in a process known as *oxy-fuel PC combustion*. The resulting flue gas consists mostly of CO_2 and can be compressed without the need for concentration, but the process has the additional cost of producing oxygen. Oxy-fuel PC combustion technology is in early commercial development, and ultimate efficiency and cost effectiveness have yet to be established.

The second approach is to gasify the coal and remove the CO_2 before burning in a technology known as *integrated gas combined cycle* (IGCC). In this process, coal is gasified to produce a mixture of hydrogen, carbon monoxide, and CO_2; known as *syngas*, this mixture is then burned in a gas turbine, and the hot turbine exhaust is used to produce steam that drives a steam turbine (thus the name combined cycle).[11]

Only a few IGCC plants are in commercial operation in the United States and Europe. The capital costs are quite high, so these plants have been established with the help of government subsidies. Nonetheless, as table 10.3 shows, IGCC has an estimated lower cost with CO_2 capture than other technologies have.

CARBON SEQUESTRATION

Carbon sequestration is the long-term (thousands of years and longer) storage of captured CO_2 in underground rock or possibly ocean reservoirs.[12] (Reforestation can be considered another form of carbon sequestration.) Carbon sequestration in underground rock reservoirs is technically and economically feasible and in fact is being carried out at several locations now. However, it has yet to be demonstrated at the scale and under the range of conditions needed to have an impact on CO_2 emissions. Given the likely continued importance of coal, the extent to which we shall be able to control CO_2 emissions may well hinge on the success of carbon capture and sequestration.

To emphasize the magnitude of the task and the reason that demonstration at scale is important, consider that sequestering a mere 1 gigaton of CO_2 per year will require capture and storage from five hundred 500-megawatt PC power plants, equivalent to all the plants currently in operation in the United States.[13] Furthermore, 1 gigaton of CO_2 in the subsurface is equivalent to more than 12.6 billion barrels of oil, or about 40 percent of the oil pumped worldwide every year.

Carbon dioxide must be stored in a liquefied form—that is, as a relatively dense fluid with properties more like liquid than like gas, which is achieved by pressurizing it. Liquefication is important because (1) it minimizes the space required, and (2) the denser the fluid, the less likely it is to escape. The appropriate depth for storage is 800 meters (2,600 feet) and more because at the pressures and temperatures encountered at this depth, CO_2 remains liquefied.[14]

There are a number of requirements for CO_2 storage, and perhaps most obvious among them is having the right rock. Most oil and other subterranean fluids are not held in otherwise empty underground caverns. Rather, the reservoirs typically consist of pore spaces and fracture networks in rocks such as sandstones. The higher the porosity and the more fractures there are, the more fluid that the rock reservoir can hold. Another important requirement is that the rock be permeable, which means that the fluid must be able to flow through it. Porosity and permeability typically decrease with depth, so good reservoir rocks cannot be too deep, but they must be deep enough so that the injected CO_2 will be dense.

Even liquefied CO_2 is less dense than water, so another requirement is that there be a cap rock above the reservoir rock to prevent the CO_2 from migrating upward and eventually escaping. Rocks with sufficiently low permeability for this purpose include shale, salt, and anhydrite (calcium sulfate). Fortunately, numerous sedimentary basins around the world offer the right conditions. Sedimentary basins are depressions in the deep, crystalline basement (the foundation of igneous or metamorphic rock on which many sedimentary rock sequences rest) in which thousands of meters of layered sedimentary rock have commonly accumulated. Total national and worldwide storage capacities appear to be enormous, albeit poorly known. For example, the estimates for U.S. storage capacity vary from 2 to 3,700 gigatons of CO_2 and for worldwide storage up to 200,000 gigatons of CO_2.[15]

Three types of reservoir rocks appear to hold the most promise for large-scale sequestration.[16] The first is abandoned oil and gas fields. Because the fields already had oil and gas in them, they are already known to be suitable as traps and will not impose any adverse environmental effects. Furthermore, most oil and gas fields have been characterized in great detail; they may already have wells and other supporting infrastructure; and in some fields, injection of CO_2 may lead to enhanced oil or gas recovery. However, these fields are also shot full of holes from exploration and production activities. The holes may or may not be well mapped, and unless they all are properly plugged, there is the potential for catastrophic leaks. A second promising reservoir type is porous sedimentary rocks saturated with salty water (brines), also known as *saline formations*. Such layers are widespread, and the brines are of no use for agriculture or human consumption. The third option for long-term CO_2 storage is coal seams that are too deep for economic mining.

When CO_2 is injected into a reservoir rock, the fluid pressure around the injection well increases, driving CO_2 into the surrounding rocks (figure 10.5). The CO_2 migrates by displacing some proportion (typically 30 to 60 percent) of whatever fluid is occupying the pore space. As the CO_2 permeates the rock, some (5 to 25 percent) will become trapped in the pore spaces. Carbon dioxide also slowly dissolves in brine, increasing its density. Therefore, CO_2-laden brine will tend to sink and remain permanently underground. Finally, over longer periods of time, CO_2 may react with the rocks to form carbonate minerals, which both immobilize the CO_2 further and seal the rock.

A real-world example illustrates how sequestration works. The Sleipner Project in the Norwegian North Sea is one of several large projects worldwide in which sequestration is being carefully monitored to determine how well it is working (figure 10.6).[17] The Norwegian state oil company, Statoil, recovers natural gas commercially from two gas fields. The gas contains 9 percent CO_2, about 70 percent of which is separated for sequestration. This recovered CO_2 is then injected into the Utsira Formation, a poorly consolidated, brine-saturated sandstone 800 to 1,000 meters

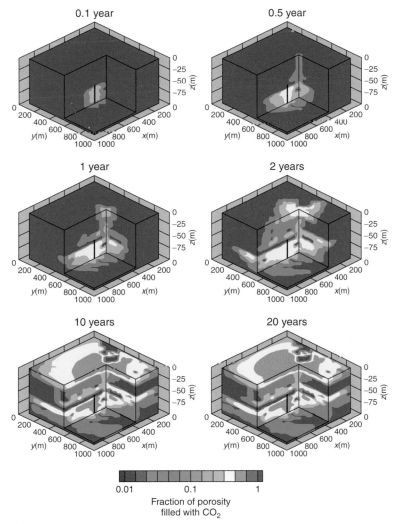

FIGURE 10.5

Simulation of the shape of a CO_2 plume as it spreads through a porous layer over a 20-year period

The porous layer is capped by an impermeable layer and contains internal zones of relatively low permeability. The horizontal and vertical scales are in meters. (After Doughty and Pruess 2004, with permission)

(2,600 to 3,300 feet) below the seabed. An 80-meter-thick (260-foot) layer of impermeable shale rests on top of the Utsira Formation and prevents the CO_2 from escaping.

The Sleipner Project commenced in October 1996 with a planned injection of 20 megatons of CO_2. By 2005, nearly 7 megatons of CO_2 had been injected with no evidence of leakage. The expectation is that the CO_2 will eventually dissolve in the saline pore water and sink. There is no shortage of storage capacity. The Utsira Formation is about 250 meters (820 feet) thick and at least 50,000 square kilometers

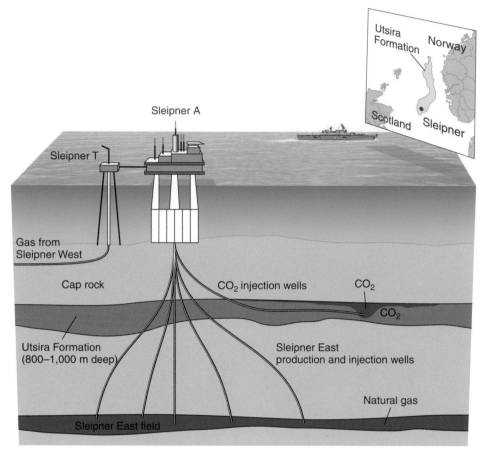

FIGURE 10.6

Schematic cross section and location of the Sleipner Project, Norwegian North Sea

The Sleipner Project is one of several large projects worldwide in which sequestration of CO_2 is being carefully monitored to determine how it is working. The CO_2 is separated from natural gas and injected into the 250-meter-thick (820-foot) Utsira Formation, a brine-saturated sandstone 800 to 1,000 meters (2,600 to 3,300 feet) below the seabed. An overlying 80-meter-thick (260-foot) layer of impermeable shale prevents the CO_2 from rising and escaping. The trapped CO_2 will eventually dissolve in the saline pore water and sink. (After Benson et al. 2005:fig. 5.4)

(19,000 square miles) in extent, and the brine-filled pores make up about 35 percent of the formation. The storage capacity is estimated at 600 gigatons of CO_2, or sufficient to hold the CO_2 emissions from all European coal- and gas-fired power stations for centuries.

There is also the possibility of storing CO_2 in the ocean.[18] Ocean storage remains largely a matter of theory, however, because there have been no demonstration projects of any size. Furthermore, some concern has been expressed about potential environmental consequences. For example, CO_2 injection would locally reduce the pH of ocean water, which may have a severe, negative impact on the creatures that live there (chapter 4). At this point, too little is known about ocean sequestration to judge its viability.

To be sure, numerous uncertainties and risks remain regarding sequestration. As noted, there have yet to be any projects of a size necessary to demonstrate that sequestration is viable at a scale necessary to offset emissions. Another uncertainty concerns the possible chemical interactions of CO_2 and the fluids and minerals that make up reservoir rocks, information that is essential for evaluating long-term storage potential. In addition, individual storage sites will have to be extensively evaluated for the possibility of leakage. Finally, there is the question of whether power plants can be located near sequestration sites, avoiding the substantial cost of transporting liquid CO_2 long distances by pipeline. Nonetheless, the existing sequestration projects have worked as hoped, and considering the necessary scale at which sequestration will have to be implemented, technical and economic evaluations have generally been positive.[19]

The Nuclear Option

The generation of electricity from nuclear power produces no CO_2 (except for relatively small amounts associated with the mining, transport, and production of uranium fuel and with the management of spent fuel) or air pollution and thus may have a significant impact on emissions. For example, a threefold increase in the current production of nuclear energy to 1,000 gigawatts by 2050 would reduce CO_2 emissions by 2.9 to 6.6 gigatons per year, depending on whether nuclear power plants displace coal- or gas-fired plants.[20] Furthermore, nuclear power is a functioning technology that already provides a significant proportion of the world's electricity. In France, 58 reactors supply 80 percent of the country's electricity needs. It is no surprise, then, that after more than two decades of dwindling support, nuclear power is enjoying a "renaissance," or, more precisely, attracting renewed interest as global CO_2 emissions mount.

Significant hurdles exist to the expansion of nuclear power, however: cost, safety, storage of high-level radioactive waste, and proliferation of nuclear weapons. One study argues that given the likely technical and economic challenges necessary to overcome these hurdles, there is little point in developing nuclear power unless it can done in a big way so as to lead to substantial reduction in greenhouse-gas emissions.[21] Such an undertaking would be massive. Expanding nuclear power from its current worldwide production level of about 350 gigawatts to 1,000 gigawatts by 2050, for example, would require construction of sixteen 1,000-megawatt plants per year for the next 40 years.

THE GENERATION OF NUCLEAR POWER

To appreciate the issues involved, we must start with the basics of how nuclear power is generated. The production begins with uranium. Nearly all (99.28 percent) of uranium exists as the isotope uranium-238, with most of the remainder being uranium-235 (0.72 percent). Uranium-235 undergoes *fission*, or the process of splitting the nucleus into smaller particles, by interaction with so-called slow or thermal (low-energy) neutrons. There are several possible reactions in the fission process; one, for example, is

$$uranium\text{-}235 + slow\ neutron \rightarrow barium\text{-}144 + krypton\text{-}90 + 2\ neutrons + 200\ megavolts$$

The reaction releases neutrons and gamma rays and generates about 200 megavolts of energy per uranium atom, which is about 50 million times more energy than generated by the burning of one atom of carbon in fossil fuel.

The neutrons produced by uranium-235 decay can be slowed down to become available to split other uranium-235 nuclei. With a "critical mass" of uranium-235, this "chain reaction" will continue until the uranium-235 decays to less than the critical mass. In a light-water reactor, the most common current design for reactors, the critical mass is 3 to 5 percent uranium-235 (for weapons, the critical mass is higher than 90 percent). So to make fuel for a reactor, the proportion of uranium-235 must be increased to above its natural proportion, or "enriched."

The neutrons from uranium-235 decay also interact with uranium-238 nuclei to produce plutonium-239. This reaction itself does not produce energy, but the fission of plutonium-239 adds to the production of energy. Spent fuel consists of uranium-238, plutonium-239, residual amounts of uranium-235, and a variety of other radioactive elements. Although the waste becomes less radioactive with time, it presents a health hazard for thousands of years.

There are two ways of dealing with the spent fuel. In the *open (or once-through) fuel cycle*, which is used in the United States, the spent fuel is removed from the reactor and stored for a decade or more in pools of water, which allows the fuel to "cool" as the short-lived radioisotopes decay away.[22] The waste is then placed in transportable, air-cooled casks to await disposal in a permanent repository, which has become a controversy of its own. The open fuel cycle is relatively inexpensive and simple. More important, the waste is too radioactive to handle, so it is resistant to theft.

In the *closed fuel cycle*, which is the strategy adopted by France, Great Britain, Russia, and Japan, the residual uranium-235 and plutonium-239 are separated from the spent fuel. Some is mixed back into new uranium stock and fabricated into fuel for another cycle, and the remainder is fashioned into glass "logs" and, along with unreprocessed spent fuel, placed in interim storage. In general, only one recycle is possible. The closed fuel cycle reduces the amount of new fuel required by about 30 percent, but the reprocessing is expensive to the point that this approach will remain uncompetitive unless the price of uranium skyrockets. Reprocessing plants also represent potential safety hazards because of their large inventories of highly radioactive materials, and the residual plutonium can be redirected to the production of nuclear weapons.

IS THERE ENOUGH URANIUM?

Significant expansion of nuclear energy raises the question of the adequacy of uranium reserves (figure 10.7). At the present rate of consumption, there is probably enough mineable uranium for three or four centuries to come.[23] But whether there is enough for expansion of nuclear energy depends first on the scale of the expan-

FIGURE 10.7

Metatorbernite

The chemical formula for metatorbernite, a copper uranium phosphate ore, is $Cu(UO_2)_2(PO_4)_2 \cdot H_2O$. The sample comes from the copper and uranium Musonoi Mine in Shaba, Zaire, which is famous among mineralogists for the fine specimens of uranium minerals it has produced. This specimen is on display at the American Museum of Natural History, New York. (Photograph by J. Newman, American Museum of Natural History)

sion, which itself depends on the cost of electricity and other factors that are inherently difficult to predict. Second, whether there is enough depends on available supplies, which are unknown but can at least be reasonably estimated.

One "global growth scenario"[24] envisions 1,000 to 1,500 reactors with a total capacity of 1,500 gigawatts providing 25 percent of global electricity needs by 2050. (In comparison, today there are 443 reactors worldwide with a capacity of about 350 gigawatts.)[25] The growth scenario would require 9.45 megatons of uranium, and if the capacity remained constant from 2050 to 2100, a total of 24.5 megatons of uranium would be needed for the entire twenty-first century.[26]

As far as uranium supplies are concerned, the so-called identified resources, which include known deposits as well as deposits that can be inferred to exist based on solid geological evidence, have been estimated at more than 11 megatons. "Undiscovered resources"—meaning those that probably exist, at least based on geological inference, but about which nothing is known—constitute an additional 12 megatons, for a total of about 23 megatons.[27] So there appears to be enough uranium for nuclear power to provide 20 to 25 percent of the world's electricity needs during the twenty-first century.

HURDLES TO THE EXPANSION OF NUCLEAR POWER

The practicality of expanding nuclear power hinges on four issues: cost, operational safety, waste storage, and weapons proliferation.[28] As far as cost is concerned, improvements must be realized in construction, operation, and maintenance of nuclear power plants to make electricity generated from nuclear power competitive. Although the necessary improvements are "plausible," nuclear power is unlikely to become truly cost effective unless the social and environmental costs of CO_2 emissions are included in the cost of electricity generation.

Operational safety is a concern not only for reactors, but also for reprocessing plants and systems that handle spent fuel. A related issue is the protection of all these facilities from terrorist attack. The world has seen two serious reactor accidents in the first 50 years of commercial nuclear energy production (at Three Mile Island, Pennsylvania, in 1979, and at Chernobyl, Ukraine, in 1986). Reactor design has advanced significantly, and a safety culture has emerged among plant operators. It is estimated that the likelihood of a reactor accident can be reduced from the historic rate of 1 in 10,000 to 1 in 100,000 reactor-years of operation.[29] The safety of reprocessing plants is another matter, however. These plants, as noted, have high inventories of highly radioactive materials and have had a higher frequency of accidents;[30] moreover, the risk of terrorist attack involving them has not been adequately evaluated.

Safety also demands a properly trained and qualified workforce to manage, operate, maintain, and even construct power plants. In the developed world, the present workforce is aging and not being replaced, whereas in the developing world an adequately trained, indigenous workforce generally does not exist to begin with. A significant expansion of nuclear energy will require renewed and focused efforts at training.

The safety issues do not appear to be insurmountable,[31] so they should not be considered an obstacle to large-scale expansion of nuclear energy. The bottom line is public acceptance, which will surely require a well-based perception that nuclear power is safe.

As for waste storage, present plans are for permanent disposal in mined repositories. The idea is to build repositories in areas unlikely to be disturbed for thousands of years by volcanic eruption or earthquakes and where radioactive material, even if it does leak from its containers, will not find its way to the surface. Experts generally (but not universally) agree that geological storage is technically feasible, but after a half century of commercial nuclear power production, no permanent waste facility is in operation anywhere in the world.[32] Needless to say, no community wants a repository in its backyard, so the issue in large part involves our ability to develop political and legal mechanisms that will allow repositories to be built. In the growth scenario whereby 20 to 25 percent of the world's electricity is generated by nuclear power by 2050, a new repository like that at Yucca Mountain, Nevada, would have to come on-line somewhere in the world every three or four years, so the issue is not trivial as far as development effort is concerned.

In dealing with this issue, a possibility is to reduce the required number of new repositories, although that may not completely solve the problem. One way to do so is to store spent fuel on the surface for several decades, as is being done now in lieu of any permanent storage option. Because the spent fuel would be less radioactive and would thus produce less heat, it would require less space in a permanent repository, so fewer repositories would be needed.

Alternative, potentially more acceptable strategies to storage have also been proposed. For example, waste canisters might be placed several kilometers below the surface in boreholes, which would then be plugged by a material such as asphalt or clay. Vast, stable regions of the continents are underlain by dense, crystalline rocks suitable for such storage. Among the attractions of this idea is that even if the fuel canisters were to leak, the waste would likely remain isolated because deep groundwater is generally not plentiful and in any case is isolated from the surface. Also, there are plenty of site possibilities. Disposal facilities might be sited with power plants in relatively remote areas. Suitable disposal sites also exist in the seabed, so one might imagine artificial islands distant from human habitation dedicated to both temporary storage and permanent disposal of waste.

Of all the potential hurdles, perhaps the most complex and worrying one is the threat of nuclear weapons proliferation. It involves two separate concerns. The first is the spread of facilities for uranium enrichment. Most of the nuclear power systems operating today require enriched uranium.[33] This requirement provides a rationale to build enrichment plants for countries such as Iran, but the plants may then be redirected to the production of highly enriched, weapons-grade uranium.

The second concern involves the reprocessing of spent fuel. As noted, reprocessing results in the production of plutonium-239, which can be used for weapons.[34] Approximately 200 metric tons of plutonium-239 now exist in Europe, Russia, and Japan, enough for more than 25,000 nuclear weapons.[35] One risk is that these existing and expanding inventories are potential targets of theft. Reprocessing might also spread to countries without adequate safeguards against theft or to countries seeking to divert plutonium to weapons production covertly. The technology for reprocessing is no secret because it is published in the open literature, and so far the world has had limited success at curbing the spread of nuclear weapons. It is therefore difficult to imagine that adequate safeguards can be developed under existing nonproliferation mechanisms in today's fractured world.

Despite this grim outlook, different combinations of fuel cycle and reactor designs may lessen (but not completely eliminate) the proliferation risk and be deployed in one or two decades.[36] One approach is to mix uranium- and thorium-bearing fuels.[37] Thorium-232 is not fissile, but it does react with slow neutrons to form the fissile isotope uranium-233, which produces energy as well as more neutrons. Proliferation can be thwarted by mixing thorium with uranium fuels in a proportion such that the uranium-233 never reaches more than 20 percent of the mixture. The mixture will consist mainly of uranium-238, so enrichment would be necessary to obtain weapons-grade uranium-233. Enrichment is much more difficult than chemical separation of plutonium-239 from conventional spent fuel. In addition, with this approach much less plutonium-239 is produced than in the standard uranium fuel cycle, and the amount of plutonium-239 as a fraction of all plutonium isotopes is less, rendering the production of weapons-grade plutonium more difficult.

Another approach is to burn thorium fuel containing some plutonium. In this case, fission of plutonium-239 instead of uranium-235 provides the neutrons for the production of uranium-233 from thorium. The proliferation resistance arises from the fact that the spent fuel is highly radioactive and thus very difficult to handle.[38] This approach is also a way to get rid of some of the existing plutonium-239 stocks.

Wind and Solar Power: Icons for the Future

Wind and solar power have been called the "iconic renewable resources."[39] The reasons are clear. They do not produce greenhouse gases or otherwise pollute (at least

TABLE 10.4 **GENERATING COSTS OF WIND AND SOLAR POWER IN 2007 FOR THREE DIFFERENT AMOUNTS OF SUNLIGHT RECEIVED AND WIND VELOCITIES**

Received irradiance (watts per square meter per year)	1,700	2,000	2,300
Solar photovoltaic (cents per kilowatt hour)	29	25	21
Solar thermal (cents per kilowatt hour)	26	22	19
Wind velocity (meters per second at 50 meters above ground)	7.0–7.5	7.5–8.0	8.0–8.8
On-shore turbines (cents per kilowatt hour)	4.6	3.8	3.4
Off-shore turbines (cents per kilowatt hour)	5.3		4.5

Source: J. A. Edmonds, M. A. Wise, J. J. Dooley, S. H. Kim, S. J. Smith, P. J. Runci, L. E. Clarke, E. L. Malone, and G. M. Stokes, *Wind and Solar Energy: A Core Element of Global Energy Technology Strategy to Address Climate Change* (College Park, Md.: Joint Global Change Research Institute, 2007).

directly); they have the potential to produce vast amounts of electricity forever; they represent a path toward national energy independence and thus security; and they insulate against risks associated with the volatility of fossil-fuel costs. Various analyses suggest that together wind and solar power can practically provide a significant and growing proportion of the world's primary energy needs through the twenty-first century.

Wind power is coming of age now. The cost of producing electricity from large wind power plants is approaching the cost of producing electricity from coal-burning plants, and the readily accessible wind resources are enormous. Meanwhile, the amount of available solar power dwarfs anything else. Capturing solar power with photovoltaic (PV) cells remains relatively expensive, however, and capturing its thermal power with mirrors is also expensive, but has recently become more competitive.[40] Although the large-scale expansion of solar power lags that of wind power (table 10.4), solar power represents perhaps the greatest hope for meeting future needs while driving down CO_2 emissions.

Nevertheless, both wind and solar power are in their infancy. To emphasize this point, recall that all renewables, which include geothermal and biofuels as well as solar and wind (but exclude hydropower), provided only 2.2 percent of electricity consumed worldwide in 2005. A more specific breakdown of renewable energy resources in the United States in 2006 further illustrates this point (figure 10.8).

THE NUTS AND BOLTS OF WIND POWER

The growth of the wind power industry is reflected in the increasing size of wind turbines. Thirty or so years ago a turbine may have had a 10-meter (33-foot) roter

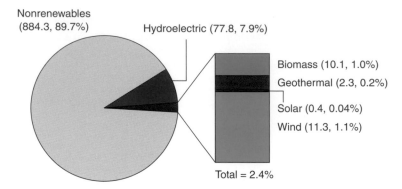

FIGURE 10.8

Renewable sources of power as proportions of total U.S. electric net summer capacity, 2006

The numbers in parentheses are gigawatts of electrical capacity with corresponding percentages of total capacity. Biomass includes wood and derived fuels, landfill gas, municipal solid waste, and agricultural by-products. (Data from Energy Information Agency, http://www.eia.doe.gov/)

(the diameter of the circle traversed by the blade). Today some turbines in the North Sea have 126-meter (413-foot) rotors and generate 5 megawatts of power in 30 mile per hour winds. Wind turbines are grouped together in "wind farms" with typical aggregate capacities of tens to hundreds of megawatts (figure 10.9). As of the beginning of 2008, the world's largest such facility was the Horse Hollow Wind Energy Center in Texas, where 421 wind turbines generate up to 735 megawatts of electricity, enough to supply the needs of 220,000 homes.[41]

The wind farm economies require an average annual wind speed of at least 6 meters per second (13 miles per hour).[42] The power that wind can theoretically provide is proportional to the cube of its velocity. Thus wind with a speed of 14 miles per hour can provide 25 percent more power than wind moving at 13 miles per hour, although turbines do not capture this entire increase. Consequently, average wind speed is an important factor in determining wind farm location.[43]

Wind speeds are on average about 90 percent stronger and more consistent over water than over land,[44] so offshore locations hold enormous potential for development of large-scale wind farms. By one estimate, the available wind offshore between Massachusetts and North Carolina might in principle provide more power than is currently being consumed by this densely populated and energy-intensive part of the world.[45] The one drawback is that up-front capital costs of offshore development are relatively large. Despite this cost, however, offshore wind farms in Europe now provide more than 1,000 megawatts of electricity annually, and new construction is proceeding rapidly.[46]

Because wind does not blow all the time, it is a variable source of energy. As a practical matter, the variability can be accommodated when wind accounts for less

FIGURE 10.9
Wind farm

The Maple Ridge Wind Farm, on the Tug Hill Plateau in New York. (Department of Energy, National Renewable Laboratory, http://www.nrel.gov/data/pix/, photograph 15236 by PPM Energy, with permission)

than about 20 percent of the electricity that a power system must deliver in any given hour because the system must be designed with enough flexibility to meet fluctuations in demand anyway.[47] In systems where wind provides more than about 20 percent of the electricity, a means of storing energy becomes necessary, which affects cost. The example of Denmark, which in 2007 obtained 21.2 percent of its energy from wind power, is telling. One major reason for Denmark's notable success in the development of wind power is that the country is connected by transmission lines to other countries, which helps to modulate the swings in wind electricity production. Nearly all of Norway's energy is produced by hydropower, so Norway in particular can absorb the swings by using excess electricity to pump water uphill to high reservoirs and then reusing that water to generate electricity when needed.

Wind power has overwhelming environmental benefits over nuclear and fossil-fuel-burning plants. Besides being nonpolluting, wind power uses almost no water (which is needed only to wash the rotor blades), an increasingly precious commodity required in large quantities in nuclear and coal-fired plants. Concerns for aesthetics and wildlife, in particular the fate of birds and bats, have nonetheless led to local opposition to some wind power projects. Wind turbines can indeed pose a significant danger to birds, the well-publicized example being a wind farm on Altamont Pass, California, home to many raptors. However, a wide-ranging study found that deaths from wind turbines account for only a minuscule fraction of the 500 million to 1 billion bird deaths in the United States annually.[48] In particular, the study estimated that 72 percent of the fatalities are caused by birds flying into windows, buildings, and power lines. Predation by cats accounts for another 11 percent, automobiles for 9 percent, pesticides for 7 percent, and communications towers for 0.5 percent, whereas wind turbines account for less than 0.01 percent. Avian death rate is site specific and can be minimized by placing wind farms where the impacts on birds will be low.

THE EXPANSION OF WIND POWER

That wind power is poised for significant expansion is suggested by the increasingly rapid rise of total global installed wind power capacity. Since 1996, that capacity has increased an average of about 28 percent per year to 94.1 gigawatts by the end of 2007 (figure 10.10) and is projected to expand to 240 gigawatts by 2012.[49] The expansion is being driven by a combination of factors. One, of course, is the world's insatiable demand for energy, which is driving up the costs of fossil fuels and making wind power increasingly more cost competitive. Another is government policies that encourage its development. Also supporting the expansion is the fact that wind power and turbine manufacturing are increasingly becoming the provenance

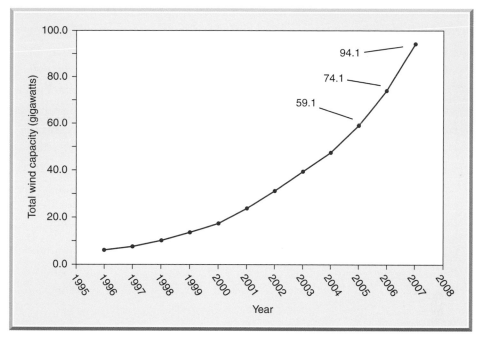

FIGURE 10.10

The growth of global installed wind power capacity, 1996–2007

Wind power, which in 2008 provided for more than 1 percent of the world's electricity demand, expanded in terms of capacity at more than 25 percent a year in 2006 and 2007, with capacity expected to reach 240 gigawatts by 2012. (Data from Global Wind Energy Council 2007 Annual Report, http://www.gwec.net/index.php?id = 90)

of large oil, utility, and manufacturing entities that can provide the significant investment necessary to increase capacity and satisfy soaring demand.

The European Union (EU) has traditionally been the largest market for wind power. As of the end of 2007, Germany remained the leader in installed capacity, followed by the United States, Spain, India, and China (table 10.5). Total EU installed capacity that year was 56.5 gigawatts (equivalent to 90 megatons of CO_2 emissions), or 3.8 percent of total electricity demand, up from 0.9 percent in 2000.[50] The EU has adopted the goal of obtaining 20 percent of its energy from renewables by 2020. If this goal is to be met, about 12 percent of all electricity will have to come from wind power.[51] Reaching the goal will require massive development, especially of offshore wind resources.

In 2007, installed wind power capacity in the United States increased 45 percent over the previous year, to 16.8 gigawatts.[52] The expansion has been encouraged by the 2 cents per kilowatt hour federal production tax credit. A high growth rate is likely to be sustained because 24 states have adopted the Renewable Portfolio Standards, a mandate that certain proportions of energy must come from renewables by specified dates. Although wind power now satisfies only about 1 percent of elec-

TABLE 10.5   **NATIONAL INSTALLED WIND POWER CAPACITIES AS OF THE END OF 2007**

	Capacity (Megawatts)	Percentage of World Capacity	Percentage of National Electricity Demand
Germany	22,247	23.6	7.0
United States	16,818	17.9	1.2[a]
Spain	15,145	16.1	11.8
India	8,000	8.5	4.0[b]
China	6,050	6.4	6.4
Denmark	3,125	3.3	21.2
Italy	2,726	2.9	1.7
France	2,454	2.6	1.2
United Kingdom	2,389	2.5	1.8
Portugal	2,150	2.3	9.3
Canada	1,846	2.0	1.1[b]
Netherlands	1,746	1.9	3.4
Japan	1,538	1.6	0.5[b]
Total Europe	57,136	60.7	3.8
World total	94,123	100.0	1.8[b]

Note: Data are for countries with capacities greater than 1,000 megawatts.

[a] For 2006.

[b] As a proportion of total national electricity generation.

Sources: Global Wind Energy Council, *Global Wind 2007 Report* (Brussels: Global Wind Energy Council, 2007), available at http://www.gwec.net/index.php?id = 90; British Petroleum, *BP Statistical Review of World Energy 2007* (London: British Petroleum, 2007), available at http://www.bp.com/ productlanding.do?categoryId = 6848&contentId = 7033471.

tricity demand in the United States, the potential for growth is enormous. In fact, by one scenario (albeit a highly uncertain one because of unknown economic and technological variables), U.S. wind resources may conceivably expand to make up about 25 percent of electricity output by 2030.[53]

Several hurdles to the massive expansion of wind power in the United States remain, however.[54] The most important is that the existing transmission infrastructure is inadequate. It is particularly a problem in the upper Midwest, where wind is a plentiful but "stranded" resource because the existing transmission system is not capable of transporting large amounts of power from potential generation sites to population centers. Another hurdle is energy storage, which, as suggested earlier, practically limits the proportion of power that wind can contribute to a local power grid.

FROM SUNLIGHT TO ELECTRICITY

The technology to turn the Sun's energy into electricity is hardly new. Spurred on by the oil crisis of the 1970s, nine solar thermal power plants with an aggregate capacity of 354 megawatts were constructed in the 1980s in the Mojave Desert, California. Interest in solar power waned as oil prices dropped, only to reawaken again as prices began to rise after 2002. In the meantime, it has gained a foothold elsewhere, especially in Japan and Germany.

There are two ways of converting solar radiation to electricity. One is by transforming solar radiation directly into electricity using PV cells; the other is to generate electricity by collecting solar heat with the use of mirrors. PV cells[55] utilize the *photoelectric effect*, the process of ejecting an electron from an atom by the atom's absorption of an incident photon, or a packet (quantum) of light. In its simplest form, the PV cell consists of layers of p and n semiconductor material, most commonly silicon, sandwiched between layers of conducting materials. Sunlight falling on the n layer produces electrons, which are collected in the p layer to create a voltage difference between the front and back faces of the cell. The voltage difference drives an electric current when the two sides are connected by a conductor. The amount of electricity produced by a PV cell is approximately proportional to the amount of solar radiation falling on it. PV cells are mounted into modules (40 cells are common), which are in turn combined to form arrays of various sizes depending on application (figure 10.11).

PV cells offer great flexibility. For example, arrays can be mounted on residential and commercial buildings. This configuration works especially well if the electricity produced locally is tied into the electrical grid, which obviates the need for expensive storage devices, such as banks of batteries, and excess energy can be sold or traded for electricity when none is to be had from the PV arrays. PV arrays are also suitable for driving devices isolated from other sources of power, such as water pumps on remote farm pastures.

In 2007, the cost of electricity from PV cells was more than 20 cents per kilowatt hour, depending on the intensity of sunlight (see table 10.4), whereas electricity from the burning of coal cost only 2 to 4 cents per kilowatt hour. Great effort, therefore, is being directed at producing PV cells with greater efficiencies at converting solar energy to electricity. Solar radiation consists of photons with a spectrum of energies, but specific semiconductor materials can absorb photons only in narrow ranges of energies, limiting their efficiencies. Therefore, one way to gain efficiency is to construct cells consisting of two or three layers of different semiconductor materials that absorb photons in different parts of the energy spectrum. The typical efficiency of commercially available PV cells is about 3 to 14 percent, but efficiencies up to 50 percent are theoretically possible. Development of more efficient PV cells

FIGURE 10.11

An array of photovoltaic panels

The Alamosa Photovoltaic Plant in southern Colorado can generate up to 8.2 megawatts of electricity. (Department of Energy, National Renewable Laboratory, http://www.nrel.gov/data/pix/, photograph 15236 by Steve Wilcox, with permission)

and less expensive ways to manufacture them are at the technological forefront and promise substantial improvements in costs.

As noted, electricity may also be generated from solar energy by using mirrors to collect the heat. This process is practical in cloud-free regions and best in those regions that receive more than 6.7 watts of solar radiation per square meter, such as in most deserts of the world and countries with the right geographical and climatic characteristics, such as Spain. There are three approaches to "concentrating solar power," as the process is known. The first is to use a linear array of parabolic troughs to focus the light onto an oil- or molten salt–filled pipe running down the center of the trough (figure 10.12). The troughs can be oriented in a north–south direction and track the sun as the day progresses. A second approach is to use a large field of sun-tracking mirrors to focus sunlight onto a receiver at the top of a central tower, where the focused light heats oil or molten salt. In either case, the hot fluid is used to boil water, which in turn drives a steam turbine.

Molten salt can also be used to store heat to provide a continuous flow of electricity at night or during cloudy days. The AndaSol project, a 150-megawatt solar thermal plant in the province of Granada, Spain, uses molten salt storage.[56] Parabolic troughs heat molten nitrate salts. This fluid is pumped into a hot storage tank at

FIGURE 10.12

Parabolic troughs

Parabolic mirrors focus sunlight on a fluid-filled pipe and track the sun at a plant located near Kramer Junction, California. This is one of a number of such plants with a combined capacity of 354 megawatts built in the 1980s after the first oil crisis. (Department of Energy, National Renewable Laboratory, http://www.nrel.gov/data/pix/, photograph 15236 by Warren Gretz, with permission).

386°C (727°F) and then run through a heat exchanger to boil water as it is transferred to a "cool" tank at 292°C (558°F). The system maintains more than seven hours of storage capacity at full output.

A third approach to concentrating solar power uses an array of mirrors mounted on a sun-tracking dish. The mirrors focus the sunlight onto a "receiver," which collects and transfers the heat to a Stirling engine.[57] The conceptual basis of the Stirling engine dates back nearly 200 years to 1816, when the Scottish inventor Rev. Dr. Robert Stirling (1790–1878)[58] conceived of the "heat economizer," a more fuel-efficient engine to replace the steam boiler. The engine works by the heating and cooling of a gas (hydrogen in the modern engine) in a sealed cylinder. As the gas expands and contracts with changing temperature, it drives the pistons of a four-cylinder reciprocating engine, which in turn drives a generator. The system is extremely efficient, having achieved a system-to-grid conversion efficiency of more than 30 percent. Two generating stations based on the Stirling engine are now under construction in southern California. Their combined initial generating capacity will be 800 megawatts, making them among the largest solar power stations to date.

THE EXPANSION OF SOLAR POWER

Like wind power, solar power is now expanding at an increasingly rapid rate. Germany, a country with only modest solar resources, accounts for much of the accelerating growth (figure 10.13). The reason for the growth there is that the government extended its "feed-in tariff" program, originally established for wind power, to all PV producers of electricity, large and small alike. Feed-in tariff programs ensure a producer of being able to connect its local system to the power grid and either sell excess power to the grid operator or trade it for electricity delivered at another time. The policy typically drives a modest increase in electricity cost because the grid operator must buy solar-generated electricity at a price above which it can otherwise produce electricity. In the case of Germany, the premium is substantial and guaranteed for a number of years. As a result, it suddenly became profitable for German home and commercial building owners to install solar panels. The policy has been wildly successful at forcing greater PV market penetration, and it has thrust Germany into a leading position as a developer of solar power technology.[59] The idea is catching on and slowly spreading beyond Europe.

Efficiency dictates that solar power plants be located in sun-soaked parts of the world, so the development of a network of high-voltage transmission lines is important to the large-scale expansion of solar power. (Wind power would also benefit from such a network, but the problem is not so severe because wind power facilities can more commonly be built closer to population centers.) Nearly all domes-

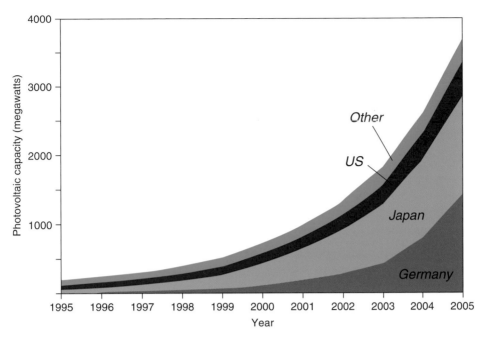

FIGURE 10.13

The worldwide growth of capacity from photovoltaic cells

Germany accounts for much of the growth because of its "feed-in tariff" program. (Data from British Petroleum 2007, http://www.bp.com/productlanding.do?categoryId=6848&contentId=7033471)

tic and industrial equipment uses alternating current (AC), so nearly all transmission lines are AC as well (98 percent by distance in United States). High-voltage AC (HVAC) transmission lines lose energy, depending on length of transmission, however. The alternative is direct current (DC) lines, which lose less energy, are more stable, and have several other advantages compared with HVAC lines. But HVDC lines require expensive converters from AC to DC and back again on the ends of the line. Consequently, HVAC lines are cheaper for distances of less than 500 to 800 kilometers (300 to 500 miles), whereas HVDC lines are more economical at greater distances.[60] The expense of converters also means that HVDC lines are not suited to providing electricity at points along a route because each point would require another converter. A logical strategy is to develop a long-distance backbone of HVDC transmission cables that feed into local and regional HVAC networks.

Several grand ideas for expanding solar power have been proposed. One argues for the construction of 120,000 square kilometers (46,000 square miles) of solar installations in the southwestern United States with an accompanying HVDC transmission network.[61] The cost would be about $400 billion over the next 40 years, but the program would provide nearly 70 percent of electricity and more than 33 percent of total energy requirements by 2050. In Europe, a plan known as Desertec has been put forward.[62] The plan is to harvest solar and wind energy with plants situated throughout North Africa and the Middle East and to distribute the energy via a "megagrid" of HVDC transmission lines connecting all of Europe, North Africa, and most of the Middle East. This ambitious plan would cost approximately €400 billion and supply about 17 percent of Europe's electrical needs by 2050. It involves numerous unanswered questions, however, including how to gauge risks from terrorism and political instability.

Solutions

Providing for the world's growing energy needs and simultaneously gaining control of and then decreasing CO_2 emissions are essential if we are to avoid potentially catastrophic climate change. We already know how to reduce emissions—the knowledge and nearly all the technology needed to achieve this goal are at hand. Yet reducing emissions to 2000 levels, the illustrative but somewhat arbitrary goal put forth at the beginning of this chapter, remains a daunting task.

To appreciate just how daunting the task is, think back to the challenges of producing clean electricity. Recall that sequestration of 1 gigaton of CO_2 per year from traditional coal-fired power plants will require capture and storage from five hundred 500-megawatt plants, and that 1 gigaton of CO_2 in the subsurface is equivalent to about 40 percent of the oil pumped worldwide every year. Recall that expanding nuclear power to 1,000 gigawatts by 2050 will require construction of sixteen 1,000-megawatt plants per year for the next 40 years and bringing on-line a large waste repository every several years. And recall that although both wind and solar power capabilities are growing fast, they still constitute but a minuscule fraction of worldwide electricity generation. Implementing solutions on the scale necessary to reduce CO_2 emissions to levels below those in 2000 will probably require the development of programs that have the scale and urgency akin to the Apollo Space Program or its mobilization at the start of World War II.[63] Furthermore, these efforts have to be global, led in particular by China, the United States, and Europe.

The shift from dirty to clean electricity will take decades to implement. In the meantime, other steps not covered in this book will be essential. Most important will be the rapid implementation of energy conservation and efficiency measures to get us through the next decade or two. These measures include moving to hybrid and electric vehicles; finding ways to reduce vehicle travel distances by expanding mass transport; retrofitting existing buildings and constructing new ones with efficient insulation, space heating, cooling, and lighting; and fitting individual buildings with PV panels to capture solar energy or devices to capture geothermal energy. Another is to substitute natural gas for coal to the maximum extent possible

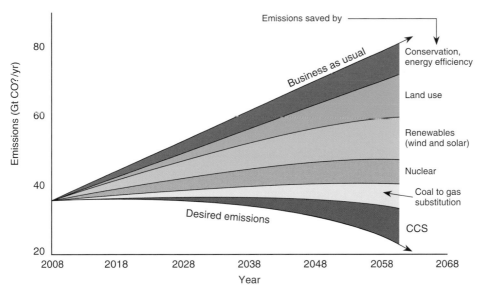

FIGURE 10.14

The stabilization triangle and wedge: a way of thinking about how to solve the emissions problem

Each stabilization wedge represents an action that reduces the CO_2 emissions rate. The actions are deployed gradually, but at an increasing rate beginning in 2008. They combine to initially maintain emissions constant at the 2008 rate, but at some point result in a decrease in emissions rate with time as new technologies are deployed.

as the fuel for electricity production. And it is not out of the question to hope that the next decade will see the implementation of promising new technologies, such as those that will directly remove CO_2 from the air.[64]

At the current time, however, there is simply no silver bullet to getting rid of the problem of greenhouse-gas emissions—efforts on a number of fronts are necessary. Various stabilization scenarios, or models of energy and economic development along with the emissions they would create, have been developed to explore the mitigation options. These scenarios assume that mitigation activities begin gradually in the near future and increase with time. A useful and widely cited conceptualization of this notion illustrates how mitigation efforts can be practically combined to achieve the desired reduction.

The conceptualization begins with a "stabilization triangle,"[65] one leg of which is the growth of CO_2 emissions projected out for the next 50 years at a business-as-usual rate, and the other leg of which represents an initially constant but then decreasing emissions rate (figure 10.14). The stabilization triangle may be divided into several "stabilization wedges," each of which represents an action that reduces the emission rate by an equal amount. The notion of a wedge implies that the action begins now and increases with time. In addition to the emissions-reducing strategies associated with energy production, possible wedges include accelerating measures to increase energy efficiency and conservation, reducing deforestation in combination with replanting campaigns, and adopting agricultural practices that limit the loss of organic material from soils by controlling erosion and practicing conservation tillage (the practice of planting seeds in boreholes in soil rather than by plowing).[66]

An increase of 1.5 percent per year in the current annual anthropogenic emission rate of 36 gigatons of CO_2 per year projects to a 2058 emission rate of 76 gigatons of CO_2 per year. (The 1.5 percent figure is the historical growth rate, but the rate of increase in CO_2 emissions is now depressingly closer to 3 percent.) Stabilization just to the 2008 emission rate thus would require mitigation actions that save 40 giga-tons a year. One can argue over which mitigation actions to take and how extensive they should be, but the point of the wedge concept is that it offers a way of thinking about practical solutions to a problem that may otherwise seem intractable due to its enormity.

This book is about how the climate system works, but it ends by emphasizing how tightly climate change is entwined with energy production. Energy production, in turn, has become a bellwether of economic development. The connection illustrates how civilization is now leading us down a path never before trodden, at least by our species. For the first time, humanity can affect the climate of the entire planet and the vast biogeochemical cycles that control the character of its surface. In a way, it is astonishing that we even know about such matters as these cycles because their scales of space and time are so far beyond those of our own lives. Think back 170 years ago to when Louis Agassiz began to understand that his world had once been covered with ice. At the time, there was hardly a conception of climate, let alone climate change, and humans were passive riders on Earth. Then think about where we are now as we peer at our planet from above with satellites, explore the deep ocean with robots, and delve into Earth's past by drilling into deep layers of rock and ice, groping for an understanding of the intricate interactions of the Sun, atmosphere, ocean, ice, life, and land—as well as of the changes forced by human activities.

And then think further: What if we had not learned the things we now know about climate and the possible dangers to civilization that climate change poses? This knowledge attests to our intelligence and innovation, just those characteristics that got us into the present predicament. We have to believe that they will also get us out of it.

NOTES

1. CLIMATE IN CONTEXT

1. The location was chosen because it is far from the disturbing influences of human activities. The CO_2 contents of the atmosphere prior to 1958 come from the compositions of bubbles of air preserved in ice.

2. U. Siegenthaler, T. F. Stocker, E. Monnin, D. Lüthi, J. Schwander, B. Stauffer, D. Raynaud, J.-M. Barnola, H. Fischer, V. Masson-Delmotte, and J. Jouzel, "Stable Carbon Cycle–Climate Relationships During the Late Pleistocene," *Science* 310 (2005): 1313–1317; D. Lüthi, M. Le Floch, B. Bereiter, T. Blunier, J.-M. Barnola, U. Siegenthaler, D. Raynaud, J. Jouzel, H. Fischer, K. Kawamura, and T. F. Stocker, "High-Resolution Carbon Dioxide Concentration Record 650,000–800,000 Years Before Present," *Nature* 453 (2008): 379–383.

3. P. Brohan, J. J. Kennedy, I. Harris, S. Tett, and P. D. Jones, "Uncertainty Estimates in Regional and Global Observed Temperature Changes: A New Data Set from 1850," *Journal of Geophysical Research* 111 (2006), doi:10.1029/2005JD006548.

4. The 11-year sunspot cycle results in a total irradiance change of about 0.1 percent between cycle maxima and minima. This change appears to be too small and the cycles too short to have any appreciable influence on surface climate. Still, the influence of changing solar output on climate remains a matter of debate. For a discussion, see J. L. Lean and D. H. Rind, "How Natural and Anthropogenic Influences Alter Global and Regional Surface Temperatures: 1889 to 2006," *Geophysical Research Letters* 35 (2008), doi:10.1029/2008GL034864, and references therein.

5. Generalizations about long-term trends cannot be based on how temperature or other conditions change from one year to the next. Similarly, they cannot be based on how temperature changes at any one location.

6. S. Levitus, J. Antonov, and T. Boyer, "Warming of the World Ocean, 1955–2003," *Geophysical Research Letters* 32 (2005), doi:10.1029/2004GL021592.

7. R. B. Alley, P. U. Clark, P. Huybrechts, and I. Joughin, "Ice-Sheet and Sea-Level Changes," *Science* 310 (2005): 456–460.

8. G. A. Meehl, T. F. Stocker, W. D. Collins, P. Friedlingstein, A. T. Gaye, J. M. Gregory, A. Kitoh, R. Knutti, J. M. Murphy, A. Noda, S. C. B. Raper, I. G. Watterson, A. J. Weaver, and Z.-C. Zhao, "Global Climate Projections," in *Climate Change 2007: The Physical Sci-*

ence Basis. Contribution of Working Group I to the Fourth Assessment Report of the Intergovernmental Panel on Climate Change, edited by S. Solomon, D. Qin, M. Manning, Z. Chen, M. Marquis, K. B. Averyt, M. Tignor, and H. L. Miller (Cambridge: Cambridge University Press, 2007), 747–846.

9. E. R. Cook, C. A. Woodhouse, C. M. Eakin, D. M. Meko, and D. W. Stahle, "Long-Term Aridity Changes in the Western United States," *Science* 306 (2004): 1015–1018.

10. L. V. Alexander, X. Zhang, T. C. Peterson, J. Caesar, B. Gleason, A. M. G. Tank, M. Haylock, D. Collins, B. Trewin, F. Rahimzadeh, A. Tagipour, K. Kumar, J. Revadekar, G. Griffiths, L. Vincent, D. B. Stephenson, J. Burn, E. Aguilar, M. Brunet, M. Taylor, M. New, P. Zhai, M. Rusticucci, and J. L. Vazquez-Aguirre, "Global Observed Changes in Daily Climate Extremes of Temperature and Precipitation," *Journal of Geophysical Research* 111 (2006), doi:10.1029/2005JD006290.

11. J. A. Patz, D. Campbell-Lendrum, T. Holloway, and J. A. Foley, "Impact of Regional Climate Change on Human Health," *Nature* 438 (2005): 310–317.

12. J. Hansen, L. Nazarenko, R. Ruedy, M. Sato, J. Willis, A. Del Genio, D. Koch, A. Lacis, K. Lo, S. Menon, T. Novakov, J. Perlwitz, G. Russell, G. A. Schmidt, and N. Tausnev, "Earth's Energy Imbalance: Confirmation and Implications," *Science* 308 (2005): 1431–1435.

13. Committee on Abrupt Climate Change, *Abrupt Climate Change: Inevitable Surprises* (Washington, D.C.: National Academy Press, 2002).

2. THE CHARACTER OF THE ATMOSPHERE

1. Good texts on the climate system are D. L. Hartmann, *Global Physical Climatology* (San Diego: Academic Press, 1994), the main source for this chapter, and the more elementary T. E. Graedel and P. J. Crutzen, *Atmosphere, Climate, and Change* (New York: Freeman, 1995).

2. The term *troposphere* is from the Greek word *tropos*, meaning "turning"; *stratosphere* is from the Latin term *stratus*, meaning "spreading out"; and *mesosphere* is from the Greek word *mesos*, meaning "middle."

3. The temperature decrease with altitude is known as the *lapse rate,* which varies with altitude, season, and latitude.

4. The aurora, a spectacular display that graces some polar nights, occurs in the thermosphere. Auroras occur when electrons streaming in from the Sun combine with ionized gases to form neutral atoms, in the process emitting radiation in the visible part of the spectrum.

5. Jet streams oscillate in waves known as *planetary* or *Rossby waves*. They are characteristic, slow-moving waves in fast-flowing fluids on rotating planetary bodies that are due to the change in the magnitude of the Coriolis force with latitude. They are named for Carl-Gustav Rossby (1898–1957), who first hypothesized their existence in the ocean. In the ocean, Rossby waves take the form of slow and large-scale meanderings of circulation patterns and are of interest because they may influence annual to decadal patterns of climate change. Analogous but more rapid meandering motions exist in the atmosphere. The winds are also deflected by permanent fixtures, such as the Himalaya and Rocky mountains, and are affected by differences in air temperature over ocean and land.

6. B. N. Goswami, V. Venugopal, D. Sengupta, M. S. Madhusoodanan, and P. K. Xavier, "Increasing Trend of Extreme Rain Events over India in a Warming Environment," *Science* 314 (2006): 1442–1445.

7. N. J. Abram, M. K. Gagan, Z. Liu, W. S. Hantoro, M. T. McCulloch, and B. W. Suwargadi, "Seasonal Characteristics of the Indian Ocean Dipole During the Holocene Epoch," *Nature* 445 (2007): 299–302. A climate record from the changing composition of corals indicates that more intense monsoons and corresponding drought conditions occurred in western Indonesia during the mid-Holocene. A worrying feature of the record is that it shows abrupt changes in monsoon activity apparently in response to a gradual shift in climate.

8. S. Weldeab, D. W. Lea, R. R. Schneider, and N. Andersen, "155,000 Years of West African Monsoon and Ocean Thermal Evolution," *Science* 316 (2007): 1303–1307.

9. J. W. Hurrell, "Decadal Trends in the North Atlantic Oscillation: Regional Temperatures and Precipitation," *Science* 269 (1995): 676–679.

10. A measure of the NAO is taken as the difference in barometric pressure at either Lisbon or the Azores or Gibraltar and at Stykkisholmur, Iceland. The NAO's positive and negative phases are characterized, respectively, by relatively large and small differences between the barometric pressures measured at the two locations. The observed pressure averaged over the winter season is divided by the standard deviation of the pressure variation over many years to obtain values that are either positive or negative, which is why the reference is to positive and negative phases of the NAO, even though the difference between the Azores and the Iceland pressures is always positive.

11. G. Beaugrand, K. M. Brander, J. A. Lindley, S. Souissi, and P. C. Reid, "Plankton Effect on Cod Recruitment in the North Sea," *Nature* 426 (2003): 661–664.

12. For example, some workers have argued that warming of sea-surface temperature in the tropical Indo-Pacific region is the root cause of the shift in wind patterns that control the NAO. In this view, the NAO is part of a hemisphere-wide fluctuation in atmospheric circulation. See M. P. Hoerling, J. W. Hurrell, and T. Xu, "Tropical Origins for Recent North Atlantic Climate Change," *Science* 292 (2001): 90–92.

13. The word *ozone* is a compound of the Greek word *ózō*, meaning "a smell."

14. But the volcanic input of ozone-destroying chemicals appears to be much less than the anthropogenic (human) input. See G. J. S. Bluth, C. C. Schnetzler, A. J. Krueger, and L. S. Walter, "The Contribution of Explosive Volcanism to Global Atmospheric Sulfur Dioxide Concentrations," *Nature* 366 (1993): 327–329; and A. Robock, "Volcanic Eruptions and Climate," *Reviews of Geophysics* 38 (2000): 191–219.

15. The reaction paths were proposed more than 20 years ago by L. T. Molina and M. J. Molina, "Production of Cl_2O_2 from the Self-Reaction of the ClO Reaction," *Journal of Physical Chemistry* 91 (1987): 433–436.

16. See, for example, J. C. Farman, B. G. Gardiner, and, J. D. Shanklin, "Large Loss of Total Ozone in Antarctica Reveal Seasonal ClO_x/NO_x Interaction," *Nature* 315 (1985): 207–210; and S. Solomon, "Stratospheric Ozone Depletion: A Review of Concepts and History," *Reviews of Geophysics* 37 (1999): 275–316.

17. See, for example, D. W. Fahey, R. S. Gao, K. S. Carslaw, J. Kettleborough, P. J. Popp, M. J. Northway, J. C. Holecek, S. C. Ciciora, R. J. McLaughlin, T. L. Thompson, R. H. Winkler, D. G. Baumgardner, B. Gandrud, P. O. Wennberg, S. Dhaniyala, K. McKinney,

T. Peter, R. J. Salawitch, T. P. Bui, J. W. Elkins, C. R. Webster, E. L. Atlas, H. Jost, J. C. Wilson, R. L. Herman, A. Kleinböhl, and M. von König, "The Detection of Large HNO₃-Containing Particles in the Winter Arctic Stratosphere," *Science* 291 (2001): 1026–1031.

18. A. E. Waibel, T. Peter, K. S. Carslaw, H. Oelhaf, G. Wetzel, P. J. Crutzen, U. Pöschl, A. Tsias, E. Reimer, and H. Fischer, "Arctic Ozone Loss Due to Denitrification," *Science* 283 (1999): 2064–2069.

19. S. A. Montzka, J. H. Butler, R. C. Myers, T. M. Thompson, T. H. Swanson, A. D. Clarke, L. T. Lock, and J. W. Elkins, "Decline in the Tropospheric Abundance of Halogen from Halocarbons: Implications for Stratospheric Ozone Depletion," *Science* 272 (1996): 1318–1322.

20. Montreal Protocol on Substances That Deplete the Ozone Layer, opened for signature on September 16, 1987, and entered into force on January 1, 1989, followed by a first meeting in Helskinki in May 1989. The treaty has been revised seven times since it entered into force.

3. THE WORLD OCEAN

1. D. L. Hartmann, *Global Physical Climatology* (San Diego: Academic Press, 1994), 171.

2. The global average sea-surface temperature is 17°C (63°F).

3. See, for example, S. Minobe, A. Kuwano-Yoshida, N. Komori, S.-P. Xie, and R. J. Small, "Influence of the Gulf Stream on the Troposphere," *Nature* 452 (2008): 206–209.

4. The primary component of this process is known as the North Atlantic Drift Current, which constitutes the warm, wind-driven surface waters that flow northwestward and spill into the Norwegian Sea. For a description of ocean currents worldwide, see http://oceancurrents.rsmas.miami.edu/index.html.

5. In theory, the Coriolis effect acting alone should drive a surface layer at 90 degrees to the wind direction. However, the water is subject to other forces. When the wind blows across the surface, it exerts a drag in the direction in which it blows. The water beneath the surface layer, however, exerts a drag in the opposite direction. These three stresses acting on the surface layer—the stress of the wind, the stress of the underlying water, and the Coriolis force—combine to cause the surface layer to deviate 45 degrees from the wind direction. But as the surface water moves, it also drags on the water immediately beneath it, propagating the wind's drag downward through the water column. The water's velocity at the surface is usually between 1 and 3 percent of the wind speed, and with depth this velocity decreases until at typically 50 to 200 meters (165 to 660 feet) depth, the effect of the wind becomes vanishingly small. With depth, the direction of water flow in this moving water column deviates more and more from the wind direction. At some depth, in fact, the water moves in exactly the opposite direction from the surface water. This structure of decreasing velocity and diverging direction of motion with depth is known as the *Ekman Spiral*.

6. W. S. Broecker, "The Great Ocean Conveyor," *Oceanography* 4 (1991): 79–89.

7. Most of the water transported from the Atlantic Ocean to the Pacific Ocean by the atmosphere is carried by moisture-laden winds that blow westward across the Isthmus of Panama.

8. See, for example, B. Lyon, "The Strength of El Niño and the Spatial Extent of Tropical Drought," *Geophysical Research Letters* 31 (2004), doi:21210.21029/22004GL020901.

9. G.-R. Walther, E. Post, P. Convey, A. Menzel, C. Parmesan, T. J. C Beebee, J.-M. Fromentin, O. Hoegh-Guldberg, and F. Bairlein, "Ecological Responses to Recent Climate Change," *Nature* 416 (2002): 389–395.

10. M. A. Cane, "The Evolution of El Niño, Past and Future," *Earth and Planetary Science Letters* 230 (2005): 227–240.

11. R W. Katz, "Sir Gilbert Walker and a Connection Between El Niño and Statistics," *Statistical Science* 17 (2002): 97–112.

12. This feedback is known as the *Bjerknes feedback* for meteorologist Jacob Bjerknes (1897–1975), who first recognized the connections among El Niño, the Southern Oscillation, and Walker circulation.

4. THE CARBON CYCLE AND HOW IT INFLUENCES CLIMATE

1. Atmospheric CO_2 is now being measured at numerous localities, including at the South Pole. An important condition is that the sites be remote to avoid variations caused by local features, such as forests or cities, that either add or remove CO_2 from the atmosphere.

2. Most magnesium is removed from seawater by the formation of magnesium-rich clays in sediments as the seawater circulates through the ocean floor, where the seawater reacts with rocks to form clay minerals.

3. This period is known as the *residence time*, which may be defined as "residence time = size of reservoir/inflow or outflow rate." The size of the atmosphere reservoir is about 760 gigatons (see table 4.1), so the residence time of carbon in the atmosphere is 760 gigatons/60 gigatons per year = 12.7 years.

4. C. L Sabine, R. A. Feely, N. Gruber, R. M. Key, K. Lee, J. L. Bullister, R. Wanninkhof, C. S. Wong, D. W. R. Wallace, B. Tilbrook, F. J. Millero, T.-H. Peng, A. Kozyr, T. Ono, and A. F. Rios, "The Oceanic Sink for Anthropogenic CO_2," *Science* 305 (2004): 367–371. Another one-fifth of the carbon has been taken up by the biosphere, and the rest is in the atmosphere.

5. D. Archer, H. Kheshgi, and E. Maier-Reimer, "Dynamics of Fossil Fuel CO_2 Neutralization by Marine $CaCO_3$," *Global Biogeochemical Cycles* 12 (1998): 259–276.

6. The reason why high-latitude waters tend to be nutrient rich is that the density of cold shallow water is not too different from that of nutrient-rich deep water, so the two tend to mix. In the tropics, however, the deep water is much denser than the warm surface layer and thus stays below it.

7. The solubility of CO_2 in seawater at 0°C (32°F) is more than twice that of CO_2 in seawater at 24°C (75°F). Even though CO_2 solubility in water increases with pressure, pressure is not a factor in determining CO_2 uptake by the ocean because the uptake always occurs at the surface.

8. The net transfer is about 100 gigatons of carbon per year in each direction.

9. In particular, pH is the negative log of the concentration of the H^+ ion in solution, so as pH decreases, the concentration of H^+ increases.

10. J. A. Kleypas, R. A. Feely, V. J. Fabry, C. Langdon, C. L. Sabine, and L. L. Robins, *Impacts of Ocean Acidification on Coral Reefs and Other Marine Calcifiers: A Guide for Future Research*, report from a workshop held April 18–20, 2005, in St. Petersburg, Fla., sponsored by the National Science Foundation (NSF), the National Oceanic and Atmo-

spheric Administration (NOAA), and the U.S. Geological Survey (USGS), 2006, available at http://www.fedworld.gov/onow.

11. CO_2 content affects the saturation levels of calcite and aragonite as follows: as CO_2 content increases, dissolved carbonate (CO_3^{2}) concentration decreases:

$$CO_3^{2-} + CO_2 + H_2O \rightarrow 2HCO_3^-$$

We can also write a dissolution reaction for calcium carbonate,

$$CaCO_3 \rightarrow Ca^{2+} + CO_3^{2-}$$
$$CaCO_3 \text{ saturation } = [Ca^{2+}][CO_3^{2-}] / K,$$

where K is the equilibrium constant for the latter reaction. The Ca^{2+} concentration of seawater is relatively constant, so the saturation levels of calcite and aragonite in seawater are dependent primarily on the concentration of CO_3^{2-}. See, for example, R. A Feely, C. L. Sabine, K. Lee, W. Berelson, J. Kleypas, V. Fabry, and F. J. Millero, "Impact of Anthropogenic CO_2 on the $CaCO_3$ System in the Oceans," *Science* 305 (2004): 362–366.

12. The depth of carbonate saturation is not the same as the carbonate compensation depth, which is the depth at which the rate of carbonate dissolution equals that of carbonate accumulation.

13. J. C. Orr, V. J. Fabry, O. Aumont, L. Bopp, S. C. Doney, R. A. Feely, A. Gnanadeskian, N. Gruber, A. Ishida, F. Joos, R. M. Key, K. Lindsay, E. Maier-Reimer, R. Matear, P. Monfray, A. Mouchet, R. G. Najjar, G.-K. Plattner, K. B. Rogers, C. L. Sabine, J. L. Sarmiento, R. Schlitzer, R. D. Slater, I. J. Totterdell, M.-F. Weirig, Y. Yamanaka, and A. Yool, "Anthropogenic Ocean Acidification over the Twenty-first Century and Its Impact on Calcifying Organisms," *Nature* 437 (2005): 681–686.

14. C. Heinze, "Simulating Oceanic $CaCO_3$ Export Production in the Greenhouse," *Geophysical Research Letters* 31 (2004), doi:10.1029/2004GL020613.

15. K. A. Denman, G. Brasseur, A. Chidthaisong, P. Ciais, P. M. Cox, R. E. Dickinson, D. Hauglustaine, C. Heinze, E. Holland, D. Jacob, U. Lohmann, S. Ramachandran, P. L. da Silva Dias, S. C. Wofsy, and X. Zhang, "Couplings Between Changes in the Climate System and Biogeochemistry," in *Climate Change 2007: The Physical Science Basis. Contribution of Working Group I to the Fourth Assessment Report of the Intergovernmental Panel on Climate Change*, edited by S. Solomon, D. Qin, M. Manning, Z. Chen, M. Marquis, K. B. Averyt, M.Tignor, and H. L. Miller (Cambridge: Cambridge University Press, 2007), 499–588. For further discussion of the uncertainty in how warming will affect ocean circulation, see also J. R. Toggweiler and J. Russell, "Ocean Circulation in a Warming Climate," *Nature* 451 (2008): 286–288.

16. Kleypas et al., *Impacts of Ocean Acidification on Coral Reefs and Other Marine Calcifiers*. Calcification rates drop even though surface waters remain supersaturated in carbonate; that is, calcifiers are sensitive to the extent of CO_3^{2-} supersaturation.

17. O. Hoegh-Guldberg, P. J. Mumby, A. J. Hooten, R. S. Steneck, P. Greenfield, E. Gomez, C. D. Harvell, P. F. Sale, A. J. Edwards, K. Caldeira, N. Knowlton, C. M. Eakin, R. Iglesias-Prieto, N. Muthiga, R. H. Bradbury, A. Dubi, and M. E. Hatziolos, "Coral

Reefs Under Rapid Climate Change and Ocean Acificiation," *Science* 318 (2007): 1737–1742, and references therein.

18. Royal Society, *Ocean Acidification Due to Increasing Atmospheric Carbon Dioxide*, Policy Document no. 12/05 (London: Royal Society, 2005).

19. U. Riebesell, K. G. Schulz, R. G. J. Bellerby, M. Botros, P. Fritsche, M. Meyerhöfer, C. Neill, G. Nondal, A. Oschlies, J. Wohlers, and E. Zöllner, "Enhanced Biological Carbon Consumption in a High CO_2 Ocean," *Nature* 450 (2007): 545–548; M. D. Iglesias-Rodriguez, P. R. Halloran, R. E. M. Rickaby, I. R. Hall, E. Colmenero-Hidalgo, J. R. Gittins, D. R. H. Green, T. Tyrrell, S. J. Gibbs, P. von Dassow, E. Rehm, E. V. Armbrust, and K. P. Boessenkool, "Phytoplankton Calcification in a High-CO_2 World," *Science* 320 (2008): 336–340.

20. Royal Society, *Ocean Acidification Due to Increasing Atmospheric Carbon Dioxide*.

21. For example, see the review in P. Falkowski, R. J. Scholes, E. Boyle, J. Canadell, D. Canfield, J. Elser, N. Gruber, K. Hibbard, P. Högberg, S. Linder, F. T. Mackenzie, B. Moore III, T. Pedersen, Y. Rosenthal, S. Seitzinger, V. Smetacek, and W. Steffen, "The Global Carbon Cycle: A Test of Our Knowledge of Earth as a System," *Science* 290 (2000): 291–296.

22. P. B. Reich, S. E. Hobbie, T. Lee, D. S. Ellsworth, J. B. West, D. Tilman, J. M. H. Knops, S. Naeem, and J. Trost, "Nitrogen Limitation Constrains Sustainability of Ecosystem Response to CO_2," *Nature* 440 (2006): 922–925.

23. A. W. King, C. A. Gunderson, W. M. Post, D. J. Weston, and S. D. Wullschleger, "Plant Respiration in a Warmer World," *Science* 312 (2006): 536–537.

24. See, for example, J. Heath, E. Ayres, M. Possell, R. D. Bardgett, H. I. J. Black, H. Grant, P. Ineson, and G. Kerstiens, "Rising Atmospheric CO_2 Reduces Sequestration of Root-Derived Soil Carbon," *Science* 309 (2005): 1711–1713; and E. A. Davidson and I. A. Janssens, "Temperature Sensitivity of Soil Carbon Decomposition and Feedbacks to Climate Change," *Nature* 440 (2006): 165–173.

5. A SCIENTIFIC FRAMEWORK FOR THINKING ABOUT CLIMATE CHANGE

1. Wavelength (λ) varies inversely as energy (E), $\lambda = hc/E$, where h is Planck's constant and c is the speed of light.

2. For further explanation, see http://edmall.gsfc.nasa.gov/inv99Project.Site/Pages/science-briefs/edstickler/edirradiance.

3. The total poleward transport of heat is about 6×10^{15} watts per year.

4. This phenomenon is described by the Stefan-Boltzmann law, which states that emitted radiation per unit area $= \sigma T^4$, where σ is a constant and T is temperature in degrees Kelvin.

5. H. Le Treut, R. Somerville, U. Cubasch, Y. Ding, C. Mauritzen, A. Mokssit, T. Peterson, and M. Prather, "Historical Overview of Climate Change," in *Climate Change 2007. The Physical Science Basis. Contribution of Working Group 1 to the Fourth Assessment Report of the Intergovernmental Panel on Climate Change*, edited by S. Solomon, D. Qin, M. Manning, Z. Chen, M. Marquis, K. B. Averyt, M. Tignor, and H. L. Miller (Cambridge: Cambridge University Press, 2007), 93–128.

6. Specifically, the boiling of water at 100°C (212°F) requires 2,260 joules per gram (heat of vaporization), and the same amount of energy is released in the reverse process

when water vapor condenses to form liquid water at that temperature. At temperatures below the boiling point, the heats of vaporization are slightly higher—for example, at 25°C (77°F) it is 2,574 joules per gram.

7. The formal definition originates from the second IPCC report: radiative forcing is "the change [relative to the year 1750] in net (down minus up) irradiance (solar plus longwave in Wm⁻²) at the tropopause after allowing for stratospheric temperatures to readjust to radiative equilibrium, but with surface and tropospheric temperatures and state held fixed at the unperturbed values" (V. Ramaswamy, O. Boucher, J. Haigh, D. Hauglustaine, J. Haywood, G. Myhre, T. Nakajima, G. Y. Shi, S. Solomon, R. Betts, R. Charlson, C. Chuang, J. S. Daniel, A. Del Genio, R. van Dorland, J. Feichter, J. Fuglestvedt, P. M. de F. Forster, S. J. Ghan, A. Jones, J. T. Kiehl, D. Koch, C. Land, J. Lean, U. Lohmann, K. Minschwaner, J. E. Penner, D. L. Roberts, H. Rodhe, G. J. Roelofs, L. D. Rotstayn, T. L. Schneider, U. Schumann, S. E. Schwartz, M. D. Schwarzkopf, K. P. Shine, S. Smith, D S. Stevenson, F. Stordal, I. Tegen, and Y. Zhang, "Radiative Forcing of Climate Change," in *Climate Change 2001: The Scientific Basis. Contribution of Working Group I to the Third Assessment Report of the Intergovernmental Panel on Climate Change,* edited by J. T. Houghton, Y. Ding, D. J. Griggs, M. Noguer, P. J. Van Der Linden, X. Dai, K. Maskell, and C. A. Johnson [Cambridge: Cambridge University Press, 2001], 353).

8. The term *greenhouse effect* is a misnomer because the process that keeps Earth warm is not the same as that which keeps a greenhouse warm. In a greenhouse, the incoming radiation warms the surfaces, the surfaces warm the air, and the greenhouse interior warms because of the restricted convective cooling of the air. The "greenhouse effect" as it pertains to the atmosphere, in contrast, involves absorption of IR radiation by certain gases. Depending on the gas, the absorption of radiation may involve:

> *Increasing the kinetic energy of gas molecules.* The molecules are made to move faster, which is equivalent to saying that temperature increases.
> *Increasing molecular rotational and vibrational energy.* The molecules rotate and vibrate such that the bonds between atoms bend or stretch around mean values. Rotational and vibrational energy is quantized, meaning that molecules exist in specific energy states. Absorption occurs when an incoming photon causes the molecule to change from a lower-energy state to a higher one. It is the mechanism by which CO_2 absorbs outgoing IR radiation.
> *Photodissociation.* This process occurs when the bond that holds a molecule together is broken. It requires incident photon wavelengths of less than one micron. For example, ozone is dissociated by radiation between 200 and 300 nanometers (millionths of a millimeter). The process is important mainly in the stratosphere.
> *Electronic excitation and photoionization.* Both require high incident energies and occur mainly in the upper atmosphere.

9. The maximum intensity is for a temperature of −18°C (0°F), which is the effective radiating temperature of Earth.

10. See, for example, Y. J. Kaufman, D. Tarné, and O. Boucher, "A Satellite View of Aerosols in the Climate System," *Nature* 419 (2002): 215–223; and Y. J. Kaufman and I. Koren, "Smoke and Pollution Aerosol Effect on Cloud Cover," *Science* 313 (2006): 655–658.

11. Kaufman, Tarné, and Boucher, "Satellite View of Aerosols."

12. D. Rosenfeld, J. Dai, X. Yu, A. Yao, X. Xu, X. Yang, and C. Du, "Inverse Relations Between Amounts of Air Pollution and Orographic Precipitation," *Science* 315 (2007): 1396–1998.

13. V. Ramanathan, M. V. Ramana, G. Roberts, D. Kim, C. Corrigan, C. Chung, and D. Winker, "Warming Trends in Asia Amplified by Brown Cloud Solar Absorption," *Nature* 448 (2007): 575–578.

14. R. T. Pinker, B. Zhang, and E. G. Dutton, "Do Satellites Detect Trends in Surface Solar Radiation?" *Science* 308 (2005): 850–854; M. Wild, H. Gilgen, A. Roesch, A. Ohmura, C. N. Long, E. G. Dutton, B. Forgan, A. Kallis, V. Russak, and A. Tsvetkov, "From Dimming to Brightening: Decadal Changes in Solar Radiation at Earth's Surface," *Science* 308 (2005): 847–850; M. I. Mishchenko, I. V. Geogdzhayev, W. B. Rossow, B. Cairns, B. E. Carlson, A. A. Lacis, L. Liu, and L. D. Travis, "Long-Term Satellite Record Reveals Likely Recent Aerosol Trend," *Science* 315 (2007): 1543.

15. D. G. Streets, Y. Wu, and M. Chin, "Two-Decadal Aerosol Trends as a Likely Explanation of the Global Dimming/Brightening Transition," *Geophysical Research Letters* 33 (2006), doi:10.1029/2006GL026471.

16. See, for example, M. O. Andreae, C. D. Jones, and P. M. Cox, "Strong Present-Day Aerosol Cooling Implies a Hot Future," *Nature* 435 (2005): 1187–1190.

17. N. Bellouin, O. Boucher, J. Haywood, and M. S. Reddy, "Global Estimate of Aerosol Direct Radiative Forcing from Satellite Measurements," *Nature* 438 (2005): 1138–1141. These authors also estimated that anthropogenic aerosols have increased cloud cover by 5 percent, which corresponds to an increase in the reflected solar flux of 5 watts per square meter. See also F.-M. Bréon, "How Do Aerosols Affect Cloudiness and Climate?" *Science* 313 (2006): 623–624.

18. J. A. Foley, R. DeFries, G. P. Asner, C. Barford, G. Bonan, S. R. Carpenter, F. S. Chapin, M. T. Coe, G. C. Daily, H. K. Gibbs, J. H. Helkowski, T. Holloway, E. A. Howard, C. J. Kucharik, C. Monfreda, J. A. Patz, I. C. Prentice, N. Ramankutty, and P. K. Snyder, "Global Consequences of Land Use," *Science* 309 (2005): 570–574. See also P. Kabat, M. Claussen, P. A. Dirmeyer, J. H. C. Gash, L. B. DeGuenni, M. Meybeck, R. A. Pielke Sr., C. J. Vörösmarty, R. W. A. Hutjes, and S. Lürkemeier, eds., *Vegetation, Water, Humans, and the Climate: A New Perspective on an Interactive System* (Berlin: Springer-Verlag, 2004).

19. J. J. Feddema, K. W. Oleson, G. B. Bonan, L. O Mearns, L. E. Bujja, G. A. Meehl, and W. M. Washington, "The Importance of Land-Cover Change in Simulating Future Climates," *Science* 310 (2005): 1674–1678.

20. P. Forster, V. Ramaswamy, P. Artaxo, T. Berntsen, R. Betts, D. W. Fahey, J. Haywood, J. Lean, D. C. Lowe, G. Myhre, J. Nganga, R. Prinn, G. Raga, M. Schulz, and R. Van Dorland, "Changes in Atmospheric Constituents and in Radiative Forcing," in Solomon et al., eds., *Climate Change 2007*, 129–234.

21. See, for example, A. Robock, "Volcanic Eruptions and Climate," *Reviews of Geophysics* 38 (2000): 191–219.

22. Ibid. The formation of sulfate aerosols preferentially heats the stratosphere over the equator compared with over the pole. The larger-than-normal equator-to-pole temperature gradient produces an unusually strong winter polar vortex, which in turn creates a characteristic pattern of circulation in the troposphere that brings relatively warm

winter conditions to regions between about 40 and 60°N latitudes in North America, western Europe, and part of Asia.

23. The radiative forcing was determined from independent climate models reported in S. Ramachandran, V. Ramaswamy, G. L. Stenchikov, and A. Robock, "Radiative Impact of the Mt. Pinatubo Volcanic Eruption: Lower Stratospheric Response," *Journal of Geophysical Research* 105 (2000): 24409–24429; and J. Hansen, M. Sato, L. Nazarenko, R. Ruedy, A. Lacis, D. Koch, I. Tegen, T. Hall, D. Shindell, B. Santer, P. Stone, T. Novakov, L. Thomason, R. Wang, Y. Wang, D. Jacob, S. Hollandsworth, L. Bishop, J. Logan, A. Thompson, R. Stolarski, J. Lean, R. Willson, S. Levitus, J. Antonov, N. Rayner, D. Parker, and J. Christy, "Climate Forcings in Goddard Institute for Space Studies SI2000 Simulations," *Journal of Geophysical Research* 107 (2002), doi:10.1029/2001JD001143. The latter paper also reported global temperature change, which has been variously reported to have cooled from 0.3 to 0.6°C (0.5 to 1.1°F).

24. See, for example, R. B. Stothers, "The Great Tambora Eruption in 1815 and Its Aftermath," *Science* 224 (1984): 1191–1198; and H. Sigurdsson and S. Carey, "The Eruption of Tambora in 1815: Environmental Effects and Eruption Dynamics," in *The Year Without a Summer? World Climate in 1816*, edited by C. R. Harington (Ottawa: Canadian Museum of Nature, 1992), 16–45.

25. J. Eddy, "Before Tambora: The Sun and Climate, 1790–1830" (abstract), in Harington, ed., *Year Without a Summer?* 9.

26. W. Baron, "1816 in Perspective: The View from the Northeastern United States," 124–144; M. K. Cleaveland, "Volcanic Effects on Colorado Plateau Douglas-Fir Tree Rings," 115–123; and J. M. Lough, "Climate in 1816 and 1811–20 as Reconstructed from Western North American Tree-Ring Chronologies," 97–114, all in Harington, ed., *Year Without a Summer?*

27. P. Foukal, C. Fröhlich, H. Spruit, and T. M. L. Wigley, "Variations in Solar Luminosity and Their Effect on the Earth's Climate," *Nature* 443 (2006): 161–166.

28. E. Bard and M. Frank, "What's New Under the Sun?" *Earth and Planetary Science Letters* 248 (2006): 1–14.

29. Forster et al., "Changes in Atmospheric Constituents and in Radiative Forcing."

30. J. L. Lean and D. H. Rind, "How Natural and Anthropogenic Influences Alter Global and Regional Surface Temperatures: 1889 to 2006," *Geophysical Research Letters* 35 (2008), doi:10.1029/2008GL034864, and references therein.

31. A. P. M. Baede, E. Ahlonsou, Y. Ding, and D. Schimel, "The Climate System: An Overview," in Houghton et al., eds., *Climate Change 2001*, 85–98.

32. There are two reasons for this sensitivity. First, there is so much water in the lower troposphere that it is essentially opaque to the portions of the IR radiation spectrum absorbed by water. Second, the strength of the feedback is dependent primarily on the fractional change in water vapor concentration, not on the change in absolute amount. The upper troposphere is cold and contains little water vapor because saturation level is dependent on temperature. As the upper troposphere warms, the fractional amount of water vapor it contains changes much more rapidly than does the amount in the water-rich lower atmosphere.

33. D. A. Randall, R. A. Wood, S. Bony, R. Colman, T. Fichefet, J. Fyfe, V. Kattsov, A. Pitman, J. Shukla, J. Srinivasan, R. J. Stouffer, A. Sumi, and K. E. Taylor, "Climate Mod-

els and Their Evaluation," in Solomon et al., eds., *Climate Change 2007*, 589–662. The uncertainty regarding the importance of the water vapor feedback originates from uncertainty in the efficiencies of processes that control relative upper troposphere humidity mainly in the tropics, such as convective transport (that is, the heating of warm, moist air near the surface causes the warm air to rise by buoyancy to the upper troposphere).

34. Committee on Abrupt Climate Change, *Abrupt Climate Change: Inevitable Surprises* (Washington, D.C.: National Academy Press, 2002), 14.

35. See, for example, J. Hansen, L. Nazarenko, R. Ruedy, M. Sato, J. Willis, A. Del Genio, D. Koch, A. Lacis, K. Lo, S. Menon, T. Novakov, J. Perlwitz, G. Russell, G. A. Schmidt, and N. Tausnev, "Earth's Energy Imbalance: Confirmation and Implications," *Science* 308 (2005): 1431–1435; and T. M. L. Wigley, "The Climate Change Commitment," *Science* 307 (2005): 1766–1769.

36. Hansen et al., "Earth's Energy Imbalance."

37. Ibid. Note that this process is not radiative forcing, which, as described previously, is the change in incoming energy minus outgoing energy in response to a factor that changes energy balance relative to the year 1750.

6. LEARNING FROM CLIMATES PAST

1. J. C. Zachos, U. Röhl, S. A. Schellenberg, A. Sluijs, D. A. Hodell, D. C. Kelly, E. Thomas, M. Nicolo, I. Raffi, L. J. Lourens, H. McCarren, and D. Kroon, "Acidification of the Ocean During the Paleocene–Eocene Thermal Maximum," *Science* 308 (2005): 1611–1615.

2. G. J. Bowen, T. J. Bralower, M. L. Delaney, G. R. Dickens, D. C. Kelly, P. L. Koch, L. R. Kump, J. Meng, L. C. Sloan, E. Thomas, S. L. Wing, and J. C. Zachos, "Eocene Hyperthermal Event Offers Insight into Greenhouse Warming," *Eos, Transactions* 87 (2006): 165.

3. W. C. Clyde and P. D. Gingrich, "Mammalian Community Response to the Latest Paleocene Thermal Maximum: An Isotaphonomic Study in the Northern Bighorn Basin, Wyoming," *Geology* 26 (1998): 1011–1014; S. L. Wing, G. J. Harrington, F. A. Smith, J. I. Bloch, D. M. Boyer, and K. H. Freeman, "Transient Floral Change and Rapid Global Warming at the Paleocene–Eocene Boundary," *Science* 310 (2005): 993–996.

4. S. J. Gibbs, P. R. Brown, J. A. Sessa, T. J. Bralower, and P. A. Wilson, "Nannoplankton Extinction and Origination Across the Paleocene–Eocene Thermal Maximum," *Science* 314 (2006): 1770–1773.

5. See, for example, G. J. Bowen, D. J. Beerling, P. L. Koch, J. C. Zachos, and T. Quattlebaum, "A Humid State During the Paleocene/Eocene Thermal Maximum," *Nature* 432 (2004): 495–499; Wing et al., "Transient Floral Change"; and Zachos et al., "Acidification of the Ocean."

6. Zachos et al., "Acidification of the Ocean."

7. G. R. Dickens, J. R. O'Neil, D. K. Rea, and R. M. Owen, "Dissociation of Oceanic Methane Hydrate as a Cause of the Carbon Isotope Excursion at the End of the Paleocene," *Paleoceanography* 10 (1995): 965–971; R. Matsumoto, "Causes of the $\delta^{13}C$ Anomalies of Carbonates and a New Paradigm 'Gas Hydrate Hypothesis,'" *Journal of the Geological Society of Japan* 11 (1995): 902–924; A. Sluijs, H. Brinkhuis, S. Schouten, S. M.

Bohaty, C. M. John, J. C. Zachos, G.-J. Reichart, J. S. Sinninghe Damsté, E. M. Crouch, and G. R. Dickens, "Environmental Precursors to Rapid Light Carbon Injection at the Paloeocene/Eocene Boundary," *Nature* 450 (2007): 1218–1221.

8. G. R. Dickens and M. S. Quinby-Hunt, "Methane Hydrate Stability in Seawater," *Geophysical Research Letters* 21 (1994): 2115–2118.

9. Sluijs et al., "Environmental Precursors to Rapid Light Carbon Injection."

10. Bowen et al., "Eocene Hyperthermal Event."

11. See, for example, Y. Yokoyama, K. Lambeck, P. De Deckker, P. Johnston, and L. K. Fifield, "Timing of the Last Glacial Maximum from Observed Sea-Level Minima," *Nature* 406 (2000): 713–716.

12. For an interesting and readable description of the ice age in Scandinavia, see B. G. Andersen and H. W. Borns Jr., *The Ice Age World* (Oslo: Scandinavian University Press, 1994).

13. E. Lurie, *A Life in Science* (Chicago: University of Chicago Press, 1960); J. Imbrie and K. P. Imbrie, *Ice Ages: Solving the Mystery*, 2d ed. (Cambridge, Mass.: Harvard University Press, 1986).

14. G. C. Bond, W. Broecker, S. Johnsen, J. McManus, L. Labeyrie, J. Jouzel, and G. Bonani, "Correlations Between Climate Records from North Atlantic Sediments and Greenland Ice," *Nature* 365 (1996): 143–147.

15. These discharges are known as Heinrich events, after their discoverer, Hartmut Heinrich (b. 1966). The rapid discharge of icebergs produced a large amount of freshwater, resulting in a reduction of salinity and a cooling of North Atlantic surface waters.

16. A. C. Mix, N. G. Pisias, W. Rugh, J. Wilson, A. Morey, and T. Hagelberg, "Benthic Foraminiferal Stable Isotope Record from Site 849, 0–5 Ma: Local and Global Climate Changes," in *Proceedings of the Ocean Drilling Program, Scientific Results*, edited by N. G. Pisias, L. Mayer, T. Janecek, A. Palmer-Julson, and T. H. Van Andel (College Station, Tex.: Ocean Drilling Program, 1995), 371–412.

17. See, for example, H. J. Dowsett, M. A. Chandler, T. M. Cronin, and G. S. Dwyer, "Middle Pliocene Sea Surface Temperature Variability," *Paleoceanography* 20 (2005), doi:10.1029/2005PA001133.

18. G. H. Denton, D. E. Sugden, D. R. Marchant, B. L. Hall, and T. I. Wilch, "East Antarctic Ice Sheet Sensitivity to Pliocene Climatic Change from a Dry Valleys Perspective," *Geografiska Annaler* 75A (1993): 155–204.

19. See, for example, K. J. Willis, A. Kleczkowski, K. M. Briggs, and C. A. Gilligan, "The Role of Sub-Milankovitch Climatic Forcing in the Initiation of the Northern Hemisphere Glaciation," *Science* 285 (1999): 568–571.

20. J. D. Hays, J. Imbrie, and N. J. Shackleton, "Variations in the Earth's Orbit: Pacemaker of the Ice Ages," *Science* 194 (1976): 1121–1132.

21. P. Huybers, "Early Pleistocene Glacial Cycles and the Integrated Summer Insolation Forcing," *Science* 313 (2006): 508–511; M. E. Raymo, L. E. Lisiecki, and K. H. Nisancioglu, "Plio-Pleistocene Ice Volume, Antarctic Climate, and the Global ^{18}O Record," *Science* 313 (2006): 492–495.

22. R. B. Alley, E. J. Brook, and S. Anandakrishnan, "A Northern Lead in the Orbital Band: North–South Phasing of Ice-Age Events," *Quaternary Science Reviews* 21 (2002):

431–441; K. Kawamura, F. Parrenin, L. Lisiecki, R. Uemura, F. Vimeux, J.P. Severing-haus, M.A. Hutterli, T. Nakazawa, S. Aoki, J. Jouzel, M.E. Raymo, K. Matsumoto, H. Nakata, H. Motoyama, S. Fujita, K. Goto-Azuma, Y. Fujii, and O. Watanabe, "North-ern Hemisphere Forcing of Climatic Cycles in Antarctica Over the Past 360,000 Years," *Nature* 448 (2007): 912–916.

23. The most extreme example was the abrupt shift 400,000 years ago from an unusu-ally cold glacial period to an unusually warm interglacial period. The latter, in addition, was warmer than any time in the past 2 million years. Much has been written about this warm interglacial, also known as *marine isotope stage 11*, because of the possibility that it is an analogue for the present interglacial. See, for example, A.W. Droxler, R.B. Alley, W.R. Howard, R.Z. Poore, and L.H. Burckle, "Unique and Exceptionally Long Intergla-cial Marine Isotope Stage 11: Window into Earth [*sic*] Warm Future Climate," in *Earth's Climate and Orbital Eccentricity: The Marine Isotope Stage 11 Question*, edited by A.W. Droxler, R.Z. Poore, and L.H. Burckle, Geophysical Monograph, no. 137 (Washington, D.C.: American Geophysical Union, 2003), 1–14.

24. P. Huybers and C. Wunsch, "Obliquity Pacing of the Late Pleistocene Glacial Ter-minations," *Nature* 343 (2005): 491–494.

25. Ibid., 493.

26. Not everyone agrees with this notion. Ice volume also reflects precession and may be sensitive to high summer insolation no matter how the cycles combine to cause it (M. Bender, personal communication, 2007).

27. J.R. Petit, J. Jouzel, D. Raynaud, N.I. Barkov, J.-M. Barnola, I. Basile, M. Bender, J. Chappellaz, M. Davis, G. Delaygue, M. Delmotte, V.M. Kotlyakov, M. Legrand, V.Y. Lipenkov, C. Lorius, L. Pépin, C. Ritz, E. Saltzman, and M. Stievenard, "Climate and Atmospheric History of the Past 420,000 Years from the Vostok Ice Core, Antarc-tica," *Nature* 399 (1999): 429–436; U. Siegenthaler, T.F. Stocker, E. Monnin, D. Lüthi, J. Schwander, B. Stauffer, D. Raynaud, J.-M. Barnola, H. Fischer, V. Masson-Delmotte, and J. Jouzel, "Stable Carbon Cycle–Climate Relationships During the Late Pleisto-cene," *Science* 310 (2005): 1313–1317; R. Spahni, J. Chappellaz, T.F. Stocker, L. Loulergue, G. Hausammann, K. Kawamura, J. Flückiger, J. Schwander, D. Raynaud, V. Masson-Delmotte, and J. Jouzel, "Atmospheric Methane and Nitrous Oxide of the Late Pleisto-cene from the Antarctic Ice Cores," *Science* 310 (2005): 1317–1321; L. Loulergue, A. Schilt, R. Spahni, V. Masson-Delmotte, T. Blunier, B. Lemieux, J.-M. Barnola, D. Raynaud, T.F. Stocker, and J. Chappellaz, "Orbital and Millennial-Scale Features of Atmospheric CH_4 over the Past 800,000 Years," *Nature* 453 (2008): 383–386; D. Lüthi, M. Le Floch, B. Be-reiter, T. Blunier, J.-M. Barnola, U. Siegenthaler, D. Raynaud, J. Jouzel, H. Fischer, K. Kawamura, and T.F. Stocker, "High-Resolution Carbon Dioxide Concentration Record 650,000–800,000 Years Before Present," *Nature* 453 (2008): 379–383.

28. The temperature is determined by the ratio of the isotopes hydrogen-2 to hydrogen-1.

29. Spahni et al., "Atmospheric Methane and Nitrous Oxide of the Late Pleistocene."

30. P.M. Grootes, M. Stuiver, J.W.C. White, S. Johnsen, and J. Jouzel, "Comparison of Oxygen Isotope Records from the GISP2 and GRIP Greenland Ice Cores," *Nature* 366 (1993): 552–554; P.A. Mayewski and M. Bender, "The GISP2 Ice Core Record—Paleoclimate Highlights," *Review of Geophysics Supplement, U.S. National Report to Inter-*

national Union of Geodesy and Geophysics, 1991–1994 (1995): 1287–1296; P. A. Mayewski, L. D. Meeker, M. S. Twickler, S. I. Whitlow, Q. Yang, W. W. Lyons, and M. Prentice, "Major Features and Forcing of High Latitude Northern Hemisphere Atmospheric Circulation Using a 110,000-Year-Long Glaciochemical Series," *Journal of Geophysical Research* 102 (1997): 26346–26366.

31. Snow currently accumulates at a rate of about 20 centimeters (8 inches) a year in Greenland. This rate is about 10 times greater than the accumulation rate at Vostok, the Russian base in Antarctica, so the annual layers in Greenland ice are thicker and can be more precisely counted and dated.

32. R. B. Alley, C. A. Shuman, D. A. Meese, A. J. Gow, K. C. Taylor, K. M. Cuffey, J. J. Fitzpatrick, P. M. Grootes, G. A. Zielinski, M. Ram, G. Spinelli, and B. Elder, "Visual-Stratigraphic Dating of the GISP2 Ice Core: Basis, Reproducibility, and Application," *Journal of Geophysical Research* 102 (1997): 26367–26381; D. A. Meese, A. J. Gow, R. B. Alley, G. A. Zielinski, P. M. Grootes, M. Ram, K. C. Taylor, P. A. Mayewski, and J. F. Bolzan, "The Greenland Ice Sheet Project 2 Depth-Age Scale: Methods and Results," *Journal of Geophysical Research* 102 (1997): 26411–26423.

33. G. A. Zielinski, P. A. Mayewski, L. D. Meeker, S. I. Whitlow, M. S. Twickler, M. C. Morrison, D. Meese, R. Alley, and A. J. Gow, "Record of Volcanism Since 7000 B.C. from the GISP2 Greenland Ice Core and Implications for the Volcano-Climate System," *Science* 264 (1994): 948–952.

34. The name Mazama age comes from C. M. Zdanowicz, G. A. Zielinski., and M. S. Germani, "Mount Mazama Eruption: Calendrical Age Verified and Atmospheric Impact Assessed," *Geology* 27 (1999): 621–624. The time assigned to the Santorini eruption has recently been questioned by G. A. Zielinski and M. S. Germani, "New Ice-Core Evidence Challenges the 1620s B.C. Age for the Santorini [Minoan] Eruption," *Journal of Archaeological Science* 25 (1998): 279–289.

35. P. A. Mayewski, W. B. Lyons, M. J. Spencer, M. S. Twickler, C. F. Buck, and S. Whitlow, "An Ice-Core Record of Atmospheric Response to Anthropogenic Sulphate and Nitrate," *Nature* 346 (1990): 554–556.

36. S. Hong, J.-P. Candelone, C. C. Patterson, and C. F. Boutron, "History of Ancient Copper Smelting Pollution During Roman and Medieval Times Recorded in Greenland Ice," *Science* 272 (1996): 246–249.

37. C. Lang, M. Leuenberger, J. Schwander, and S. Johnsen, "16°C Rapid Temperature Variation in Central Greenland 70,000 Years Ago," *Science* 286 (1999): 934–937; J. P. Severinghaus and E. J. Brook, "Abrupt Climate Change at the End of the Last Glacial Period Inferred from Trapped Air in Polar Ice," *Science* 286 (1999): 930–934.

38. The name Younger Dryas comes from the tundra shrub dryas. Fossil dryas plants are present in three stratigraphic intervals in sediments in northern Europe. These periods became known as the Oldest, Older, and Younger Dryas intervals.

39. W. S. Broecker and G. H. Denton, "The Role of Ocean–Atmosphere Reorganizations in Glacial Cycles," *Geochimica et Geochemica Acta* 53 (1989): 2465–2501; W. S. Broecker and G. H. Denton, "What Drives Glacial Cycles?" *Scientific American*, January 1990, 48–56.

40. C. R. W. Ellison, M. R. Chapman, and I. R. Hall, "Surface and Deep Ocean Interactions During the Cold Climate Event 8200 Years Ago," *Science* 312 (2006): 1929–1932; H. F. Kleiven, C. Kissel, C. Laj, U. S. Ninnemann, T. O. Richter, and E. Cortijo, "Reduced

North Atlantic Deep Water Coeval with the Glacial Lake Agassiz Freshwater Outburst," *Science* 319 (2008): 60–64.

41. See, for example, T. Blunier and E. J. Brook, "Timing of Millennial-Scale Climate Change in Antarctica and Greenland During the Last Glacial Period," *Science* 291 (2001): 109–112.

42. EPICA Community Members, "One-to-One Coupling of Glacial Climate Variability in Greenland and Antarctica," *Nature* 444 (2006): 195–198. This is in the Atlantic sector of East Antarctica, so the EDML core is more directly related to climate over the Atlantic Ocean than are other Antarctic ice cores. In addition, snow accumulated two to three times faster in this sector than at other drill sites in Antarctica. Thus, the temporal resolution of the EDML core, which extends back about 150,000 years, is comparable to the resolution of the cores taken in Greenland.

43. T. F. Stocker, D. G. Wright, and W. S. Broecker, "The Influence of High-Latitude Surface Forcing on the Global Thermocline Circulation," *Paleoceanography* 7 (1992): 529–541.

44. T. F. Stocker and S. J. Johnsen, "A Minimum Thermodynamic Model for the Bipolar Seesaw," *Paleoceanography* 18 (2003), doi:1010.1029/2003PA000920.

45. B. Martrat, J. O. Grimalt, N. J. Shackleton, L. de Abreu, M. A. Hutterli, and T. F. Stocker, "Four Climate Cycles of Recurring Deep Surface Water Destabilizations on the Iberian Margin," *Science* 317 (2007): 502–507.

46. T. Corrège, M. K. Gagan, J. W. Beck, G. S. Burr, G. Cabloch, and F. Le Cornec, "Interdecadal Variation in the Extent of South Pacific Tropical Waters During the Younger Dryas Event," *Nature* 428 (2004): 927–929.

47. T. V. Lowell, C. J. Heusser, B. G. Andersen, P. I. Moreno, A. Hauser, L. E. Heusser, C. Schlüchter, D. R. Marchant, and G. H. Denton, "Interhemispheric Correlation of Late Pleistocene Glacial Events," *Science* 269 (1995): 1541–1549. This correspondence remains uncertain because of lack of precision in the timing of the events in the Andes.

48. D. W. Lea, D. K. Pak, L. C. Peterson, and K. A. Hughen, "Synchroneity of Tropical and High-Latitude Atlantic Temperatures over the Last Glacial Termination," *Science* 301 (2003): 1361–1364.

49. T. T. Barrows, S. J. Lehman, L. K. Fifield, and P. De Deckker, "Absence of Cooling in New Zealand and the Adjacent Ocean During the Younger Dryas Chronozone," *Science* 318 (2007): 86–89.

50. Milankovitch cycles predict that 60,000 years from now Earth will be in a full glacial period. What actually happens, however, may depend on what humanity does to influence climate. For recent thinking on this matter, see A. Berger and M. F. Loutre, "An Exceptionally Long Interglacial Ahead?" *Science* 297 (2002): 1287–1288.

51. See, for example, H. C. Fritts, "Tree-Ring Analysis," in *The Encyclopedia of Climatology*, edited by J. E. Oliver and R. W. Fairbridge (New York: Van Nostrand Reinhold, 1987), 858–875. The long chronology was established with both live and dead trees, mainly the long-lived bristlecone pine in North America and the waterlogged oaks found in Ireland and Germany.

52. The term *varve* comes from the Swedish word *varv*, meaning "layer."

53. Corals build skeletons of the calcium carbonate mineral aragonite. The process is complex, so there is some uncertainty regarding how closely the skeletons' compositional characteristics reflect environmental conditions.

54. See, for example, J. W. Beck, R. L. Edwards, E. Ito, F. W. Taylor, J. Recy, F. Rougerie, P. Joannot, and C. Henin, "Sea-Surface Temperature from Coral Skeletal Strontium/Calcium Ratios," *Science* 31 (1992): 644–647. Depending on location, they may also be sensitive to salinity.

55. J. E. Cole, G. T. Shen, R. G. Fairbanks, and M. Moore, "Cora Monitors of El Niño/Southern Oscillation Dynamics Across the Equatorial Pacific," in *El Niño*, edited by H. F. Diaz and V. Markgraf (Cambridge: Cambridge University Press, 1992), 349–375.

56. Y. Wang, H. Cheng, R. L. Edwards, X. Kong, X. Shao, S. Chen, J. Wu, X. Jiang, X. Wang, and Z. An, "Millennial- and Orbital-Scale Changes in the East Asian Monsoon over the Past 224,000 Years," *Nature* 451 (2008): 1090–1093.

57. M. E. Mann, R. S. Bradley, and M. K. Hughes, "Global-Scale Temperature Patterns and Climate Forcing over the Past Six Centuries," *Nature* 392 (1998): 779–787; M. E. Mann, R. S. Bradley, and M. K. Hughes, "Northern Hemisphere Temperatures During the Past Millennium: Inferences, Uncertainties, and Limitations," *Geophysical Research Letters* 26 (1999): 759–762.

58. W. S. Broecker, "Was the Medieval Warm Period Global?" *Science* 291 (2001): 1497–1499.

59. G. C. Bond, W. Showers, M. Elliot, M. Evans, R. Lotti, I. Hajdas, G. Bonani, and S. Johnson, "The North Atlantic's 1–2 kyr Climate Rhythm: Relation to Heinrich Events, Dansgaard/Oeschger Cycles, and the Little Ice Age," in *Mechanisms of Global Climate Change at Millennial Time-Scales*, edited by P. U. Clark, R. S. Webb, and L. D. Keigwin (Washington, D.C.: American Geophysical Union, 1999), 35–58.

60. P. D. Jones and M. E. Mann, "Climate over Past Millennia," *Reviews of Geophysics* 42 (2004), doi:10.1029/2003RG000143.

61. R. D'Arrigo, R. Wilson, and G. Jacoby, "On the Long-Term Context for Late Twentieth Century Warming," *Journal of Geophysical Research* 111 (2006), doi:10.1029/2005JD006352.

7. A CENTURY OF WARMING AND SOME CONSEQUENCES

1. See, for example, S. Levitus, J. Antonov, and T. Boyer, "Warming of the World Ocean, 1955–2003," *Geophysical Research Letters* 32 (2005), doi:10.1029/2004GL021592.

2. P. Forster, V. Ramaswamy, P. Artaxo, T. Berntsen, R. Betts, D. W. Fahey, J. Haywood, J. Lean, D. C. Lowe, G. Myhre, J. Nganga, R. Prinn, G. Raga, M. Schulz, and R. Van Dorland, "Changes in Atmospheric Constituents and in Radiative Forcing," in *Climate Change 2007: The Physical Science Basis. Contribution of Working Group I to the Fourth Assessment Report of the Intergovernmental Panel on Climate Change*, edited by S. Solomon, D. Qin, M. Manning, Z. Chen, M. Marquis, K. B. Averyt, M.Tignor, and H. L. Miller (Cambridge: Cambridge University Press, 2007), 129–234.

3. J. Hansen, M. Sato, R. Ruedy, K. Lo, D. W. Lea, and M. Medina-Elizade, "Global Temperature Change," *Proceedings of the National Academy of Science* 103 (2006): 14288–14293.

4. P. Brohan, J. J. Kennedy, I. Harris, S. Tett, and P. D. Jones, "Uncertainty Estimates in Regional and Global Observed Temperature Changes: A New Data Set from 1850," *Journal of Geophysical Research* 111 (2006), doi:10.1029/2005JD006548.

5. Ibid. At the 95 percent confidence interval, the decadal errors are about 0.1°C (0.18°F) for nineteenth-century and early-twentieth-century warming and 0.05°C (0.09°F) for late-twentieth-century warming.

6. J. Oerlemans, "Extracting a Climate Signal from 169 Glacier Records," *Science* 308 (2005): 675–677. The temperature record was based on observations from 169 glaciers.

7. See, for example, H. N. Pollack and S. Huang, "Climate Reconstruction from Subsurface Temperatures," *Annual Reviews of Earth and Planetary Sciences* 28 (2000): 339–365. Thermal profiles in ice through the Greenland Ice Sheet also yield records of past temperature.

8. Thermal profiles are governed by (1) conditions at the surface and (2) the heat flowing up from Earth's interior. At several kilometers depth, the rocks are warmer than at the surface due to Earth's internal heat. The long-term temperature perturbations at the surface impose themselves on thermal gradient due to outward heat flow.

9. J. R. Lanzante, T. C. Peterson, F. J. Wentz, and K. Y. Vinnikov, "What Do Observations Indicate About the Change of Temperatures in the Atmosphere and at the Surface Since the Advent of Measuring Temperatures Vertically?" in *Temperature Trends in the Lower Atmosphere: Steps for Understanding and Reconciling Differences*, edited by T. R. Karl, S. J. Hassol, C. D. Miller, and W. L. Murray (Washington, D.C.: Climate Change Science Program, Subcommittee on Global Change Research, 2006), 47–70.

10. V. Ramaswamy, M. D. Schwarzkopf, W. J. Randel, B. D. Santer, B. J. Soden, and G. L. Stenchikov, "Anthropogenic and Natural Influences in the Evolution of Lower Stratospheric Cooling," *Science* 311 (2006): 1138–1141.

11. J. Laštovička, R. A. Akmaev, G. Beig, J. Bremer, and J. T. Emmert, "Global Change in the Upper Atmosphere," *Science* 314 (2006): 1253–1254.

12. R. S. Vose, D. R. Easterling, and B. Gleason, "Maximum and Minimum Temperature Trends for the Globe: An Update Through 2004," *Geophysical Research Letters* 32 (2005), doi:10.1029/2004GL024379.

13. L. V. Alexander, X. Zhang, T. C. Peterson, J. Caesar, B. Gleason, A. M. G. Tank, M. Haylock, D. Collins, B. Trewin, F. Rahimzadeh, A. Tagipour, K. Kumar, J. Revadekar, G. Griffiths, L. Vincent, D. B. Stephenson, J. Burn, E. Aguilar, M. Brunet, M. Taylor, M. New, P. Zhai, M. Rusticucci, and J. L. Vazquez-Aguirre, "Global Observed Changes in Daily Climate Extremes of Temperature and Precipitation," *Journal of Geophysical Research* 111 (2006), doi:10.1029/2005JD006290.

14. K. E. Trenberth, P. D. Jones, P. Ambenje, R. Bojariu, D. Easterling, A. Klein Tank, D. Parker, F. Rahimzadeh, J. A. Renwick, M. Rusticucci, B. Soden, and P. Zhai, "Observations: Surface and Atmospheric Climate Change," in Solomon et al., eds., *Climate Change 2007*, 235–336.

15. Levitus, Antonov, and Boyer, "Warming of the World Ocean."

16. See, for example, T. P. Barnett, D. W. Pierce, K. M. AchutaRao, P. J. Gleckler, B. D. Santer, J. M. Gregory, and W. M. Washington, "Penetration of Human-Induced Warming into the World's Oceans," *Science* 309 (2005): 284–287; and D. W. Pierce, T. P. Barnett, K. M. AchutaRao, P. J. Gleckler, J. M. Gregory, and W. M. Washington, "Anthropogenic Warming of the Oceans: Observations and Model Results," *Journal of Climate* 19 (2006): 1873–1900.

17. See, for example, L. Bengtsson, K. I. Hodges, and E. Roeckner, "Storm Tracks and Climate Change," *Journal of Climate* 19 (2006): 3518–3543.

18. P. Y. Groisman, R. W. Knight, T. R. Karl, D. R. Easterling, B. Sun, and J. H. Lawrimore, "Contemporary Changes of the Hydrological Cycle over the Contiguous United States: Trends Derived from in Situ Observations," *Journal of Hydrometeorology* 5 (2004): 64–85.

19. Trenberth et al., "Observations." Curiously, they also report that there is no obvious decadal trend in cloud cover.

20. F. J. Wentz, L. Ricciardulli, K. Hilburn, and C. Mears, "How Much More Rain Will Global Warming Bring?" *Science* 317 (2007): 233–235.

21. Trenberth et al., "Observations."

22. A. Dai, K. E. Trenberth, and T. Qian, "A Global Data Set of Palmer Drought Severity Index for 1870–2002: Relationship with Soil Moisture and Effects of Surface Warming," *Journal of Hydrometeorology* 5 (2004): 1117–1130.

23. K. E. Trenberth and D. J. Shea, "Relationships Between Precipitation and Surface Temperature," *Geophysical Research Letters* 32 (2005), doi:10.1029/2005GL022760.

24. An enormous archive of photographs is available online from the Library of Congress at http://www.loc.gov/rr/print/coll/052_fsa.html.

25. An even more severe drought than that of the 1930s occurred in the 1950s, although it was not accompanied by much dust.

26. For long-term analyses of the climate records, see C. A. Woodhouse and J. T. Overpeck, "2000 Years of Drought Variability in the Central United States," *Bulletin of the American Meteorological Society* 79 (1998): 2693–2714; and A. J. Cook, A. J. Fox, D. G. Vaughn, and J. G. Ferrigno, "Retreating Glacier Fronts on the Antarctic Peninsula over the Past Half-Century," *Science* 308 (2005): 541–544. Megadroughts of the Medieval Warm Period are also recorded in the presence of tree trunks rooted in present-day lakes in the Sierra Nevada of California. See S. Stine, "Extreme and Persistent Drought in California and Patagonia During Mediaeval Time," *Nature* 369 (1994): 546–549.

27. M. P. Hoerling and A. Kumar, "The Perfect Ocean for Drought," *Science* 299 (2003): 691–694; S. D. Schubert, M. J. Suarez, P. J. Pegion, R. D. Koster, and J. T. Bachmeister, "Causes of Long-Term Drought in the U.S. Great Plains," *Journal of Climate* 17 (2004): 485–503; G. J. McCabe, M. A. Palecki, and J. L. Betancourt, "Pacific and Atlantic Ocean Influences on Multidecadal Drought Frequency in the United States," *Proceedings of the National Academy of Science* 101 (2004): 4136–4141. La Niña conditions also seem to have prevailed during a severe and long drought that struck the region between 4,100 and 4,300 years ago. See R. K. Booth, S. T. Jackson, S. L. Forman, J. E. Kutzbach, E. A. Bettis III, J. Kreig, and D. K. Wright, "A Severe Centennial-Scale Drought in Mid-continental North America 4200 Years Ago and Apparent Global Linkages," *Holocene* 15 (2005): 321–328.

28. Schubert et al., "Causes of Long-Term Drought in the U.S. Great Plains."

29. S. D. Schubert, M. J. Suarez, P. J. Pegion, R. D. Koster, and J. T. Bachmeister, "On the 1930s Dust Bowl," *Science* 303 (2004): 1855–1859.

30. E. R. Cook, C. A. Woodhouse, C. M. Eakin, D. M. Meko, and D. W. Stahle, "Long-Term Aridity Changes in the Western United States," *Science* 306 (2004): 1015–1018, and references therein. Indeed, the previous great dust bowl in the American West appears to have occurred during the Medieval Warm Period. See V. Sridhar, D. B. Loopoe, J. B. Swinehart, J. A. Mason, R. J. Oglesby, and C. M. Rowe, "Large Wind Shift on the Great Plains During the Medieval Warm Period," *Science* 313 (2006): 345–347.

31. R. Seager, M. Ting, I. Held, Y. Kushnir, J. Lu, G. Vecchi, H.-P. Huang, N. Harnik, A. Leetmaa, N.-C. Lau, C. Li, J. Velez, and N. Naik, "Model Projections of an Imminent Transition to a More Arid Climate in Southwestern North America," *Science* 316 (2007): 1181–1184.

32. P. M. Cox, P. P. Harris, C. Huntingford, R. A. Betts, M. Collings, C. D. Jones, T. E. Jupp, J. A. Marengo, and C. A. Nobre, "Increasing Risk of Amazonian Drought Due to Decreasing Aerosol Pollution," *Nature* 453 (2008): 212–215.

33. See, for example, A. Giannini, R. Saravanan, and P. Chang, "Oceanic Forcing of Sahel Rainfall on Interannual to Interdecadal Time Scales," *Science* 302 (2003): 1027–1030; and G. A. Meehl and A. Hu, "Megadroughts in the Indian Monsoon Region and Southwest North America and a Mechanism for Associated Multidecadal Pacific Sea Surface Temperature Anomalies," *Journal of Climate* 19 (2006): 1605–1623.

34. This discussion is based on the review in T. P. Barnett, J. C. Adam, and D. P. Lettenmaier, "Potential Impacts of a Warming Climate on Water Availability in Snow-Dominated Regions," *Nature* 438 (2005): 303–309, where references to the numerous original works may be found.

35. P. W. Mote, A. F. Hamlet, M. P. Clark, and D. P Lettenmaier, "Declining Mountain Snowpack in Western North America," *Bulletin of the American Meteorological Society* 86 (2005): 39–49; T. P. Barnett, D. W. Pierce, H. G. Hidalgo, C. Bonfils, B. D. Santer, T. Das, G. Bala, A. W. Wood, T. Nozawa, A. A. Mirim, D. R. Cayan, and M. D. Dettinger, "Human-Induced Changes in the Hydrology of the Western United States," *Science* 319 (2008): 1080–1083.

36. Barnett, Adam, and Lettenmaier, "Potential Impacts of a Warming Climate on Water Availability."

37. Ibid., 306.

38. See, for example, G. A. Vecchi and B. J. Soden, "Effect of Remote Sea Surface Temperature Change on Tropical Cyclone Potential Intensity," *Nature* 450 (2007): 1066–1070; and M. A. Saunders and A. S. Lea, "Large Contribution of Sea Surface Warming to Recent Increase in Atlantic Hurricane Activity," *Nature* 451 (2008): 557–560, and the references in both.

39. Alexander et al., "Global Observed Changes in Daily Climate Extremes."

40. G. C. Hegerl, F. W. Zwiers, P. Braconnot, N. P. Gillett, Y. Luo, J. A. Marengo Orsini, N. Nicholls, J. E. Penner, and P. A. Stott, "Understanding and Attributing Climate Change," in Solomon et al., eds., *Climate Change 2007*, 663–746.

41. J. A. Patz, D. Campbell-Lendrum, T. Holloway, and J. A. Foley, "Impact of Regional Climate Change on Human Health," *Nature* 438 (2005): 310–317.

42. C. Schär, P. L. Vidale, D. Lüthi, C. Frei, C. Häberli, M. A. Liniger, and C. Appenzeller, "The Role of Increasing Temperature Variability in European Summer Heatwaves," *Nature* 427 (2004): 332–337.

43. This conclusion is based on the assumption that the distribution is Gaussian. The standard deviation of the 137-year record is 0.94°C (1.7°F) ; the 2003 average summer temperature (22.3°C) (72.1°F) is 5.4 standard deviations (5.1°C) (9.2°F) above the mean (17.2°C) (63.0°F) of these data. Another way to deduce the probability is to run a climate model under the same conditions multiple times. Such an approach suggests that the probability of the 2003 event was 1 in 1,000.

44. This account is summarized from D. D. Breshears, N. S. Cobb, P. M. Rich, K. P. Price, C. D. Allen, R. G. Balice, W. H. Romme, J. H. Kastens, M. L. Floyd, J. Belnap, J. J. Anderson, O. B. Myers, and C. W. Meyer, "Regional Vegetation Die-off in Response to Global-Change-Type Drought," *Proceedings of the National Academy of Sciences* 102 (2005): 15144–15148.

45. This account is summarized from A. Woods, K. D. Coates, and A. Hamann, "Is an Unprecedented Dothistroma Needle Blight Epidemic Related to Climate Change?" *BioScience* 55 (2005): 761–769. The forests in British Columbia are also suffering from infestations of other insects—for example, mountain pine beetle.

46. A. L. Westerling, H. G. Hidalgo, D. R. Cayan, and T. W. Swetnam, "Warming and Earlier Spring Increase Western U.S. Forest Wildfire Activity," *Science* 313 (2006): 940–943.

47. C. Wiedinmyer and J. C. Neff, "Estimates of CO_2 from Fires in the United States: Implications for Carbon Management," *Carbon Balance and Management* 2 (2007), doi:10.1186/1750-0680-2-10.

8. THE SENSITIVE ARCTIC AND SEA-LEVEL RISE

1. L. D. Hinzman, N. D. Bettez, W. R. Bolton, F. S. Chapin, M. B. Dyurgerov, C. L. Fastie, B. Griffith, R. D. Hollister, A. Hope, H. P. Huntington, A. M. Jensen, G. J. Jia, T. Jorgenson, D. L. Kane, D. R. Klein, G. Kofinas, A. H. Lynch, A. H. Lloyd, A. D. McGuire, F. E. Nelson, W. C. Oechel, T. E. Osterkamp, C. H. Racine, V. E. Romanovsky, R. S. Stone, D. A. Stow, M. Sturm, C. E. Tweedie, G. L. Vourlitis, M. D. Walker, D. A. Walker, P. J. Webber, J. M. Welker, K. S. Winker, and K. Yoshikawa, "Evidence and Implications of Recent Climate Change in Northern Alaska and Other Arctic Regions," *Climatic Change* 72 (2005), 256.

2. See, for example, D. W. J. Thompson and J. M. Wallace, "The Arctic Oscillation Signature in the Wintertime Geopotential Height and Temperature Fields," *Geophysical Research Letters* 25 (1998): 1297–1300; and I. G. Rigor, J. M. Wallace, and R. L. Colony, "Response of Sea Ice to the Arctic Oscillation," *Journal of Climate* 15 (2002): 2648–2663. The Arctic Oscillation is also referred to as the Northern Annual Mode. See, for example, M. C. Serreze, M. M. Holland, and J. Stroeve, "Perspectives on the Arctic's Shrinking Sea-Ice Cover," *Science* 315 (2007): 1533–1536. There are different interpretations of what drives the Arctic Oscillation.

3. A. Y. Proshutinsky and M. A. Johnson, "Two Circulation Regimes of the Wind-Driven Arctic Ocean," *Journal of Geophysical Research* 102 (1997): 12493–12514.

4. J. Richter-Menge, J. Overland, A. Proshutinsky, V. Romanovsky, L. Bengtsson, L. Brigham, M. Dyurgerov, J. C. Gascard, S. Gerland, R. Graversen, C. Haas, M. Karcher, P. Kuhry, J. Maslanik, H. Melling, W. Maslowski, J. Morison, D. Perovich, R. Przybylak, V. Rachold, I. Rigor, A. Shiklomanov, J. Stroeve, D. Walker, and J. Walsh, *State of the Arctic Report*, NOAA/OAR Special Report (Seattle: National Oceanic and Atmospheric Administration, Office of Oceanic and Atmospheric Research, and Pacific Marine Environmental Laboratory, 2006). See also the review in R. E. Moritz, C. M. Bitz, and E. J. Steig, "Dynamics of Recent Climate Change in the Arctic," *Science* 297 (2002): 1497–1502.

5. See, for example, D. K. Peroich, B. Light, H. Eicken, K. F. Jones, K. Runciman, and S. V. Nghiem, "Increasing Solar Heating of the Arctic Ocean and Adjacent Seas, 1979–2005: Attribution and Role in the Ice-Albedo Feedback," *Geophysical Research Letters* 34 (2007), doi:10.1029/2007GL031480.

6. Stephanie Pfirman, personal communication, 2008.

7. S. Pfirman, W. F. Haxby, R. Colony, and I. Rigor, "Variability in Arctic Sea Ice Drift," *Geophysical Research Letters* 31 (2004), doi:10.1029/2004GL020063; I. Rigor and J. M. Wallace, "Variations in the Age of Arctic Sea-Ice and Summer Sea-Ice Extent," *Geophysical Research Letters* 31 (2004), doi:10.1029/2004GL019492.

8. S. V. Nghiem, I. G. Rigor, D. K. Perovich, P. Clemente-Colón, J. W. Weatherly, and G. Neumann, "Rapid Reduction of Arctic Perennial Sea Ice," *Geophysical Research Letters* 34 (2007), doi:10.1029/2007GL031138. See also the review in Serreze, Holland, and Stroeve, "Perspectives on the Arctic's Shrinking Sea-Ice Cover."

9. See, for example, Serreze, Holland, and Stroeve, "Perspectives on the Arctic's Shrinking Sea-Ice Cover."

10. Ibid.

11. For a detailed summary of changes in arctic climate and the environmental, ecological, and human consequences of those changes, see Hinzman et al., "Evidence and Implications of Recent Climate Change in Northern Alaska."

12. S. Zhang, J. E. Walsh, J. Zhang, U. S. Bhatt, and M. Ikeda, "Climatology and Interannual Variability of Arctic Cyclone Activity: 1948–2002," *Journal of Climate* 17 (2004): 2300–2317. The average is based on ground-temperature measurements made at 17 stations. The depths ranged from 1.2 to 2.3 meters (3.9 to 7.5 feet).

13. F. E. Nelson, "(Un)Frozen in Time," *Science* 299 (2003): 1673–1675; S. A. Zimov, E. A. G. Schuur, and F. S. Chapin III, "Permafrost and the Global Carbon Budget," *Science* 312 (2006): 1612–1613.

14. V. E. Romanovsky, T. S. Sazonova, V. T. Balobaev, N. I. Shender, and D. O. Sergueev, "Past and Recent Changes in Air and Permafrost Temperatures in Eastern Siberia," *Global and Planetary Change* 56 (2007): 399–413. The corresponding increase in ground temperature was 0.8°C (1.4°F) at 1.6 meters (5.2 feet) depth.

15. D. M. Lawrence and A. G. Slater, "A Projection of Severe Near-Surface Permafrost Degradation During the 21st Century," *Geophysial Research Letters* 32 (2005), doi:10.1029/2005GL025080.

16. Zimov, Schuur, and Chapin, "Permafrost and the Global Carbon Budget," 1612.

17. Ibid.

18. K. M. Walter, S. A. Zimov, J. P. Chanton, D. Verbyla, and F. S. Chapin III, "Methane Bubbling from Siberian Thaw Lakes as a Positive Feedback to Climate Warming," *Nature* 443 (2006): 71–75.

19. R. S. Stone, E. G. Dutton, J. M. Harris, and D. Longenecker, "Earlier Spring Snowmelt in Northern Alaska as an Indicator of Climate Change," *Journal of Geophysical Research* 107 (2002), doi:10.1029/2000JD000286.

20. L. C. Smith, Y. Sheng, G. M. MacDonald, and L. D. Hinzman, "Disappearing Arctic Lakes," *Science* 308 (2005): 1429.

21. B. J. Peterson, J. McClelland, R. Curry, R. M. Holmes, J. E. Walsh, and K. Aagaard, "Trajectory Shifts in the Arctic and Subarctic Freshwater Cycle," *Science* 313 (2006): 1061–1066.

22. Hinzman et al., "Evidence and Implications of Recent Climate Change in Northern Alaska." These authors report that in 41 years, surface water decreased 2 millimeters (0.08 inch) per year for the Alaskan coastal plain and 5.5 millimeters (0.22 inch) per year for interior regions.

23. W. C. Oechel, G. L. Vourlitis, S. J. Hastings, R. C. Zulueta, L. Hinzman, and D. Kane, "Acclimation of Ecosystem CO_2 Exchange in the Alaskan Arctic in Response to Decadal Climate Warming," *Nature* 406 (2000): 978–981.

24. G. J. Jia, H. E. Epstein, and D. A. Walker, "Greening of Arctic Alaska, 1981–2001," *Geophysical Research Letters* 30 (2003), doi:10.1029/2003GL018268; S. J. Goetz, A. G. Bunn, G. J. Fiske, and R. A. Houghton, "Satellite-Observed Photosynthetic Trends Across Boreal North America Associated with Climate and Fire Disturbance," *Proceedings of the National Academy of Science* 102 (2005): 13521–13525.

25. A. G. Bunn, S. J. Goetz, J. S. Kimball, and K. Zhang, "Northern High-Latitude Ecosystems Respond to Climate Change," *Eos, Transactions* 88 (2007): 333–335.

26. See, for example, Y. Yokoyama, K. Lambeck, P. De Deckker, P. Johnston, and L. K. Fifield, "Timing of the Last Glacial Maximum from Observed Sea-Level Minima," *Nature* 406 (2000): 713–716.

27. See, for example, L. Miller and B. C. Douglas, "Mass and Volume Contributions to Twentieth-Century Global Sea Level Rise," *Nature* 428 (2004): 406–409.

28. For a concise review of the current state of understanding, see A. Shepherd and D. Wingham, "Recent Sea-Level Contribution of the Antarctic and Greenland Ice Sheets," *Science* 315 (2007): 1529–1532.

29. G. A. Meehl, T. F. Stocker, W. D. Collins, P. Friedlingstein, A. T. Gaye, J. M. Gregory, A. Kitoh, R. Knutti, J. M. Murphy, A. Noda, S. C. B. Raper, I. G. Watterson, A. J. Weaver, and Z.-C. Zhao, "Global Climate Projections," in *Climate Change 2007: The Physical Science Basis. Contribution of Working Group I to the Fourth Assessment Report of the Intergovernmental Panel on Climate Change*, edited by S. Solomon, D. Qin, M. Manning, Z. Chen, M. Marquis, K. B. Averyt, M. Tignor, and H. L. Miller (Cambridge: Cambridge University Press, 2007), 747–846.

30. Based on a "semiempirical" relationship between global temperature and sea level, Stefan Rahmstorf estimated a sea level increase of 0.5 to 1.4 meters (1.6 to 4.6 feet) by 2100. See S. Rahmstorf, "A Semi-empirical Approach to Projecting Future Sea-Level Rise," *Science* 315 (2007): 368–370.

31. One-tenth of the world's population—634 million people—live within 10 meters (33 feet) of sea level; 75 percent of them are in Asia.

32. J. M. Gregory, P. Huybrechts, and S. C. B. Raper, "Threatened Loss of the Greenland Ice-Sheet," *Nature* 428 (2004): 616.

33. G. Ekström, M. Nettles, and V. C. Tsai, "Seasonality and Increasing Frequency of Greenland Glacial Earthquakes," *Science* 311 (2006): 1756–1758.

34. E. Rignot and P. Kanagaratnam, "Changes in the Velocity Structure of the Greenland Ice Sheet," *Science* 311 (2006): 986–990.

35. In a project known as the Gravity Recovery and Climate Experiment (GRACE), two satellites measure the subtle variations in Earth's gravity field, one closely following the other in their orbit around the planet. As the satellites approach a region of relatively high gravity, the lead satellite speeds up slightly so that the distance between the two increases, and as the latter passes over the region of high gravity, the rear satellite catches up. The gravity field is mapped by keeping precise track of the changing distance between the satellites. Large masses of ice sitting high on the solid Earth are regions where gravity is unusually high. As the ice mass changes, so does the local gravity, which is what the satellites detect.

36. I. Velicogna and J. Wahr, "Acceleration of Greenland Ice Mass Loss in Spring 2004," *Nature* 443 (2006): 329–331.

37. These estimates are from J. L. Chen, C. R. Wilson, and B. D. Tapley, "Satellite Gravity Measurements Confirm Accelerated Melting of Greenland Ice Sheet," *Science* 313 (2006): 1958–1960; and S. B. Luthcke, H. J. Zwally, W. Abdalati, D. D. Rowlands, R. D. Ray, R. S. Nerem, F. G. Lemoine, J. J. McCarthy, and D. S. Chinn, "Recent Greenland Ice Mass Loss by Drainage System from Satellite Gravity Observations," *Science* 314 (2006): 1286–1289. The differences reflect different methodologies and illustrate the inherent large uncertainties.

38. See, for example, G. H. Denton, D. E. Sugden, D. R. Marchant, B. L. Hall, and T. I. Wilch, "East Antarctic Ice Sheet Sensitivity to Pliocene Climatic Change from a Dry Valleys Perspective," *Geografiska Annaler* 75A (1993): 155–204.

39. E. Domack, D. Duran, A. Leventer, S. Ishman, S. Doane, S. McCallum, D. Amblas, J. Ring, R. Gilbert, and M. Prentice, "Stability of the Larsen B Ice Shelf on the Antarctic Peninsula During the Holocene Epoch," *Nature* 436 (2005): 681–685.

40. E. Rignot, G. Casassa, P. Gogineni, W. Krabill, A. Rivera, and R. Thomas, "Accelerated Ice Discharge from the Antarctic Peninsula Following the Collapse of Larsen B Ice Shelf," *Geophysical Research Letters* 31 (2004), doi:10.1029/2004GL020697.

41. A. J. Cook, A. J. Fox, D. G. Vaughn, and J. G. Ferrigno, "Retreating Glacier Fronts on the Antarctic Peninsula over the Past Half-Century," *Science* 308 (2005): 541–544.

42. A. Shepherd, D. Wingham, and E. Rignot, "Warm Ocean Is Eroding West Antarctic Ice Sheet," *Geophysical Research Letters* 31 (2004), doi:10.1029/2004GL021106.

43. I. Velicogna and J. Wahr, "Measurements of Time-Variable Gravity Show Mass Loss in Antarctica," *Science* 311 (2006): 1754–1756. At the same time, the gravity data suggest that the East Antarctic Ice Sheet has not experienced significant ice loss, which may be because the increased loss of ice along the margins is approximately balanced by increased precipitation. Indeed, based on satellite radar altimeter measurements, C. H. Davis and colleagues report that the interior of the East Antarctic Ice Sheet increased in mass by 45 ± 7 billion tons per year from 1992 to 2003. See C. H. Davis, Y. Li, J. R. McConnell, M. M. Frey, and E. Hanna, "Snowfall-Driven Growth in East Antarctic Ice Sheet Mitigates Recent Sea-Level Rise," *Science* 308 (2005): 1898–1901. However, a longer time series, derived mainly from ice cores, indicated no "statistically significant" change in snowfall for the past 50 years, as reported in A. J. Monaghan, D. H. Bromwich, R. L. Fogt, S.-H. Wang, P. A. Mayewski, D. A. Dixon, A. Ekaykin, M. Frezzotti, I. Goodwin,

E. Isaksson, S. D. Kaspari, V. I. Morgan, H. Oerter, T. D. Van Ommen, C. J. Van der Veen, and J. Wen, "Insignificant Change in Antarctic Snowfall Since the International Geophysical Year," *Science* 313 (2006): 827–831.

44. K. Muhs, R. Simmons, and B. Steinke, "Timing and Warmth of the Last Interglacial Period: New U-Series Evidence from Hawaii and Bermuda and a New Fossil Compilation for North America," *Quaternary Science Reviews* 21 (2002): 1355–1383.

45. B. L. Otto-Bliesner, S. J. Marshall, J. T. Overpeck, G. H. Miller, A. Hu, and CAPE Last Interglacial Project Members, "Simulating Arctic Climate Warmth and Icefield Retreat in the Last Interglacial," *Science* 311 (2006): 1751–1753.

46. J. T. Overpeck, B. L. Otto-Bliesner, G. H. Miller, D. R. Muhs, R. B. Alley, and J. T. Kiehl, "Paleoclimatic Evidence for Future Ice-Sheet Instability and Rapid Sea-Level Rise," *Science* 311 (2006): 1747–1750.

47. R. B. Alley, P. U. Clark, P. Huybrechts, and I. Joughin, "Ice-Sheet and Sea-Level Changes," *Science* 310 (2005): 456–460.

48. M. T. McCulloch and T. Esat, "The Coral Record of Last Interglacial Sea Levels and Sea Surface Temperatures," *Chemical Geology* 169 (2000): 107–129. A more recent, unpublished estimate, based on study of past sea-level rise, suggests a somewhat lower rate of less than 1.5 meters (5 feet) a century (Mark Siddall, personal communication).

9. CLIMATE MODELS AND THE FUTURE

1. For brief descriptions of climate models, see A. Scaife, C. Folland, and J. Mitchell, "A Model Approach to Climate Change," *Physics World* 20 (2007): 20–25; and G. A. Schmidt, "The Physics of Climate Modeling, 2007," *Physics Today* 60 (2007): 72–73.

2. Schmidt, "Physics of Climate Modeling."

3. But the real climate system may be chaotic, at least on short timescales of decades or less. In any case, climate models are not chaotic because they are "a statistical description of the mean state and variability of a system, not an individual path through [it]" (ibid., 72).

4. B. J. McAvaney, C. Covey, S. Joussaume, V. Kattsov, A. Kitoh, W. Ogana, A. J. Pitman, A. J. Weaver, R. A. Wood, and Z.-C. Zhao, "Model Evaluation," in *Climate Change 2001: The Scientific Basis. Contributions of Working Group I to the Third Assessment Report of the Intergovernmental Panel on Climate Change*, edited by J. T. Houghton, Y. Ding, D. J. Griggs, M. Noguer, P. J. Van Der Linden, X. Dai, K. Maskell, and C. A. Johnson (Cambridge: Cambridge University Press, 2001), 471–523.

5. See, for example, B. D. Santer, T. M. L. Wigley, C. Doutriaux, J. S. Boyle, J. E. Hansen, P. D. Jones, G. A. Meehl, E. Roeckner, S. Sengupta, and K. E. Taylor, "Accounting for the Effects of Volcanoes and ENSO in Comparisons of Modeled and Observed Temperature Trends," *Journal of Geophysical Research* 106 (2001): 28033–28059.

6. D. A. Randall, R. A. Wood, S. Bony, R. Colman, T. Fichefet, J. Fyfe, V. Kattsov, A. Pitman, J. Shukla, J. Srinivasan, R. J. Stouffer, A. Sumi, and K. E. Taylor, "Climate Models and Their Evaluation," in *Climate Change 2007: The Physical Science Basis. Contribution of Working Group I to the Fourth Assessment Report of the Intergovernmental Panel on Climate Change*, edited by S. Solomon, D. Qin, M. Manning, Z. Chen, M. Mar-

quis, K. B. Averyt, M. Tignor, and H. L. Miller (Cambridge: Cambridge University Press, 2007), 589–662.

7. G. C. Hegerl, F. W. Zwiers, P. Braconnot, N. P. Gillett, Y. Luo, J. A. Marengo Orsini, N. Nicholls, J. E. Penner, and P. A. Stott, "Understanding and Attributing Climate Change," in Solomon et al., eds., *Climate Change 2007*, 663–746.

8. See, for example, D. J. Karoly and Q. Wu, "Detection of Regional Surface Temperature Trends," *Journal of Climate* 18 (2005): 4337–4343; and R. R. Knutson, R. L. Delworth, K. W. Dixon, I. M. Held, J. Lu, V. Ramaswamy, M. D. Schwarzkopf, G. Stenchikov, and R. J. Stouffer, "Assessment of Twentieth-Century Regional Surface Temperature Trends Using the GFDL CM2 Coupled Models," *Journal of Climate* 19 (2006): 1624–1651.

9. K. Braganza, D. J. Karoly, A. C. Hirst, M. E. Mann, P. Stott, R. J. Stouffer, and S. F. B. Tett, "Simple Indices of Global Climate Variability and Change: Part I—Variability and Correlation Structure," *Climate Dynamics* 20 (2003): 491–502; K. Braganza, D. J. Karoly, A. C. Hirst, P. Stott, R. J. Stouffer, and S. F. B. Tett, "Simple Indices of Global Climate Variability and Change: Part II—Attribution of Climate Change During the Twentieth Century," *Climate Dynamics* 22 (2004): 823–838.

10. Hegerl et al., "Understanding and Attributing Climate Change."

11. G. A. Meehl and C. Tebaldi, "More Intense, More Frequent, and Longer Lasting Heat Waves in the 21st Century," *Science* 305 (2004): 994–997; C. Schär, P. L. Vidale, D. Lüthi, C. Frei, C. Häberli, M. A. Liniger, and C. Appenzeller, "The Role of Increasing Temperature Variability in European Summer Heatwaves," *Nature* 427 (2004): 332–337; P. A. Stott, D. A. Stone, and M. R. Allen, "Human Contribution to the European Heatwave of 2003," *Nature* 432 (2004): 610–614.

12. N. Nakićenović and R. Swart, "Background and Overview," in *Special Report on Emissions Scenarios: A Special Report of Working Group III of the Intergovernmental Panel on Climate Change*, edited by N. Nakićenović and R. Swart (Cambridge: Cambridge University Press 2000).

13. All the scenarios assume that income gaps around the world narrow and that the world becomes more affluent as gross world product increases by 10 to 26 times. The scenarios mentioned here have been summarized:

A2 (*high-emissions scenario*): "A very heterogeneous world. The underlying theme is self-reliance and preservation of local identities. Fertility patterns across regions converge very slowly, which results in continuously increasing global population. Economic development is primarily regionally oriented and per capita economic growth and technological change are more fragmented and slower than in other storylines."

A1B (*medium-emissions scenario*): "A future world of very rapid economic growth, global population that peaks in mid-century and declines thereafter, and the rapid introduction of new and more efficient technologies. Major underlying themes are convergence among regions, capacity building, and increased cultural and social interactions, with a substantial reduction in regional differences in per capita income." The world relies on energy from all sources, with improvements in all energy sectors and end-use technologies.

B1 (*low-emissions scenario*): "A convergent world with the same global population that peaks in mid-century and declines thereafter, as in the A1 storyline, but with rapid changes in economic structures toward a service and information economy, with reductions in material intensity, and the introduction of clean and resource-efficient technologies. The emphasis is on global solutions to economic, social, and environmental sustainability, including improved equity, but without additional climate initiatives." (Ibid., 4–5)

14. The 2007 IPCC also produced projections for certain idealized situations, such as a 1 percent per year CO_2 increase, in order to understand what happens when, say, the atmospheric CO_2 content doubles.

15. G. A. Meehl, T. F. Stocker, W. D. Collins, P. Friedlingstein, A. T. Gaye, J. M. Gregory, A. Kitoh, R. Knutti, J. M. Murphy, A. Noda, S. C. B. Raper, I. G. Watterson, A. J. Weaver, and Z.-C. Zhao, "Global Climate Projections," in Solomon et al., eds., *Climate Change 2007*, 747–846.

16. The ranges include the error of individual models.

17. Here I use temperature increase as a proxy for all changes in climate.

18. M. D. Mastrandrea and S. H. Schneider, "Probabilistic Integrated Assessment of 'Dangerous' Climate Change," *Science* 304 (2004): 571–575.

10. ENERGY AND THE FUTURE

1. But note the uncertainties in table 9.1, which derive from the uncertainty in the sensitivity of climate to atmospheric CO_2 buildup and translate to uncertainty in the target emissions levels needed to limit warming to a desired level. For more discussion on this matter, see K. Caldeira, A. K. Jain, and M. I. Hoffert, "Climate Sensitivity Uncertainty and the Need for Energy Without CO_2 Emission," *Science* 299 (2003): 2052–2054.

2. Yogi Berra supposedly said, "It's tough to make predictions, especially about the future." This point is worth bearing in mind when it comes to energy. Indeed, Vaclav Smil shows that practically all long-term forecasts of energy demand, cost, intensity, and source, along with intrinsically more speculative forecasts of technical advance, have been abysmal failures. See V. Smil, *Energy at the Crossroads: Global Perspectives and Uncertainties* (Cambridge, Mass.: MIT Press, 2003). This finding should not be too surprising; after all, most matters of energy are tightly bound to world economic and social development and thus to the numerous, entwined, and complex factors on which development depends.

3. Energy Information Agency, http://www.eia.doe.gov/.

4. Massachusetts Institute of Technology, *The Future of Geothermal Energy: Impact of Enhanced Geothermal Systems (EGS) on the United States in the 21st Century* (Cambridge, Mass.: MIT, 2006), available at http://geothermal.inel.gov/.

5. British Petroleum, *BP Statistical Review of World Energy 2007* (London: British Petroleum, 2007), available at http://www.bp.com/productlanding.do?categoryId = 6848&contentId = 7033471.

6. Massachusetts Institute of Technology, *The Future of Coal: Options for a Carbon-Constrained World* (Cambridge, Mass.: MIT, 2007), available at http://web.mit.edu/coal/.

7. Unless otherwise stated, the figures are from Energy Information Agency, *International Energy Outlook* (Washington, D.C.: Department of Energy, 2007), available at http://www.eia.doe.gov/oiaf/ieo/index.html. But see the discussion by David Rutledge, California Institute of Technology, concerning reserve estimates, at http://rutledge.caltech.edu/.

8. The data and discussion in this section are based on MIT, *Future of Coal*, in which more detailed information may be found.

9. The term *supercritical* refers to steam at temperatures and pressures higher than 550°C (1,022°F) and 22.0 megapascals, respectively. Under supercritical conditions, steam and liquid water become one and the same.

10. The limestone breaks down to lime (CaO), which reacts with the sulfur gases to produce calcium sulfate.

11. Most modern power plants that burn natural gas already employ a combined cycle. A future design combines IGCC with a fuel cell, which might boost efficiency to 70 to 80 percent and simplify CO_2 capture. See, for example, http://www.fossil.energy.gov/programs/powersystems/gasification/index.html.

12. The information for this section comes mainly from S. Benson, P. Cook, J. Anderson, S. Bachu, H. B. Nimir, B. Basu, J. Bradshaw, G. Deguchi, J. Gale, G. von Goerne, W. Heidug, S. Holloway, R. Kamal, D. Keith, P. Lloyd, P. Rocha, W. Senior, J. Thomson, T. Torp, T. Wildenborg, M. Wilson, F. Zarlenga, and D. Zhou, "Underground Geological Storage," in *IPCC Special Report on Carbon Dioxide Capture and Storage*, edited by B. Metz, O. Davidson, H. C. de Coninck, M. Loos, and L. A. Meyer (Cambridge: Cambridge University Press, 2005), 195–276; and MIT, *Future of Coal*. See also the summaries in K. S. Lackner, "Carbonate Chemistry for Sequestering Fossil Carbon," *Annual Review of Energy and the Environment* 27 (2002): 1521–1527; K. S. Lackner, "A Guide to CO_2 Sequestration," *Science* 300 (2003): 1677–1678; and S. J. Friedmann, "Geological Carbon Sequestration," *Elements* 3 (2007): 179–184.

13. MIT, *Future of Coal*.

14. Carbon dioxide is supercritical at temperatures higher than 31°C (88°F) and pressures greater than 73 times atmospheric pressure.

15. As compiled in MIT, *Future of Coal*.

16. Another possibility is to sequester CO_2 in the seabed at water depths of several thousand meters and beneath several hundred meters of sediment. See, for example, K. Z. House, D. P. Schrag, C. F. Harvey, and K. S. Lackner, "Permanent Carbon Dioxide Storage in Deep-Sea Sediments," *Proceedings of the National Academy of Science* 103 (2006): 12291–12295.

17. Benson et al., "Underground Geological Storage."

18. Ibid.

19. See, for example, MIT, *Future of Coal*.

20. R. C. Ewing, "The Nuclear Fuel Cycle: A Role for Mineralogy and Geochemistry," *Elements* 2 (2007): 331–334, and references therein.

21. Massachusetts Institute of Technology, *The Future of Nuclear Power: An Interdisciplinary MIT Study* (Cambridge, Mass.: MIT, 2003), available at http://web.mit.edu/nuclearpower/.

22. The radioactivity diminishes from about 3 million curies per metric ton at one year after removal to 0.6 million curies per metric ton after ten years.

23. J. A. Plant, P. R. Simpson, B. Smith, and B. Windley, "Uranium Ore Deposits— Products of the Radioactive Earth," in *Uranium: Mineralogy, Geochemistry, and the Environment*, edited by P. C. Burns and R. Finch, Reviews in Mineralogy, no. 38 (Chantilly, Va.: Mineralogical Society of America, 1999), 255–319. For a concise summary, see also A. M. Macfarlane and M. Miller, "Nuclear Energy and Uranium Resources," *Elements* 3 (2007): 185–192. According to the latter source, uranium production in 2004 was dominated by Canada (29 percent), Australia (22 percent), Kazakhstan (9 percent), Russia (8 percent), Namibia (8 percent), and Uzbekistan (5 percent).

24. MIT, *Future of Nuclear Power*.

25. Energy Information Agency, *World Net Nuclear Electric Power Generation, 1980– 2005*, available at http://www.eia.doe.gov/fuelnuclear.html.

26. Macfarlane and Miller, "Nuclear Energy and Uranium Resources."

27. The resource data are for 2005 and come from the Organization for Economic Co-operation and Development, as cited in ibid. Total resources are likely to be considerably greater than these data indicate. First, certain unconventional resources are not included in this estimate. Second, Macfarlane and Miller state that "as is the case with other ores, there is no economic incentive for mining companies to prove out resources or convert them to reserves many years before they can be sold. Thus, predictions of the future availability of uranium based on current cost, price, and geological data . . . are likely to be extremely conservative" (189).

28. These issues are assessed in MIT, *Future of Nuclear Power*, the primary source for this discussion.

29. Ibid.

30. Four such accidents worldwide are cited in MIT, *Future of Nuclear Power*, which points out that a complete inventory of accidents at reprocessing plants does not exist.

31. Ibid.

32. The two planned repositories are Yucca Mountain in Nevada and Onkalo in southern Finland. The latter should be operational in 2020. The opening of Yucca Mountain has been delayed, and it is not clear when it will happen.

33. The exception is the so-called CANDU (Canada Duterium Uranium) system, which uses natural uranium to heat deuterium-rich water. The CANDU system was the basis for India's entry into the club of nuclear nations.

34. For further discussion, see E. C. Ewing, "Environmental Impact of the Nuclear Fuel Cycle," in *Energy, Waste, and the Environment: A Geochemical Perspective*, edited by R. Gieré and P. Stille, Special Publication, no. 236 (London: Geological Society, 2004), 7–23.

35. MIT, *Future of Nuclear Power*.

36. For example, an organization known as the Generation IV International Forum has been established as a mechanism for international collaboration in the development of the next generation of nuclear reactors slated for deployment by 2030. The organization has defined six designs that offer improvements in proliferation resistance, safety, and economy. See http://www.gen-4.org/index.html.

37. For a brief summary of the thorium fuel cycle, see Macfarlane and Miller, "Nuclear Energy and Uranium Resources." For a detailed account of the benefits and challenges

associated with the use of the thorium fuels and the fuel cycle, see International Atomic Energy Agency, *Thorium Fuel Cycle—Potential Benefits and Challenges* (Vienna: IAEA, 2005), available at http://www-pub.iaea.org/MTCD/publications/PDF/TE_1450_web .pdf, TECDOC-1450. The latter also describes practical implementation scenarios.

38. In reactors, thorium fuels produce uranium-232. This isotope decays with a half-life of 73.6 years and produces highly radioactive daughter products, resulting in spent fuel with high radiotoxicity for hundreds of years.

39. J. A. Edmonds, M. A. Wise, J. J. Dooley, S. H. Kim, S. J. Smith, P. J. Runci, L. E. Clarke, E. L. Malone, and G. M. Stokes, *Wind and Solar Energy: A Core Element of Global Energy Technology Strategy to Address Climate Change* (College Park, Md.: Joint Global Change Research Institute, 2007), 4.

40. The issue is not just present-day cost. Solar (and wind) power are becoming more attractive because there is more certainty about their future costs than the future costs of coal and natural gas.

41. FPL Energy, Plant Fact Sheet, available at http://www.fplenergy.com/portfolio/ wind/plantfactsheet.shtml.

42. American Wind Energy Association, http://www.awea.org/.

43. Maps of U.S. and international wind resources are available at http://rredc.nrel.gov/ wind/pubs/atlas/ and http://www.nrel.gov/wind/international_wind_resources.html.

44. C. L. Archer and M. Z. Jacobson, "Evaluation of Global Wind Power," *Journal of Geophysical Research* 110 (2005), doi:10.1029/2004JD005462.

45. W. Kempton, C. L. Archer, A. Dhanju, R. W. Garvine, and M. Z. Jacobson, "Large CO_2 Reductions via Offshore Wind Power Matched to Inherent Storage in Energy End-Uses," *Geophysial Research Letters* 34 (2007), doi:10.1029/2006GL028016.

46. This new construction notably includes a project known as the London Array, which will install 431 turbines 20 kilometers (12 miles) off the coast of southeastern England and generate another 1,000 megawatts of electricity, enough for 750,000 English homes. See http://www.londonarray.com/.

47. As summarized by the American Wind Energy Association at http://www.awea .org/. For a technical study, see B. Parsons, M. Milligan, J. C. Smith, E. DeMeo, B. Oakleaf, K. Wolf, M. Schuerger, R. Zavadil, M. Ahlstrom, and D. Yen Nakafuji, *Grid Impacts of Wind Power Variability: Recent Assessments from a Variety of Utilities in the United States*, Report no. NREL/CP-500–39955 (Washington, D.C.: Department of Energy, 2006), available at http://www.osti.gov/bridge.

48. W. P. Erickson, G. D. Johnson, and D. P. Young Jr., "A Summary and Comparison of Bird Mortality from Anthropogenic Causes with an Emphasis on Collisions," in *USDA Forest Service General Technical Report PSW-GTR-191* (Washington, D.C.: Forest Service, Department of Agriculture, 2005), 1029–1042. However, wind turbines claim a much higher proportion of deaths of large, high-flying birds, such as raptors and geese, than of small birds, which tend to fly much lower to the ground.

49. Global Wind Energy Council, *Global Wind 2007 Report* (Brussels: Global Wind Energy Council, 2007), available at http://www.gwec.net/index.php?id = 90.

50. The figures are from the European Wind Energy Association, http://www.ewea .org/.

51. European Wind Energy Association, *2007 Annual Report* (Brussels: European Wind Energy Association, 2007), available at http://www.ewea.org/.

52. American Wind Energy Association, http://www.awea.org/.

53. G. P. Kyle, J. P Lurz, S. J. Smith, D. Barrie, and M. A. Wise, *Long-Term Modeling of Wind Energy in the United States*, Report no. PNNL-16316 (College Park, Md.: Joint Global Change Research Institute. 2007), available at http://www.pnl.gov/publications/. It is important to understand that this description is not a prediction, but a model-based scenario that combines various assumptions of economic conditions and technological advances.

54. For a fuller discussion, see L. Flowers and P. Dougherty, "Toward a 20% Wind Electricity Supply in the United States," in *2007 European Wind Energy Conference and Exhibition*, Conference Paper no. NREL/CP-500-41579 (Washington, D.C.: Department of Energy, 2007), available at http://www.osti.gov/bridge.

55. A good source of information on PV cells is http://www1.eere.energy.gov/solar/photovoltaics.html.

56. The project is being constructed in three stages. AndaSol 1 began operating in November 2008, ultimately will provide 50 megawatts of capacity, and will be followed by AndaSol 2 and AndaSol 3. The salt is composed of a mixture of 60 percent sodium nitrate and 40 percent potassium nitrate. See http://www.flagsol.com/andasol_projects.htm.

57. See, for example, http://www.stirlingenergy.com/ and http://www.sandia.gov/news/resources/releases/2008/solargrid.html.

58. See http://www.stirlingengines.org.uk/pioneers/pion2.html.

59. In a way, the idea has been too successful. The demand for solar PV cells has far outstripped the supply, with the result that it has driven their cost up substantially.

60. K. Meah and S. Ula, "Comparative Evaluation of HVDC and HVAC Transmission Systems," *Power Engineering Society General Meeting* 1–5 (2007), doi:10.1109/PES.2007.385993.

61. K. Zweibel, J. Mason, and V. Fthenakis, "A Solar Grand Plan," *Scientific American*, January 2008, 64–73.

62. The plan comes from an initiative known as the Trans-Mediterranean Renewable Energy Cooperation (TREC), http://www.desertec.org/index.html. An important part of the Desertec vision is to combine solar power and desalination plants, which are desperately needed in some desert countries. For example, Sana'a, the capital of Yemen at 2,200 meters (7,200 feet) elevation, will likely exhaust its groundwater in about 15 years. The alternative to displacing more than 2 million people, which TREC estimates will cost €30 billion, is to build solar and desalination plants and the necessary pipeline, all of which will cost approximately €5 billion.

63. M. I. Hoffert, K. Caldeira, A. K. Jain, E. F. Haites, L. D. Danny Harvey, S. D. Potter, M. E. Schlesinger, S. H. Schneider, R. G. Watts, T. M. L. Wigley, and D. J. Wuebbles, "Energy Implications of Future Stabilization of Atmospheric CO_2," *Nature* 395 (1998): 881–884; Martin Hoffert, personal communication, 2008.

64. Lackner, "Guide to CO_2 Sequestration"; F. S. Zeman and K. S. Lackner, "Capturing Carbon Dioxide Directly from the Atmosphere," *World Resource Review* 16 (2004): 157–172.

65. S. Pacala and R. Socolow, "Stabilization Wedges: Solving the Climate Problem for the Next 50 Years with Current Technologies," *Science* 305 (2004): 968–972.

66. Ibid.

GLOSSARY

ABYSSAL PLAINS   The featureless, flat expanses of the deep oceans.

ACTIVE LAYER   The soil layer above permafrost that freezes in winter and thaws in summer.

AEROSOLS   Particles, at least partly solid and typically less than 2 microns across, suspended in the air.

ALBEDO   The fraction of the total incident electromagnetic radiation that is reflected.

ANTARCTIC CIRCUMPOLAR CURRENT   The eastward-flowing current in the Southern Ocean circling Antarctica.

ANTHROPOGENIC   Relating to the influence of human beings.

ANTICYCLONIC FLOW   The flow of wind in the clockwise direction in the Northern Hemisphere and the counterclockwise direction in the Southern Hemisphere.

ARAGONITE   An orthorhombic mineral with the chemical formula $CaCO_3$. The composition is the same as that of calcite, but the mineral structure is different.

ARCTIC OSCILLATION   The five- to seven-year oscillation between two natural, different, and distinctive patterns of Arctic atmosphere and ocean circulation.

BENTHIC   Referring to the deepest level of a body of water, usually the ocean. For example, benthic marine organisms are those that live in the deep ocean.

BICARBONATE   The ion HCO_3^-, the principal carbon species dissolved in seawater.

BIOLOGICAL PUMP (FOR OCEAN CARBON)   The growth of organisms in the shallow ocean and the transfer of their dead remains to the deep ocean, thus transporting carbon from the shallow to the deep ocean and removing CO_2 from the atmosphere.

BIOMASS   The total mass of living organisms in a certain area.

BIOSPHERE   All of Earth's life.

CALCITE   The common, hexagonal, rock-forming mineral with the chemical formula $CaCO_3$. The composition is the same as that of aragonite, but the mineral structure is different.

CARBONATE COMPENSATION DEPTH (CCD)   The depth of the ocean floor below which calcareous sediments, such as chalk and limestone, do not form because carbonate shells of dead organisms dissolve more rapidly than they accumulate.

CARBONATE OOZE Fine-grained, carbonate-rich mud that accumulates on the ocean floor.

CARBONATE PUMP (FOR OCEAN CARBON) The removal of bicarbonate from the ocean when organisms build calcium carbonate ($CaCO_3$) shells and die and the shells settle to the deeper ocean to accumulate as sediment. Calcite formation, however, also increases ocean CO_2 content, causing CO_2 to be pumped from the ocean into the atmosphere.

CARBON CYCLE The cycling of carbon among the lithosphere (including accumulations of oil and gas), hydrosphere (ocean), atmosphere, and biosphere, which are the primary reservoirs for carbon.

CARBON SEQUESTRATION The long-term (thousands of years and longer) storage of captured CO_2 in underground rock or the deep ocean.

CLIMATE The average weather for a particular region for a significant period of time, typically several years.

CLIMATE SYSTEM The system of combined chemical and physical interactions among the atmosphere, hydrosphere, cryosphere, biosphere, and lithosphere that result in climate.

CLOSED FUEL CYCLE (IN A NUCLEAR REACTOR) The cycle in which spent nuclear fuel is removed from the reactor, residual uranium-235 and plutonium-239 are separated from the spent fuel, and some of the uranium-235 is mixed back into new uranium stock and fabricated into fuel for another cycle.

COCCOLITH The microscopic, calcareous, skeletal plates excreted by certain planktonic organisms. Coccoliths may accumulate on the ocean floor to make the sedimentary rock chalk.

COMMITTED WARMING OR "WARMING IN THE PIPELINE" The warming of the atmosphere over several decades that will occur without additional rise in atmospheric greenhouse gases.

CONCENTRATING SOLAR POWER The process by which mirrors are used to collect and concentrate solar energy.

CONTINENTAL SHELF The shallow edges of the continents extending into the ocean.

CONTINENTAL SLOPE The slope extending from the continental shelf to the deep ocean.

CONVECTION The transfer of heat by transfer of matter, most commonly involving the rise of warm, less dense matter and the sinking of cool, more dense matter.

CORIOLIS FORCE The deflective component of centrifugal force on a rotating sphere, which deflects Earth's winds and ocean currents. It is named for the French mathematician Gaspard-Gustave de Coriolis (1792–1843), who first explained how Earth's rotation influences the movement of fluid.

COSMIC RAYS High-energy subatomic particles from the Sun and elsewhere from outer space. Approximately 90 percent of cosmic rays are protons.

CRITICAL MASS (NUCLEAR REACTIONS) The mass required to sustain a nuclear chain reaction until the decay of the fissile isotope brings it to a concentration below the critical mass.

CRYOSPHERE The components of Earth that are perennially frozen, including glaciers, ice sheets, sea ice, and permafrost.

CYCLONIC FLOW The flow of wind in the quasi-circular clockwise direction in the Northern Hemisphere and the counter-clockwise direction in the Southern Hemisphere.

DALTON MINIMUM   A time of relatively low solar irradiance that approximately corresponded to the cold period from 1790 to 1830. It is named for the English meteorologist John Dalton (1766–1854).

DANSGAARD-OESCHGER (D-O) EVENTS   Cycles of rapid warming followed by slow cooling that characterized the most recent glacial period in the North Atlantic. The cycles occurred at approximately 1,500-year intervals and are best recorded in ice cores from Greenland.

DEEP ZONE   The part of the ocean below 500 to 900 meters (1,600 to 3,000 feet) to depth where salinity and temperature vary only slightly. The deep zone constitutes about 65 percent of ocean water and at its deepest and coldest is about 2°C (35.6°F).

DETRITAL   Describing a sediment formed from detritus, or grains, from other rocks.

EKMAN TRANSPORT   The flow of surface water at ideally 45 degrees to the wind direction as a consequence of the Coriolis effect. The flow is to the right of the wind direction in the Northern Hemisphere and to the left in the Southern Hemisphere.

EL NIÑO   The warming of the eastern equatorial Pacific Ocean that takes place typically every three to seven years, usually beginning around Christmas and persisting into May or June.

EL NIÑO-SOUTHERN OSCILLATION (ENSO)   The coupled oscillations of ocean and wind currents across the equatorial Pacific Ocean, creating El Niño and La Niña states.

EMISSION SCENARIO   A model of how greenhouse-gases emissions will change based on specific assumptions of how economic and social conditions will change.

ENERGY BALANCE   The condition whereby the amount of energy entering the Earth system is equal to that exiting it.

EOLIAN SEDIMENT   A sediment deposited by wind.

EVAPORATION   The change of state of matter from a liquid to a vapor.

EVAPOTRANSPIRATION   The combined processes of evaporation and plant transpiration of water.

FEEDBACK   A phenomenon that amplifies or diminishes a force that acts to change climate.

FEED-IN TARIFF PROGRAM   A program in which a producer of energy is allowed to connect to the power grid and either sell excess power to the grid operator or trade it for electricity delivered at another time.

FERREL CELLS   The global atmospheric circulation patterns between about the 30 and 60° latitudes characterized by the rise of cold air at about the 50 to 60° latitudes and the descent of that air at about the 30 to 40° latitudes. The Ferrel cells are caused by eddy circulations, or small swirls in the opposite direction from the main flow, that transport heat and moisture poleward and result in southwesterly Northern Hemisphere and northwesterly Southern Hemisphere surface winds. They are named for the American meteorologist William Ferrel (1817–1891).

FISSION   The process of splitting of the nucleus of isotope into smaller particles, with the liberation of energy.

FORAMINIFERA   Protozoans of the subclass Sarcodina, order Foraminifera, characterized by a test composed of secreted calcite or, more rarely, aragonite or silica.

FORCING FACTORS   Influences external to, or not part of, the climate system that can cause climate to change.

FOSSIL FUEL   Coal, oil, natural gas, or any other organic material from Earth that may be used for fuel.

GENERATING EFFICIENCY (IN POWER PRODUCTION)   The amount of electricity produced per unit of thermal energy in the fuel.

GREENHOUSE EFFECT   The phenomenon of warming of the atmosphere whereby greenhouse gases, principally water vapor and CO_2 but also several other constituents of the atmosphere, absorb infrared radiation originating mostly from Earth's surface.

GROUNDING LINE   The line that demarcates ice resting on land and ice floating on water.

GULF STREAM   The western boundary current of the North Atlantic Ocean.

GYRE   A spinning current of water or air.

HADLEY CELLS   The global atmospheric circulation pattern characterized by convective rise of air near the equator, its flow in the upper troposphere toward the poles, descent at about the 30° latitudes, and return at the surface toward the equator. Hadley circulation is deflected by the Coriolis force, resulting in the trade winds. They are named for the English mathematician George Hadley (1685–1768), who first explained the phenomenon.

HOLOCENE   The geological epoch comprising the past 11,600 years, during which climate has been relatively stable and temperate.

HYDROLOGIC CYCLE   The flow of water through the ocean, atmosphere, and land reservoirs.

HYDROSPHERE   The waters of Earth, including the ocean, lakes, rivers, groundwater, soil water, and water vapor in the atmosphere.

ICE AGE   A time of extensive glaciation.

ICE SHEET   A continuous sheet of ice of more than 50,000 square kilometers (19,000 square miles) in extent on land.

ICE SHELF   A sheet of ice, typically hundreds to thousands of square kilometers in extent, that floats on water and is attached on at least one side to land.

INDO-AUSTRALIAN CONVERGENCE ZONE (IACZ)   The zone of low pressure due to the rise of warm, moist air over a region approximately centered on New Guinea in the western equatorial Pacific Ocean. The convergence occurs because the rising air is replaced by easterly winds from the Pacific and westerly winds from Australasia.

INFRARED (IR) RADIATION   The radiation in the electromagnetic spectrum with wavelengths between 750 nanometers and 1 millimeter.

INSOLATION   The amount of radiation received per unit time at any one location. The common units are watts per square meter per day.

INTEGRATED GAS COMBINED CYCLE (IGCC)   The process by which coal is gasified to produce a mixture of hydrogen and carbon monoxide (known as syngas) and in which the syngas is burned in a gas turbine and the hot turbine exhaust used to produce steam that drives a steam turbine.

INTERGLACIAL   Pertaining to a time interval of relative warmth between two glacial intervals.

INTERTROPICAL CONVERGENCE ZONE (ITCZ)   The equatorial band where the Northern and Southern Hemisphere trade winds converge. The large-scale upward motion of air results in low surface air pressure, high evaporation rate of water from the ocean, high rainfall, and transfer to the atmosphere of substantial amounts of heat from the surface.

ISOTOPE　Any of the different atomic masses of the same element.

JET STREAM　High-speed, horizontal rivers of wind hundreds of kilometers wide and several kilometers thick. The polar jet streams undulate back and forth near the polar fronts at 10 kilometers (32,000 feet) altitude and are in essence the upper tropospheric manifestations of the polar fronts. The subtropical jet streams exist at the 30° latitudes at an altitude of about 12 kilometers (40,000 feet), corresponding to the downward limbs of the Hadley cells. The polar jet streams affect weather patterns between about 45 and 60° latitudes, but the subtropical jet streams have little influence on weather because of their high altitude.

KEELING CURVE　The change in atmospheric CO_2 content with time, as determined by direct measurement of the atmosphere. Direct measurement was begun by Charles Keeling (1928–2005) in 1958 and has been made since then.

LA NIÑA　The state whereby winds blow westward across the eastern equatorial Pacific Ocean, causing upwelling of cold water along the equator. La Niña states alternate with El Niño states. The name is commonly reserved for times when the westward winds are unusually strong.

LAPSE RATE　The rate at which temperature decreases with height in the atmosphere.

LIMESTONE　A sedimentary rock composed mainly of calcium carbonate minerals, generally calcite. Limestone may contain lesser amounts of the magnesium carbonate mineral dolomite.

LITHOSPHERE　The rigid outer part of Earth, including sediments, extending to about 100 kilometers (62 miles) depth.

LITTLE ICE AGE　A period of unusual cold that was apparently global, but is best documented in Europe, where it lasted from about 1350 to the mid-nineteenth century.

LOESS　A fine-grained, homogeneous, buff to yellowish brown, semiconsolidated deposit of silt representing the accumulation of windblown dust, mostly formed during the Pleistocene.

MAUNDER MINIMUM　A period of low sunspot activity between 1645 and 1715, corresponding to unusual cold conditions. It is named for the English astronomer E. W. Maunder (1851–1928).

MEDIEVAL WARM PERIOD　A period of warmth that was apparently global and lasted from about 850 to 1200.

MERIDIONAL　Referring to north–south directed flow, normally of ocean water or of air.

MESOPAUSE　The top boundary of the mesosphere. At the mesopause, the temperature of the atmosphere reaches of minimum of about –100°C (–148°F).

MESOSPHERE　The layer of atmosphere above the stratosphere, from about 50 to 85 kilometers (30 to 50 miles) altitude, where temperature slowly decreases with height.

METHANOGENESIS　The process of bacterial production of methane gas from organic materials.

MILANKOVITCH CYCLES　The collective cycles of Earth's changing precession, obliquity, and eccentricity that yield cycles of changing Northern Hemisphere insolation. Milankovitch theory holds that these changes in insolation account for the Pleistocene glacial–interglacial cycles. They are named for the Serbian civil engineer and mathematician Milutin Milankovitch (1879–1958), who quantified Earth's orbital characteristics to explain glaciation.

MIXED LAYER The thin, warm surface layer of the ocean. The layer is 20 to 200 meters (65 to 650 feet) deep, with an average of 70 meters (230 feet).

MONSOON A tropical wind system that changes direction with the season, bringing seasonal rains.

NORTH ATLANTIC DEEP WATER (NADW) The deep, cold water formed in the Norwegian-Greenland and Labrador seas during the winter by the sinking of cold surface water.

NORTH ATLANTIC OSCILLATION (NAO) The wintertime oscillation of atmospheric pressure between the polar and the subtropical North Atlantic. The oscillations occur over the course of a weeks, but there are also decadal oscillations in which one or the other state predominates. The NAO influences the weather and climates of Europe, western Asia, parts of the Mediterranean basin, and certain regions of eastern North America.

OBLIQUITY The angle between the rotational axis of a planet and the perpendicular to its orbital plane.

OCEAN CONVEYOR SYSTEM The global circulation of water from the Atlantic Ocean through the Indian and Southern oceans to the Pacific Ocean and back. A full cycle takes about 1,000 years. The circulation is driven by variations in water salinity and temperature. This system is also known as *thermohaline circulation*.

OPEN (OR ONCE-THROUGH) FUEL CYCLE (IN A NUCLEAR REACTOR) The cycle in which spent fuel is removed from a reactor and stored for ultimate disposal rather than being reprocessed.

ORBITAL FORCING The changes in Earth's precession, obliquity, or eccentricity that influence radiation balance.

OXY-FUEL PC COMBUSTION The burning of pulverized coal (PC) in an oxygen-rich gas.

OZONE The gaseous molecule O_3. Ozone exists naturally in the stratosphere, where it absorbs ultraviolet radiation, but it is also a pollutant in the troposphere.

OZONE HOLE The stratospheric region of low ozone concentration that develops in a seasonal cycle over Antarctica in the Southern Hemisphere spring.

PALEOCENE–EOCENE THERMAL MAXIMUM (PETM) An approximately 120,000-year period of unusual warmth that occurred about 55 million years ago. The PETM was characterized by an abrupt increase in atmospheric CO_2, a rise in average global surface temperature of 5 to 9°C (9 to 16°F), and a shallowing of the ocean carbonate compensation depth—changes that dramatically affected terrestrial and marine ecosystems.

PALEOCLIMATE Past climate, the climate of a specific interval of time in the past.

PALEOCLIMATOLOGY The study of past climate, its variations, and their causes. Paleoclimatology involves mainly interpretation of climate records represented by ice, ocean and lake sediments, corals, cave deposits, trees, and fossils.

PALMER DROUGHT SEVERITY INDEX A measure of soil moisture calculated from a combination of precipitation records and estimates of evaporation based on temperature.

PARAMETERIZATION An equation relating subgrid-scale phenomena in climate models based on knowledge of how the phenomena are related on the grid scale.

PERMAFROST Soil or rock that has been below 0°C (32°F) and frozen for at least two years.

PHENOLOGY   The study of the timing of organisms' life-cycle events, such as egg laying, flower blooming, and spring migration.

PHOTIC ZONE   The upper layer of the ocean, usually 50 to 150 meters (165 to 490 feet) deep, where photosynthesis occurs.

PHOTOELECTRIC EFFECT   The process of ejection of an electron from an atom by absorption of an incident photon, or a packet (quantum) of light.

PHOTOSYNTHESIS   The process by which living matter uses sunlight to convert CO_2 and water into organic matter. Oxygen is produced as a by-product.

PHOTOVOLTAIC (PV) CELLS   Devices that collect sunlight and transform it into electricity utilizing the photoelectric effect.

PHYTOPLANKTON   Single-celled organisms that engage in photosynthesis and live in the photic zone of the ocean.

PLANKTON   Drifting or weakly swimming aquatic organisms.

PLEISTOCENE   The geological epoch encompassing the period from 1.8 million to 11,600 years ago.

PLIOCENE   The geological epoch encompassing the period from 5.3 million to 1.8 million years ago.

POLAR CELLS   The atmospheric circulation near the poles characterized by the descent of cold, dry air in vortices at the poles and diverging surface flow toward the equator.

POLAR FRONT   The location at which the polar cell easterlies encounter the Ferrel cell westerlies, creating a zone of unstable weather between about the 50 and 60° latitudes.

PRECESSION   The change in the direction of the axis of rotation of a body.

PULVERIZED COAL (PC) POWER PLANT   A coal-fired power plant in which PC and air are injected into the combustion chamber.

RADIATIVE FORCING   The change, relative to the year 1750, in energy entering the Earth system minus energy exiting the Earth system in response to a factor that changes energy balance.

RADIOLARIAN   An actinopod of the subclass Radiolaria characterized by a siliceous skeleton.

SAHEL   The southern borderland of the Sahara Desert, stretching from Senegal in the west to Sudan in the east.

SALINE FORMATIONS   Porous sedimentary rocks, typically sandstones, in which the fluid is saline water.

SALINITY   Grams of dissolved salt per kilogram of water.

SHALE   A laminated sedimentary rock consisting of particles of which more than two-thirds are clay-size (less than 1/256 millimeter in diameter).

SHORT-TERM CARBON CYCLE   The flow of carbon through its surface reservoirs, composed mainly of the ocean, atmosphere, and biosphere.

SOLAR CONSTANT   The energy flux (energy per unit time) received by Earth from the Sun. The solar constant is the cross-sectional value of 1,367 watts per square meter. Averaged over the entire surface area of the upper atmosphere of Earth, this amount is equivalent to 342 watts per square meter.

SOLAR IRRADIANCE   The amount of the Sun's radiant energy reaching Earth.

SOLUBILITY PUMP (FOR OCEAN CARBON)   The process by which cold, carbon-rich surface water sinks and transports carbon to the deep ocean, thus removing CO_2 from the atmosphere.

SOUTHERN OSCILLATION   The swings in barometric pressure between the eastern and western parts of the equatorial Pacific Ocean.

SOUTHERN OSCILLATION INDEX   The deviation from the mean difference in atmospheric pressure between Tahiti and Darwin, Australia.

SPECIFIC HEAT   The amount of energy required to raise a unit mass of matter 1°C.

SPELEOTHERMS   Deposits in caves formed by the precipitation of calcium carbonate from groundwater.

STRATOSPHERE   The layer of the atmosphere between the troposphere and the overlying mesosphere. The base of the stratosphere varies in altitude from 16 to 18 kilometers (52,000 to 59,000 feet) in the tropics to 8 to 10 kilometers (26,000 to 33,000 feet) in the polar regions; the top of the stratosphere is at about 50 kilometers (30 miles) altitude. The temperature in the stratosphere slowly increases from about –60°C (–76°F) at its base to 0°C (32°F) at its top.

STRATOSPHERIC CLOUDS   Clouds of microscopic ice particles that form in the stratosphere when temperatures dip below –80°C (–112°F). The ice particles lead to the formation of ozone-destroying molecules of chlorine.

SUNSPOTS   A region on the Sun's surface of much lower temperature than the surrounding surface. The average number of sunspots on the Sun changes in an approximately 11-year cycle.

THERMOCLINE   The layer below the mixed layer in the ocean. In the thermocline, temperature decreases and salinity increases rapidly with depth. The base of the thermocline is typically 500 to 900 meters (1,600 to 3,000 feet) below the sea surface and at a temperature of 5°C (41°F).

THERMOHALINE CIRCULATION   See *ocean conveyor system*.

THERMOKARST   A pit and depression formed by subsidence as underlying permafrost melts.

THERMOSPHERE   The layer of the atmosphere from about 85 kilometers (50 miles) altitude to the outer reaches of the atmosphere. In the thermosphere, temperature increases with altitude.

TIPPING POINT   A large, abrupt shift in climate in response to a factor, such as the buildup of greenhouse gases in the atmosphere, that had been causing climate to change gradually.

TRADE WINDS   The system of subtropical winds that blow from the subtropical belts of high atmospheric pressure to the equatorial belt of low atmospheric pressure, representing the return limb of Hadley circulation. The trade winds blow to the southwest in the Northern Hemisphere and to the northwest in the Southern Hemisphere.

TRANSPOLAR DRIFT STREAM   The rapid current that carries ice through the central Arctic Ocean, across the pole, and out through the Fram Strait.

TROPOPAUSE   The boundary between the troposphere and the overlying stratosphere. The tropopause, at about –60°C (–76°F), represents the temperature minimum between the troposphere and the stratosphere. It is stable and tends to isolate the two layers—for example, by restricting water vapor and weather to the troposphere.

TROPOSPHERE   The lowest layer of the atmosphere, extending from the surface to an elevation of 16 to 18 kilometers (52,0000 to 59,000 feet) in the tropics and 8 to 10 kilometers (26,000 to 33,000 feet) in the polar regions. In the troposphere, temperature decreases with height.

ULTRAVIOLET (UV) RADIATION   The radiation in the electromagnetic spectrum with wavelengths between 40 and 400 nanometers.

VARVES   Sediment sequences consisting of alternating dark and light layers. In lakes sealed under ice, dead microscopic organisms and fine clay accumulate on the lake bottom in the winter to produce the dark layer; sand and silt deposited from spring runoff form the light layer.

WALKER CIRCULATION   The equatorial convective circulation of the atmosphere characterized by the rise of air in the vicinity of eastern Indonesia–New Guinea and the descent of that air to the east and west.

WEATHER   The temperature, precipitation, humidity, wind, atmospheric pressure, and cloudiness at any one time at any one place.

WEATHERING (OF ROCK)   The in situ physical and chemical disintegration of rock by air, water, biological compounds, and other chemical compounds in the environment at or near Earth's surface.

WESTERN BOUNDARY CURRENTS   Strong poleward currents flowing along the eastern coastlines of continents, corresponding to the western sides of the ocean basins.

YEDOMA   Frozen loess underlying vast areas of Siberia. Yedoma commonly contains 2 to 5 percent carbon.

YOUNGER DRYAS   The interval of extreme cold that abruptly descended on the northern Atlantic Ocean and neighboring regions beginning about 12,900 years ago and ending just as abruptly 1,300 years later. The name comes from the tundra wildflower white dryas (*Dryas octopetala*).

ZENITH ANGLE   The angle between the vertical and a line from Earth's surface to the Sun. Zenith angle changes with time of day, season, and latitude.

ZOOPLANKTON   Animal plankton that consume other plankton.

BIBLIOGRAPHY

Abram, N. J., M. K. Gagan, Z. Liu, W. S. Hantoro, M. T. McCulloch, and B. W. Suwargadi. "Seasonal Characteristics of the Indian Ocean Dipole During the Holocene Epoch." *Nature* 445 (2007): 299–302.

Alexander, L. V., X. Zhang, T. C. Peterson, J. Caesar, B. Gleason, A. M. G. Tank, M. Haylock, D. Collins, B. Trewin, F. Rahimzadeh, A. Tagipour, K. Kumar, J. Revadekar, G. Griffiths, L. Vincent, D. B. Stephenson, J. Burn, E. Aguilar, M. Brunet, M. Taylor, M. New, P. Zhai, M. Rusticucci, and J. L. Vazquez-Aguirre. "Global Observed Changes in Daily Climate Extremes of Temperature and Precipitation." *Journal of Geophysical Research* 111 (2006), doi:10.1029/2005JD006290.

Alley, R. B., T. Berntsen, N. L. Bindoff, Z. Chen, A. Chidthaisong, P. Friedlingstein, J. Gregory, G. Hegerl, M. Heimann, B. Hewitson, B. Hoskins, F. Joos, J. Jouzel, V. Kattsov, U. Lohmann, M. Manning, T. Matsuno, M. Molina, N. Nicholls, J. Overpeck, D. Qin, G. Raga, V. Ramaswamy, J. Ren, M. Rusticucci, S. Solomon, R. Somerville, T. F. Stocker, P. Stott, R. J. Stouffer, P. Whetton, R. A. Wood, and D. Wratt. "Summary for Policymakers." In *Climate Change 2007: The Physical Science Basis. Contribution of Working Group I to the Fourth Assessment Report of the Intergovernmental Panel on Climate Change*, edited by S. Solomon, D. Qin, M. Manning, Z. Chen, M. Marquis, K. B. Averyt, M. Tignor, and H. L. Miller, 1–18. Cambridge: Cambridge University Press, 2007.

Alley, R. B., E. J. Brook, and S. Anandakrishnan. "A Northern Lead in the Orbital Band: North–South Phasing of Ice-Age Events." *Quaternary Science Reviews* 21 (2002): 431–441.

Alley, R. B., P. U. Clark, P. Huybrechts, and I. Joughin. "Ice-Sheet and Sea-Level Changes." *Science* 310 (2005): 456–460.

Alley, R. B., C. A. Shuman, D. A. Meese, A. J. Gow, K. C. Taylor, K. M. Cuffey, J. J. Fitzpatrick, P. M. Grootes, G. A. Zielinski, M. Ram, G. Spinelli, and B. Elder. "Visual-Stratigraphic Dating of the GISP2 Ice Core: Basis, Reproducibility, and Application." *Journal of Geophysical Research* 102 (1997): 26367–26381.

Andersen, B. G., and H. W. Borns Jr. *The Ice Age World*. Oslo: Scandinavian University Press, 1994.

Andreae, M. O., C. D. Jones, and P. M. Cox. "Strong Present-Day Aerosol Cooling Implies a Hot Future." *Nature* 435 (2005): 1187–1190.

Archer, C. L., and M. Z. Jacobson. "Evaluation of Global Wind Power." *Journal of Geophysical Research* 110 (2005), doi:10.1029/2004JD005462.

Archer, D., H. Kheshgi, and E. Maier-Reimer. "Dynamics of Fossil Fuel CO_2 Neutralization by Marine $CaCO_3$." *Global Biogeochemical Cycles* 12 (1998): 259–276.

Baede, A. P. M., E. Ahlonsou, Y. Ding, and D. Schimel. "The Climate System: An Overview." In *Climate Change 2001: The Scientific Basis. Contributions of Working Group I to the Third Assessment Report of the Intergovernmental Panel on Climate Change*, edited by J. T. Houghton, T. Y. Ding, D. J. Griggs, M. Noguer, P. J. Van Der Linden, X. Dai, K. Maskell, and C. A. Johnson, 85–98. Cambridge: Cambridge University Press, 2001.

Bard, E., and M. Frank. "What's New Under the Sun?" *Earth and Planetary Science Letters* 248 (2006): 1–14.

Barker, T., I. Bashmakov, L. Bernstein, J. E. Bogner, P. R. Bosch, R. Dave, O. R. Davidson, B. S. Fisher, S. Gupta, K. Halsnæs, G. J. Heij, S. Kahn Ribeiro, S. Kobayashi, M. D. Levine, D. L. Martino, O. Masera, B. Metz, L. A. Meyer, G.-J. Nabuurs, A. Najam, N. Nakićenović, H.-H. Rogner, J. Roy, J. Sathaye, R. Schock, P. Shukla, R. E. H. Sims, P. Smith, D. A. Tirpak, D. Urge-Vorsatz, and D. Zhou. "Technical Summary." In *Climate Change 2007: Mitigation of Climate Change. Contribution of Working Group III to the Fourth Assessment Report of the Intergovernmental Panel on Climate Change*, edited by B. Metz, O. R. Davidson, P. R. Bosch, R. Dave, and L. A. Meyer, 25–94. Cambridge: Cambridge University Press, 2007.

Barnett, T. P., J. C. Adam, and D. P. Lettenmaier. "Potential Impacts of a Warming Climate on Water Availability in Snow-Dominated Regions." *Nature* 438 (2005): 303–309.

Barnett, T. P., D. W. Pierce, K. M. Achuta Rao, P. J. Gleckler, B. D. Santer, J. M. Gregory, and W. M. Washington. "Penetration of Human-Induced Warming into the World's Oceans." *Science* 309 (2005): 284–287.

Barnett, T. P., D. W. Pierce, H. G. Hidalgo, C. Bonfils, B. D. Santer, T. Das, G. Bala, A. W. Wood, T. Nozawa, A. A. Mirim, D. R. Cayan, and M. D. Dettinger. "Human-Induced Changes in the Hydrology of the Western United States." *Science* 319 (2008): 1080–1083.

Baron, W. "1816 in Perspective: The View from the Northeastern United States." In *The Year Without a Summer? World Climate in 1816*, edited by C. R. Harington, 124–144. Ottawa: Canadian Museum of Nature, 1992.

Barrows, T. T., S. J. Lehman, L. K. Fifield, and P. De Deckker. "Absence of Cooling in New Zealand and the Adjacent Ocean During the Younger Dryas Chronozone." *Science* 318 (2007): 86–89.

Beaugrand, G., K. M. Brander, J. A. Lindley, S. Souissi, and P. C. Reid. "Plankton Effect on Cod Recruitment in the North Sea." *Nature* 426 (2003): 661–664.

Beck, J. W., R. L. Edwards, E. Ito, F. W. Taylor, J. Recy, F. Rougerie, P. Joannot, and C. Henin. "Sea-Surface Temperature from Coral Skeletal Strontium/Calcium Ratios." *Science* 31 (1992): 644–647.

Bellouin, N., O. Boucher, J. Haywood, and M. S. Reddy. "Global Estimate of Aerosol Direct Radiative Forcing from Satellite Measurements." *Nature* 438 (2005): 1138–1141.

Bengtsson, L., K. I. Hodges, and E. Roeckner. "Storm Tracks and Climate Change." *Journal of Climate* 19 (2006): 3518–3543.

Benson, S., P. Cook, J. Anderson, S. Bachu, H. B. Nimir, B. Basu, J. Bradshaw, G. Deguchi, J. Gale, G. von Goerne, W. Heidug, S. Holloway, R. Kamal, D. Keith, P. Lloyd, P. Rocha, W. Senior, J. Thomson, T. Torp, T. Wildenborg, M. Wilson, F. Zarlenga, and D. Zhou. "Underground Geological Storage." In *IPCC Special Report on Carbon Dioxide Capture and Storage*, edited by B. Metz, O. Davidson, H. C. de Coninck, M. Loos, and L. A. Meyer, 195–276. Cambridge: Cambridge University Press, 2005.

Berger, A., and M. F. Loutre. "An Exceptionally Long Interglacial Ahead?" *Science* 297 (2002): 1287–1288.

Bigg, G. R. *The Oceans and Climate*. Cambridge: Cambridge University Press, 1996.

Blunier, T., and E. J. Brook. "Timing of Millennial-Scale Climate Change in Antarctica and Greenland During the Last Glacial Period." *Science* 291 (2001): 109–112.

Bluth, G. J. S., C. C. Schnetzler, A. J. Krueger, and L. S. Walter. "The Contribution of Explosive Volcanism to Global Atmospheric Sulfur Dioxide Concentrations." *Nature* 366 (1993): 327–329.

Bond, G. C., W. Broecker, S. Johnsen, J. McManus, L. Labeyrie, J. Jouzel, and G. Bonani. "Correlations Between Climate Records from North Atlantic Sediments and Greenland Ice." *Nature* 365 (1996): 143–147.

Bond, G. C., W. Showers, M. Elliot, M. Evans, R. Lotti, I. Hajdas, G. Bonani, and S. Johnson. "The North Atlantic's 1–2 kyr Climate Rhythm: Relation to Heinrich Events, Dansgaard/Oeschger Cycles, and the Little Ice Age." In *Mechanisms of Global Climate Change at Millennial Time-Scales*, edited by P. U. Clark, R. S. Webb, and L. D. Keigwin, 35–58. Washington, D.C.: American Geophysical Union, 1999.

Booth, R. K., S. T. Jackson, S. L. Forman, J. E. Kutzbach, E. A. Bettis III, J. Kreig, and D. K. Wright. "A Severe Centennial-Scale Drought in Mid-continental North America 4200 Years Ago and Apparent Global Linkages." *Holocene* 15 (2005): 321–328.

Bowen, G. J., D. J. Beerling, P. L. Koch, J. C. Zachos, and T. Quattlebaum. "A Humid State During the Paleocene/Eocene Thermal Maximum." *Nature* 432 (2004): 495–499.

Bowen, G. J., T. J. Bralower, M. L. Delaney, G. R. Dickens, D. C. Kelly, P. L. Koch, L. R. Kump, J. Meng, L. C. Sloan, E. Thomas, S. L. Wing, and J. C. Zachos. "Eocene Hyperthermal Event Offers Insight into Greenhouse Warming." *Eos, Transactions* 87 (2006): 165.

Braganza, K., D. J. Karoly, A. C. Hirst, M. E. Mann, P. Stott, R. J. Stouffer, and S. F. B. Tett. "Simple Indices of Global Climate Variability and Change: Part I—Variability and Correlation Structure." *Climate Dynamics* 20 (2003): 491–502.

Braganza, K., D. J. Karoly, A. C. Hirst, P. Stott, R. J. Stouffer, and S. F. B. Tett. "Simple Indices of Global Climate Variability and Change: Part II—Attribution of Climate Change During the Twentieth Century." *Climate Dynamics* 22 (2004): 823–838.

Bréon, F.-M. "How Do Aerosols Affect Cloudiness and Climate?" *Science* 313 (2006): 623–624.

Breshears, D. D., N. S. Cobb, P. M. Rich, K. P. Price, C. D. Allen, R. G. Balice, W. H. Romme, J. H. Kastens, M. L. Floyd, J. Belnap, J. J. Anderson, O. B. Myers, and C. W. Meyer. "Regional Vegetation Die-off in Response to Global-Change-Type Drought." *Proceedings of the National Academy of Sciences* 102 (2005): 15144–15148.

British Petroleum. *BP Statistical Review of World Energy 2007*. London: British Petroleum, 2007. Available at http://www.bp.com/productlanding.do?categoryId = 6848&contentId = 7033471.

Broecker, W. S. "The Great Ocean Conveyor." *Oceanography* 4 (1991): 79–89.

———. "Was the Medieval Warm Period Global?" *Science* 291 (2001): 1497–1499.

Broecker, W. S., and G. H. Denton. "The Role of Ocean–Atmosphere Reorganizations in Glacial Cycles." *Geochimica et Cosmochemica Acta* 53 (1989): 2465–2501.

———. "What Drives Glacial Cycles?" *Scientific American*, January 1990, 48–56.

Brohan, P., J. J. Kennedy, I. Harris, S. Tett, and P. D. Jones. "Uncertainty Estimates in Regional and Global Observed Temperature Changes: A New Data Set from 1850." *Journal of Geophysical Research* 111 (2006), doi:10.1029/2005JD006548.

Bunn, A. G., S. J. Goetz, J. S. Kimball, and K. Zhang. "Northern High-Latitude Ecosystems Respond to Climate Change." *Eos, Transactions* 88 (2007): 333–335.

Caldeira, K., A. K. Jain, and M. I. Hoffert. "Climate Sensitivity Uncertainty and the Need for Energy Without CO_2 Emission." *Science* 299 (2003): 2052–2054.

Cane, M. A. "The Evolution of El Niño, Past and Future." *Earth and Planetary Science Letters* 230 (2005): 227–240.

Cane, M. A., G. Eshel, and R. W. Buckland. "Forecasting Zimbabwean Maize Yield Using Eastern Equatorial Pacific Sea Surface Temperature." *Nature* 370 (1994): 204–205.

Chen, J. L., C. R. Wilson, and B. D. Tapley. "Satellite Gravity Measurements Confirm Accelerated Melting of Greenland Ice Sheet." *Science* 313 (2006): 1958–1960.

Christy, J. R., D. J. Seidel, and S. C. Sherwood. "What Kinds of Atmospheric Temperature Variations Can the Current Observing Systems Detect and What Are Their Strengths and Limitations, Both Spatially and Temporally?" In *Temperature Trends in the Lower Atmosphere: Steps for Understanding and Reconciling Differences*, edited by T. R. Karl, S. J. Hassol, C. D. Miller, and W. L. Murray, 29–46. Washington, D.C.: Climate Change Science Program, Subcommittee on Global Change Research, 2006.

Cleaveland, M. K. "Volcanic Effects on Colorado Plateau Douglas-Fir Tree Rings." In *The Year Without a Summer? World Climate in 1816*, edited by C. R. Harington, 115–123. Ottawa: Canadian Museum of Nature, 1992.

Clyde, W. C., and P. D. Gingrich. "Mammalian Community Response to the Latest Paleocene Thermal Maximum: An Isotaphonomic Study in the Northern Bighorn Basin, Wyoming." *Geology* 26 (1998): 1011–1014.

Cole J. E., G. T. Shen, R. G. Fairbanks, and M. Moore. "Cora Monitors of El Niño/Southern Oscillation Dynamics Across the Equatorial Pacific." In *El Niño*, edited by H. F. Diaz and V. Markgraf, 349–375. Cambridge: Cambridge University Press, 1992.

Committee on Abrupt Climate Change. *Abrupt Climate Change: Inevitable Surprises*. Washington, D.C.: National Academy Press, 2002.

Cook, A. J., A. J. Fox, D. G. Vaughn, and J. G. Ferrigno. "Retreating Glacier Fronts on the Antarctic Peninsula over the Past Half-Century." *Science* 308 (2005): 541–544.

Cook, E. R., C. A. Woodhouse, C. M. Eakin, D. M. Meko, and D. W. Stahle. "Long-Term Aridity Changes in the Western United States." *Science* 306 (2004): 1015–1018.

Corrège, T., M. K. Gagan, J. W. Beck, G. S. Burr, G. Cabloch, and F. Le Cornec. "Interdecadal Variation in the Extent of South Pacific Tropical Waters During the Younger Dryas Event." *Nature* 428 (2004): 927–929.

Cox, P. M., P. P. Harris, C. Huntingford, R. A. Betts, M. Collings, C. D. Jones, T. E. Jupp, J. A. Marengo, and C. A. Nobre. "Increasing Risk of Amazonian Drought Due to Decreasing Aerosol Pollution." *Nature* 453 (2008): 212–215.

Dai, A., K. E. Trenberth, and T. Qian. "A Global Data Set of Palmer Drought Severity Index for 1870–2002: Relationship with Soil Moisture and Effects of Surface Warming." *Journal of Hydrometeorology* 5 (2004): 1117–1130.

D'Arrigo, R., R. Wilson, and G. Jacoby. "On the Long-Term Context for Late Twentieth Century Warming." *Journal of Geophysical Research* 111 (2006), doi:10.1029/2005JD006352.

Davidson, E. A., and I. A. Janssens. "Temperature Sensitivity of Soil Carbon Decomposition and Feedbacks to Climate Change." *Nature* 440 (2006): 165–173.

Davis, C. H., Y. Li, J. R. McConnell, M. M. Frey, and E. Hanna. "Snowfall-Driven Growth in East Antarctic Ice Sheet Mitigates Recent Sea-Level Rise." *Science* 308 (2005): 1898–1901.

Denman, K. L., G. Brasseur, A. Chidthaisong, P. Ciais, P. M. Cox, R. E. Dickinson, D. Hauglustaine, C. Heinze, E. Holland, D. Jacob, U. Lohmann, S. Ramachandran, P. L. da Silva Dias, S. C. Wofsy, and X. Zhang. "Couplings Between Changes in the Climate System and Biogeochemistry." In *Climate Change 2007: The Physical Science Basis. Contribution of Working Group I to the Fourth Assessment Report of the Intergovernmental Panel on Climate Change*, edited by S. Solomon, D. Qin, M. Manning, Z. Chen, M. Marquis, K. B. Averyt, M.Tignor, and H. L. Miller, 499–588. Cambridge: Cambridge University Press, 2007.

Denman, K. L., and M. Miyake. "Upper Layer Modification at Ocean Station *Papa*: Observations and Simulations." *Journal of Physical Oceanography* 3 (1973): 185–196.

Denton, G. H., D. E. Sugden, D. R. Marchant, B. L. Hall, and T. I. Wilch. "East Antarctic Ice Sheet Sensitivity to Pliocene Climatic Change from a Dry Valleys Perspective." *Geografiska Annaler* 75A (1993): 155–204.

Dickens, G. R., J. R. O'Neil, D. K. Rea, and R. M. Owen. "Dissociation of Oceanic Methane Hydrate as a Cause of the Carbon Isotope Excursion at the End of the Paleocene." *Paleoceanography* 10 (1995): 965–971.

Dickens, G. R., and M. S. Quinby-Hunt. "Methane Hydrate Stability in Seawater." *Geophysical Research Letters* 21 (1994): 2115–2118.

Domack, E., D. Duran, A. Leventer, S. Ishman, S. Doane, S. McCallum, D. Amblas, J. Ring, R. Gilbert, and M. Prentice. "Stability of the Larsen B Ice Shelf on the Antarctic Peninsula During the Holocene Epoch." *Nature* 436 (2005): 681–685.

Doughty, C., and K. Pruess. "Modeling Supercritical Carbon Dioxide Injection in Heterogeneous Porous Media." *Vadose Zone Journal* 3 (2004): 837–847.

Dowsett, H. J., M. A. Chandler, T. M. Cronin, and G. S. Dwyer. "Middle Pliocene Sea Surface Temperature Variability." *Paleoceanography* 20 (2005), doi:10.1029/2005PA001133.

Droxler, A. W., R. B. Alley, W. R. Howard, R. Z. Poore, and L. H. Burckle. "Unique and Exceptionally Long Interglacial Marine Isotope Stage 11: Window into Earth [*sic*] Warm Future Climate." In *Earth's Climate and Orbital Eccentricity: The Marine Isotope Stage 11 Question*, edited by A. W. Droxler, R. Z. Poore, and L. H. Burckle, 1–14. Geophysical Monograph, no. 137. Washington, D.C.: American Geophysical Union, 2003.

Eddy, J. "Before Tambora: The Sun and Climate, 1790–1830" (abstract). In *The Year Without a Summer? World Climate in 1816*, edited by C. R. Harington, 9. Ottawa: Canadian Museum of Nature, 1992.

Edmonds, J. A., M. A. Wise, J. J. Dooley, S. H. Kim, S. J. Smith, P. J. Runci, L. E. Clarke, E. L. Malone, and G. M. Stokes. *Wind and Solar Energy: A Core Element of Global Energy Technology Strategy to Address Climate Change*. College Park, Md.: Joint Global Change Research Institute, 2007.

Ekström, G., M. Nettles, and V. C. Tsai. "Seasonality and Increasing Frequency of Greenland Glacial Earthquakes." *Science* 311 (2006): 1756–1758.

Ellison, C. R. W., M. R. Chapman, and I. R. Hall. "Surface and Deep Ocean Interactions During the Cold Climate Event 8200 Years Ago." *Science* 312 (2006): 1929–1932.

Energy Information Agency. *International Energy Outlook*. Washington, D.C.: Department of Energy, 2007. Available at http://www.eia.doe.gov/oiaf/ieo/index.html.

———. *World Net Nuclear Electric Power Generation, 1980–2005*. Washington, D.C.: Department of Energy, 2007. Available at http://www.eia.doe.gov/fuelnuclear.html.

EPICA Community Members. "One-to-One Coupling of Glacial Climate Variability in Greenland and Antarctica." *Nature* 444 (2006): 195–198.

Erickson, W. P., G. D. Johnson, and D. P. Young Jr. "A Summary and Comparison of Bird Mortality from Anthropogenic Causes with an Emphasis on Collisions." In *USDA Forest Service General Technical Report PSW-GTR-191*, 1029–1042. Washington, D.C.: Forest Service, Department of Agriculture, 2005.

European Wind Energy Association. *2007 Annual Report*. Brussels: European Wind Energy Association, 2007. Available at http://www.ewea.org/.

Ewing, R. C. "Environmental Impact of the Nuclear Fuel Cycle." In *Energy, Waste, and the Environment: A Geochemical Perspective*, edited by R. Gieré and P. Stille, 7–23. Special Publication, no. 236. London: Geological Society, 2004.

———. "The Nuclear Fuel Cycle: A Role for Mineralogy and Geochemistry." *Elements* 2 (2007): 331–334.

Fahey, D. W., R. S. Gao, K. S. Carslaw, J. Kettleborough, P. J. Popp, M. J. Northway, J. C. Holecek, S. C. Ciciora, R. J. McLaughlin, T. L. Thompson, R. H. Winkler, D. G. Baumgardner, B. Gandrud, P. O. Wennberg, S. Dhaniyala, K. McKinney, T. Peter, R. J. Salawitch, T. P. Bui, J. W. Elkins, C. R. Webster, E. L. Atlas, H. Jost, J. C. Wilson, R. L. Herman, A. Kleinböhl, and M. von König. "The Detection of Large HNO_3-Containing Particles in the Winter Arctic Stratosphere." *Science* 291 (2001): 1026–1031.

Falkowski, P., R. J. Scholes, E. Boyle, J. Canadell, D. Canfield, J. Elser, N. Gruber, K. Hibbard, P. Högberg, S. Linder, F. T. Mackenzie, B. Moore III, T. Pedersen, Y. Rosenthal, S. Seitzinger, V. Smetacek, and W. Steffen. "The Global Carbon Cycle: A Test of Our Knowledge of Earth as a System." *Science* 290 (2000): 291–296.

Farman, J. C., B. G. Gardiner, and, J. D. Shanklin. "Large Losses of Total Ozone in Antarctica Reveal Seasonal ClO_x/NO_x Interaction." *Nature* 315 (1985): 207–210.

Feddema, J. J., K. W. Oleson, G. B. Bonan, L. O Mearns, L. E. Bujja, G. A. Meehl, and W. M. Washington. "The Importance of Land-Cover Change in Simulating Future Climates." *Science* 310 (2005): 1674–1678.

Feely, R. A., C. L. Sabine, K. Lee, W. Berelson, J. Kleypas, V. Fabry, and F. J. Millero. "Impact of Anthropogenic CO_2 on the $CaCO_3$ System in the Oceans." *Science* 305 (2004): 362–366.

Flowers, L., and P. Dougherty. "Toward a 20% Wind Electricity Supply in the United States." In *2007 European Wind Energy Conference and Exhibition*. Conference Paper no. NREL/CP-500-41579. Washington, D.C.: Department of Energy, 2007. Available at http://www.osti.gov/bridge.

Foley, J. A., R. DeFries, G. P. Asner, C. Barford, G. Bonan, S. R. Carpenter, F. S. Chapin, M. T. Coe, G. C. Daily, H. K. Gibbs, J. H. Helkowski, T. Holloway, E. A. Howard, C. J. Kucharik, C. Monfreda, J. A. Patz, I. C. Prentice, N. Ramankutty, and P. K. Snyder. "Global Consequences of Land Use." *Science* 309 (2005): 570–574.

Forster, P., V. Ramaswamy, P. Artaxo, T. Berntsen, R. Betts, D. W. Fahey, J. Haywood, J. Lean, D. C. Lowe, G. Myhre, J. Nganga, R. Prinn, G. Raga, M. Schulz, and R. Van Dorland. "Changes in Atmospheric Constituents and in Radiative Forcing." In *Climate Change 2007: The Physical Science Basis. Contribution of Working Group I to the Fourth Assessment Report of the Intergovernmental Panel on Climate Change*, edited by S. Solomon, D. Qin, M. Manning, Z. Chen, M. Marquis, K. B. Averyt, M. Tignor, and H. L. Miller, 129–234. Cambridge: Cambridge University Press, 2007.

Foukal, P., C. Fröhlich, H. Spruit, and T. M. L. Wigley. "Variations in Solar Luminosity and Their Effect on the Earth's Climate." *Nature* 443 (2006): 161–166.

Friedmann, S. J. "Geological Carbon Sequestration." *Elements* 3 (2007): 179–184.

Fritts, H. C. "Tree-Ring Analysis." In *The Encyclopedia of Climatology*, edited by J. E. Oliver and R. W. Fairbridge, 858–875. New York: Van Nostrand Reinhold, 1987.

Giannini, A., R. Saravanan, and P. Chang. "Oceanic Forcing of Sahel Rainfall on Interannual to Interdecadal Time Scales." *Science* 302 (2003): 1027–1030.

Gibbs, S. J., P. R. Brown, J. A. Sessa, T. J. Bralower, and P. A. Wilson. "Nannoplankton Extinction and Origination Across the Paleocene–Eocene Thermal Maximum." *Science* 314 (2006): 1770–1773.

Global Wind Energy Council. *Global Wind 2007 Report*. Brussels: Global Wind Energy Council, 2007. Available at http://www.gwec.net/index.php?id = 90.

Goetz, S. J., A. G. Bunn, G. J. Fiske, and R. A. Houghton. "Satellite-Observed Photosynthetic Trends Across Boreal North America Associated with Climate and Fire Disturbance." *Proceedings of the National Academy of Science* 102 (2005): 13521–13525.

Goody, R. M., and Y. L. Yung. *Atmospheric Radiation*. New York: Oxford University Press, 1989.

Goswami, B. N., V. Venugopal, D. Sengupta, M. S. Madhusoodanan, and P. K. Xavier. "Increasing Trend of Extreme Rain Events over India in a Warming Environment." *Science* 314 (2006): 1442–1445.

Graedel, T. E., and P. J. Crutzen. *Atmosphere, Climate, and Change*. New York: Freeman, 1995.

Gregory, J. M., P. Huybrechts, and S. C. B. Raper. "Threatened Loss of the Greenland Ice-Sheet." *Nature* 428 (2004): 616.

Groisman, P. Y., R. W. Knight, T. R. Karl, D. R. Easterling, B. Sun, and J. H. Lawrimore. "Contemporary Changes of the Hydrological Cycle over the Contiguous United States: Trends Derived from in Situ Observations." *Journal of Hydrometeorology* 5 (2004): 64–85.

Grootes, P. M., M. Stuiver, J. W. C. White, S. Johnsen, and J. Jouzel. "Comparison of Oxygen Isotope Records from the GISP2 and GRIP Greenland Ice Cores." *Nature* 366 (1993): 552–554.

Hansen, J., L. Nazarenko, R. Ruedy, M. Sato, J. Willis, A. Del Genio, D. Koch, A. Lacis, K. Lo, S. Menon, T. Novakov, J. Perlwitz, G. Russell, G. A. Schmidt, and N. Tausnev. "Earth's Energy Imbalance: Confirmation and Implications." *Science* 308 (2005): 1431–1435.

Hansen, J., M. Sato, L. Nazarenko, R. Ruedy, A. Lacis, D. Koch, I. Tegen, T. Hall, D. Shindell, B. Santer, P. Stone, T. Novakov, L. Thomason, R. Wang, Y. Wang, D. Jacob, S. Hollandsworth, L. Bishop, J. Logan, A. Thompson, R. Stolarski, J. Lean, R. Willson, S. Levitus, J. Antonov, N. Rayner, D. Parker, and J. Christy. "Climate Forcings in Goddard Institute for Space Studies SI2000 Simulations." *Journal of Geophysical Research* 107 (2002), doi:10.1029/2001JD001143.

Hansen, J., M. Sato, R. Ruedy, K. Lo, D. W. Lea, and M. Medina-Elizade. "Global Temperature Change." *Proceedings of the National Academy of Science* 103 (2006): 14288–14293.

Hartmann, D. L. *Global Physical Climatology*. San Diego: Academic Press, 1994.

Hays, J. D., J. Imbrie, and N. J. Shackleton. "Variations in the Earth's Orbit: Pacemaker of the Ice Ages." *Science* 194 (1976): 1121–1132.

Heath, J., E. Ayres, M. Possell, R. D. Bardgett, H. I. J. Black, H. Grant, P. Ineson, and G. Kerstiens. "Rising Atmospheric CO_2 Reduces Sequestration of Root-Derived Soil Carbon." *Science* 309 (2005): 1711–1713.

Hegerl, G. C., F. W. Zwiers, P. Braconnot, N. P. Gillett, Y. Luo, J. A. Marengo Orsini, N. Nicholls, J. E. Penner, and P. A. Stott. "Understanding and Attributing Climate Change." In *Climate Change 2007: The Physical Science Basis. Contribution of Working Group I to the Fourth Assessment Report of the Intergovernmental Panel on Climate Change*, edited by S. Solomon, D. Qin, M. Manning, Z. Chen, M. Marquis, K. B. Averyt, M. Tignor, and H. L. Miller, 663–746. Cambridge: Cambridge University Press, 2007.

Heinze, C. "Simulating Oceanic $CaCO_3$ Export Production in the Greenhouse." *Geophysical Research Letters* 31 (2004), doi:10.1029/2004GL020613.

Hinzman, L. D., N. D. Bettez, W. R. Bolton, F. S. Chapin, M. B. Dyurgerov, C. L. Fastie, B. Griffith, R. D. Hollister, A. Hope, H. P. Huntington, A. M. Jensen, G. J. Jia, T. Jorgenson, D. L. Kane, D. R. Klein, G. Kofinas, A. H. Lynch, A. H. Lloyd, A. D. McGuire, F. E. Nelson, W. C. Oechel, T. E. Osterkamp, C. H. Racine, V. E. Romanovsky, R. S. Stone, D. A. Stow, M. Sturm, C. E. Tweedie, G. L. Vourlitis, M. D. Walker, D. A. Walker, P. J. Webber, J. M. Welker, K. S. Winker, and K. Yoshikawa. "Evidence and Implications of Recent Climate Change in Northern Alaska and Other Arctic Regions." *Climatic Change* 72 (2005): 251–298.

Hoegh-Guldberg, O., P. J. Mumby, A. J. Hooten, R. S. Steneck, P. Greenfield, E. Gomez, C. D. Harvell, P. F. Sale, A. J. Edwards, K. Caldeira, N. Knowlton, C. M. Eakin, R. Iglesias-Prieto, N. Muthiga, R. H. Bradbury, A. Dubi, and M. E. Hatziolos. "Coral Reefs Under Rapid Climate Change and Ocean Acificiation." *Science* 318 (2007): 1737–1742.

Hoerling, M. P., J. W. Hurrell, and T. Xu. "Tropical Origins for Recent North Atlantic Climate Change." *Science* 292 (2001): 90–92.

Hoerling, M. P., and A. Kumar. "The Perfect Ocean for Drought." *Science* 299 (2003): 691–694.

Hoffert, M. I., K. Caldeira, A. K. Jain, E. F. Haites, L. D. Danny Harvey, S. D. Potter, M. E. Schlesinger, S. H. Schneider, R. G. Watts, T. M. L. Wigley, and D. J. Wuebbles. "Energy Implications of Future Stabilization of Atmospheric CO_2." *Nature* 395 (1998): 881–884.

Hong, S., J.-P. Candelone, C. C. Patterson, and C. F. Boutron. "History of Ancient Copper Smelting Pollution During Roman and Medieval Times Recorded in Greenland Ice." *Science* 272 (1996): 246–249.

House, K. Z., D. P. Schrag, C. F. Harvey, and K. S. Lackner. "Permanent Carbon Dioxide Storage in Deep-Sea Sediments." *Proceedings of the National Academy of Science* 103 (2006): 12291–12295.

Hurrell, J. W. "Decadal Trends in the North Atlantic Oscillation: Regional Temperatures and Precipitation." *Science* 269 (1995): 676–679.

Huybers, P. "Early Pleistocene Glacial Cycles and the Integrated Summer Insolation Forcing." *Science* 313 (2006): 508–511.

Huybers, P., and C. Wunsch. "Obliquity Pacing of the Late Pleistocene Glacial Terminations." *Nature* 343 (2005): 491–494.

Iglesias-Rodriguez, M. D., P. R. Halloran, R. E. M. Rickaby, I. R. Hall, E. Colmenero-Hidalgo, J. R. Gittins, D. R. H. Green, T. Tyrrell, S. J. Gibbs, P. von Dassow, E. Rehm, E. V. Armbrust, and K. P. Boessenkool. "Phytoplankton Calcification in a High-CO_2 World." *Science* 320 (2008): 336–340.

Imbrie, J., and K. P. Imbrie. *Ice Ages: Solving the Mystery.* 2d ed. Cambridge, Mass.: Harvard University Press, 1986.

Intergovernmental Panel on Climate Change. "Summary for Policymakers." In *Climate Change 2007: The Physical Science Basis. Contribution of Working Group I to the Fourth Assessment Report of the Intergovernmental Panel on Climate Change,* edited by S. Solomon, D. Qin, M. Manning, Z. Chen, M. Marquis, K. B. Averyt, M. Tignor and H. L. Miller, 1–18. Cambridge: Cambridge University Press, 2007.

International Atomic Energy Agency. *Thorium Fuel Cycle—Potential Benefits and Challenges.* Vienna: IAEA, 2005. Available at http://www-pub.iaea.org/MTCD/publications/PDF/TE_1450_web.pdf, TECDOC-1450.

Jia, G. J., H. E. Epstein, and D. A. Walker. "Greening of Arctic Alaska, 1981–2001." *Geophysical Research Letters* 30 (2003), doi:10.1029/2003GL018268.

Jones, P. D., and M. E. Mann. "Climate over Past Millennia." *Reviews of Geophysics* 42 (2004), doi:10.1029/2003RG000143.

Kabat, P., M. Claussen, P. A. Dirmeyer, J. H. C. Gash, L. B. DeGuenni, M. Meybeck, R. A. Pielke Sr., C. J. Vörösmarty, R. W. A. Hutjes, and S. Lürkemeier, eds. *Vegetation, Water, Humans, and the Climate: A New Perspective on an Interactive System.* Berlin: Springer-Verlag, 2004.

Karoly, D. J., and Q. Wu. "Detection of Regional Surface Temperature Trends." *Journal of Climate* 18 (2005): 4337–4343.

Katz, R. W. "Sir Gilbert Walker and a Connection Between El Niño and Statistics." *Statistical Science* 17 (2002): 97–112.

Kaufman, Y. J., and I. Koren. "Smoke and Pollution Aerosol Effect on Cloud Cover." *Science* 313 (2006): 655–658.

Kaufman, Y. J., D. Tarné, and O. Boucher. "A Satellite View of Aerosols in the Climate System." *Nature* 419 (2002): 215–223.

Kawamura, K., F. Parrenin, L. Lisiecki, R. Uemura, F. Vimeux, J. P. Severinghaus, M. A. Hutterli, T. Nakazawa, S. Aoki, J. Jouzel, M. E. Raymo, K. Matsumoto, H. Nakata, H. Motoyama, S. Fujita, K. Goto-Azuma, Y. Fujii, and O. Watanabe. "Northern Hemisphere Forcing of Climatic Cycles in Antarctica over the Past 360,000 Years." *Nature* 448 (2007): 912–916.

Kempton, W., C. L. Archer, A. Dhanju, R. W. Garvine, and M. Z. Jacobson. "Large CO_2 Reductions via Offshore Wind Power Matched to Inherent Storage in Energy End-Uses." *Geophysial Research Letters* 34 (2007), doi:10.1029/2006GL028016.

King, A. W., C. A. Gunderson, W. M. Post, D. J. Weston, and S. D. Wullschleger. "Plant Respiration in a Warmer World." *Science* 312 (2006): 536–537.

Kleiven, H. F., C. Kissel, C. Laj, U. S. Ninnemann, T. O. Richter, and E. Cortijo. "Reduced North Atlantic Deep Water Coeval with the Glacial Lake Agassiz Freshwater Outburst." *Science* 319 (2008): 60–64.

Kleypas, J. A., R. A. Feely, V. J. Fabry, C. Langdon, C. L. Sabine, and L. L. Robins. *Impacts of Ocean Acidification on Coral Reefs and Other Marine Calcifiers: A Guide for Future Research*. Report of a workshop held April 18–20, 2005, in St. Petersburg, Fla., sponsored by the National Science Foundation (NSF), the National Oceanic and Atmospheric Administration (NOAA), and the U.S. Geological Survey (USGS), 2006. Available at http://www.fedworld.gov/onow.

Knutson, R. R., R. L. Delworth, K. W. Dixon, I. M. Held, J. Lu, V. Ramaswamy, M. D. Schwarzkopf, G. Stenchikov, and R. J. Stouffer. "Assessment of Twentieth-Century Regional Surface Temperature Trends Using the GFDL CM2 Coupled Models." *Journal of Climate* 19 (2006): 1624–1651.

Kump, L. R., J. F. Kasting, and R. G. Crane. *The Earth System*. 2d ed. Upper Saddle River, N.J.: Prentice Hall, 2004.

Kyle, G. P., J. P Lurz, S. J. Smith, D. Barrie, and M. A. Wise. *Long-Term Modeling of Wind Energy in the United States*. Report no. PNNL-16316. College Park, Md.: Joint Global Change Research Institute, 2007. Available at http://www.pnl.gov/publications/.

Lackner, K. S. "Carbonate Chemistry for Sequestering Fossil Carbon." *Annual Review of Energy and the Environment* 27 (2002): 1521–1527.

——. "A Guide to CO_2 Sequestration." *Science* 300 (2003): 1677–1678.

Lang, C., M. Leuenberger, J. Schwander, and S. Johnsen. "16°C Rapid Temperature Variation in Central Greenland 70,000 Years Ago." *Science* 286 (1999): 934–937.

Lanzante, J. R., T. C. Peterson, F. J. Wentz, and K. Y. Vinnikov. "What Do Observations Indicate About the Change of Temperatures in the Atmosphere and at the Surface Since the Advent of Measuring Temperatures Vertically?" In *Temperature Trends in the Lower Atmosphere: Steps for Understanding and Reconciling Differences*, edited by T. R. Karl, S. J. Hassol, C. D. Miller, and W. L. Murray, 47–70. Washington, D.C.: Climate Change Science Program, Subcommittee on Global Change Research, 2006.

Laštovička, J., R. A. Akmaev, G. Beig, J. Bremer, and J. T. Emmert. "Global Change in the Upper Atmosphere." *Science* 314 (2006): 1253–1254.

Lawrence, D. M., and A. G. Slater. "A Projection of Severe Near-Surface Permafrost Degradation During the 21st Century." *Geophysial Research Letters* 32 (2005), doi:10.1029/2005GL025080.

Lea, D. W., D. K. Pak, L. C. Peterson, and K. A. Hughen. "Synchroneity of Tropical and High-Latitude Atlantic Temperatures over the Last Glacial Termination." *Science* 301 (2003): 1361–1364.

Lean, J. L., and D. H. Rind. "How Natural and Anthropogenic Influences Alter Global and Regional Surface Temperatures: 1889 to 2006." *Geophysical Research Letters* 35 (2008), doi:10.1029/2008GL034864.

Le Treut, H., R. Somerville, U. Cubasch, Y. Ding, C. Mauritzen, A. Mokssit, T. Peterson, and M. Prather. "Historical Overview of Climate Change." In *Climate Change 2007: The Physical Science Basis. Contribution of Working Group I to the Fourth Assessment Report of the Intergovernmental Panel on Climate Change*, edited by S. Solomon, D. Qin, M. Manning, Z. Chen, M. Marquis, K. B. Averyt, M. Tignor, and H. L. Miller, 93–128. Cambridge: Cambridge University Press, 2007.

Levitus, S., J. Antonov, and T. Boyer. "Warming of the World Ocean, 1955–2003." *Geophysical Research Letters* 32 (2005), doi:10.1029/2004GL021592.

Lough, J. M. "Climate in 1816 and 1811–20 as Reconstructed from Western North American Tree-Ring Chronologies." In *The Year Without a Summer? World Climate in 1816*, edited by C. R. Harington, 97–114. Ottawa: Canadian Museum of Nature, 1992.

Loulergue, L., A. Schilt, R. Spahni, V. Masson-Delmotte, T. Blunier, B. Lemieux, J.-M. Barnola, D. Raynaud, T. F. Stocker, and J. Chappellaz. "Orbital and Millennial-Scale Features of Atmospheric CH_4 over the Past 800,000 Years." *Nature* 453 (2008): 383–386.

Lowell, T. V., C. J. Heusser, B. G. Andersen, P. I. Moreno, A. Hauser, L. E. Heusser, C. Schlüchter, D. R. Marchant, and G. H. Denton. "Interhemispheric Correlation of Late Pleistocene Glacial Events." *Science* 269 (1995): 1541–1549.

Lurie, E. *A Life in Science*. Chicago: University of Chicago Press, 1960.

Luthcke, S. B., H. J. Zwally, W. Abdalati, D. D. Rowlands, R. D. Ray, R. S. Nerem, F. G. Lemoine, J. J. McCarthy, and D. S. Chinn. "Recent Greenland Ice Mass Loss by Drainage System from Satellite Gravity Observations." *Science* 314 (2006): 1286–1289.

Lüthi, D., M. Le Floch, B. Bereiter, T. Blunier, J.-M. Barnola, U. Siegenthaler, D. Raynaud, J. Jouzel, H. Fischer, K. Kawamura, and T. F. Stocker. "High-Resolution Carbon Dioxide Concentration Record 650,000–800,000 Years Before Present." *Nature* 453 (2008): 379–383.

Lyon, B. "The Strength of El Niño and the Spatial Extent of Tropical Drought." *Geophysical Research Letters* 31 (2004), doi:21210.21029/22004GL020901.

Macfarlane, A. M., and M. Miller. "Nuclear Energy and Uranium Resources." *Elements* 3 (2007): 185–192.

Mann, M. E., R. S. Bradley, and M. K. Hughes. "Global-Scale Temperature Patterns and Climate Forcing over the Past Six Centuries." *Nature* 392 (1998): 779–787.

———. "Northern Hemisphere Temperatures During the Past Millennium: Inferences, Uncertainties, and Limitations." *Geophysical Research Letters* 26 (1999): 759–762.

Martrat, B., J. O. Grimalt, N. J. Shackleton, L. de Abreu, M. A. Hutterli, and T. F. Stocker. "Four Climate Cycles of Recurring Deep Surface Water Destabilizations on the Iberian Margin." *Science* 317 (2007): 502–507.

Massachusetts Institute of Technology. *The Future of Coal: Options for a Carbon-Constrained World*. Cambridge, Mass.: MIT, 2007. Available at http://web.mit.edu/coal/.

———. *The Future of Geothermal Energy: Impact of Enhanced Geothermal Systems (EGS) on the United States in the 21st Century*. Cambridge, Mass.: MIT, 2006. Available at http://geothermal.inel.gov/.

———. *The Future of Nuclear Power: An Interdisciplinary MIT Study*. Cambridge, Mass.: MIT, 2003. Available at http://web.mit.edu/nuclearpower/.

Mastrandrea, M. D., and S. H. Schneider. "Probabilistic Integrated Assessment of 'Dangerous' Climate Change." *Science* 304 (2004): 571–575.

Matsumoto, R. "Causes of the $\delta^{13}C$ Anomalies of Carbonates and a New Paradigm 'Gas Hydrate Hypothesis.'" *Journal of the Geological Society of Japan* 11 (1995): 902–924.

Mayewski, P. A., and M. Bender. "The GISP2 Ice Core Record—Paleoclimate Highlights." *Review of Geophysics Supplement, U.S. National Report to International Union of Geodesy and Geophysics, 1991–1994* (1995): 1287–1296.

Mayewski, P. A., W. B. Lyons, M. J. Spencer, M. S. Twickler, C. F. Buck, and S. Whitlow. "An Ice-Core Record of Atmospheric Response to Anthropogenic Sulphate and Nitrate." *Nature* 346 (1990): 554–556.

Mayewski, P. A., L. D. Meeker, M. S. Twickler, S. I. Whitlow, Q. Yang, W. W. Lyons, and M. Prentice. "Major Features and Forcing of High Latitude Northern Hemisphere Atmospheric Circulation Using a 110,000-Year-Long Glaciochemical Series." *Journal of Geophysical Research* 102 (1997): 26346–26366.

McAvaney, B. J., C. Covey, S. Joussaume, V. Kattsov, A. Kitoh, W. Ogana, A. J. Pitman, A. J. Weaver, R. A. Wood, and Z.-C. Zhao. "Model Evaluation." In *Climate Change 2001: The Scientific Basis. Contributions of Working Group I to the Third Assessment Report of the Intergovernmental Panel on Climate Change*, edited by J. T. Houghton, Y. Ding, D. J. Griggs, M. Noguer, P. J. Van Der Linden, X. Dai, K. Maskell, and C. A. Johnson, 471–523. Cambridge: Cambridge University Press, 2001.

McCabe, G. J., M. A. Palecki, and J. L. Betancourt. "Pacific and Atlantic Ocean Influences on Multidecadal Drought Frequency in the United States." *Proceedings of the National Academy of Science* 101 (2004): 4136–4141.

McCulloch, M. T., and T. Esat. "The Coral Record of Last Interglacial Sea Levels and Sea Surface Temperatures." *Chemical Geology* 169 (2000): 107–129.

McPhaden, M. J., S. E. Zebiak, and M. H. Glantz. "ENSO as an Integrating Concept in Earth Science." *Science* 314 (2006): 1740–1745.

Meah, K., and S. Ula. "Comparative Evaluation of HVDC and HVAC Transmission Systems." *Power Engineering Society General Meeting* 1–5 (2007), doi:10.1109/ PES.2007.385993.

Meehl, G. A., and A. Hu. "Megadroughts in the Indian Monsoon Region and Southwest North America and a Mechanism for Associated Multidecadal Pacific Sea Surface Temperature Anomalies." *Journal of Climate* 19 (2006): 1605–1623.

Meehl, G. A., T. F. Stocker, W. D. Collins, P. Friedlingstein, A. T. Gaye, J. M. Gregory, A. Kitoh, R. Knutti, J. M. Murphy, A. Noda, S. C. B. Raper, I. G. Watterson, A. J. Weaver, and Z.-C. Zhao. "Global Climate Projections." In *Climate Change 2007: The Physical Science Basis. Contribution of Working Group I to the Fourth Assessment Report of the Intergovernmental Panel on Climate Change*, edited by S. Solomon, D. Qin, M. Manning, Z. Chen, M. Marquis, K. B. Averyt, M. Tignor, and H. L. Miller, 747–846. Cambridge: Cambridge University Press, 2007.

Meehl, G. A., and C. Tebaldi. "More Intense, More Frequent, and Longer Lasting Heat Waves in the 21st Century." *Science* 305 (2004): 994–997.

Meese, D. A., A. J. Gow, R. B. Alley, G. A. Zielinski, P. M. Grootes, M. Ram, K. C. Taylor, P. A. Mayewski, and J. F. Bolzan. "The Greenland Ice Sheet Project 2 Depth-

Age Scale: Methods and Results." *Journal of Geophysical Research* 102 (1997): 26411–26423.

Ménot, G., E. Bard, F. Rostek, J. W. H. Weijers, E. C. Hopmans, S. Schouten, and J. S. Sinninghe Damste. "Early Reactivation of European Rivers During the Last Deglaciation." *Science* 313 (2006): 1623–1625.

Miller, L., and B. C. Douglas. "Mass and Volume Contributions to Twentieth-Century Global Sea Level Rise." *Nature* 428 (2004): 406–409.

Minobe, S., A. Kuwano-Yoshida, N. Komori, S.-P. Xie, and R. J. Small. "Influence of the Gulf Stream on the Troposphere." *Nature* 452 (2008): 206- 209.

Mishchenko, M. I., I. V. Geogdzhayev, W. B. Rossow, B. Cairns, B. E. Carlson, A. A. Lacis, L. Liu, and L. D. Travis. "Long-Term Satellite Record Reveals Likely Recent Aerosol Trend." *Science* 315 (2007): 1543.

Mix, A. C., N. G. Pisias, W. Rugh, J. Wilson, A. Morey, and T. Hagelberg. "Benthic Foraminiferal Stable Isotope Record from Site 849, 0–5 Ma: Local and Global Climate Changes." In *Proceedings of the Ocean Drilling Program, Scientific Results*, edited by N. G. Pisias, L. Mayer, T. Janecek, A. Palmer-Julson, and T. H. Van Andel, 138: 371–412. College Station, Tex.: Ocean Drilling Program, 1995.

Molina, L. T., and M. J. Molina. "Production of Cl_2O_2 from the Self-Reaction of the ClO Reaction." *Journal of Physical Chemistry* 91 (1987): 433–436.

Monaghan, A. J., D. H. Bromwich, R. L. Fogt, S.-H. Wang, P. A. Mayewski, D. A. Dixon, A. Ekaykin, M. Frezzotti, I. Goodwin, E. Isaksson, S. D. Kaspari, V. I. Morgan, H. Oerter, T. D. Van Ommen, C. J. Van der Veen, and J. Wen. "Insignificant Change in Antarctic Snowfall Since the International Geophysical Year." *Science* 313 (2006): 827–831.

Montzka, S. A., J. H. Butler, R. C. Myers, T. M. Thompson, T. H. Swanson, A. D. Clarke, L. T. Lock, and J. W. Elkins. "Decline in the Tropospheric Abundance of Halogen from Halocarbons: Implications for Stratospheric Ozone Depletion." *Science* 272 (1996): 1318–1322.

Moritz, R. E., C. M. Bitz, and E. J. Steig. "Dynamics of Recent Climate Change in the Arctic." *Science* 297 (2002): 1497–1502.

Mote, P. W., A. F. Hamlet, M. P. Clark, and D. P. Lettenmaier. "Declining Mountain Snowpack in Western North America." *Bulletin of the American Meteorological Society* 86 (2005): 39–49.

Muhs, K., R. Simmons, and B. Steinke. "Timing and Warmth of the Last Interglacial Period: New U-Series Evidence from Hawaii and Bermuda and a New Fossil Compilation for North America." *Quaternary Science Reviews* 21 (2002): 1355–1383.

Nakićenović, N., and R. Swart, eds. *Special Report on Emissions Scenarios: A Special Report of Working Group III of the Intergovernmental Panel on Climate Change.* Cambridge: Cambridge University Press, 2000.

Nelson, F. E. "(Un)Frozen in Time." *Science* 299 (2003): 1673–1675.

Nghiem, S. V., I. G. Rigor, D. K. Perovich, P. Clemente-Colón, J. W. Weatherly, and G. Neumann. "Rapid Reduction of Arctic Perennial Sea Ice." *Geophysical Research Letters* 34 (2007), doi:10.1029/2007GL031138.

Occhel, W. C., G. L. Vourlitis, S. J. Hastings, R. C. Zulueta, L. Hinzman, and D. Kane. "Acclimation of Ecosystem CO_2 Exchange in the Alaskan Arctic in Response to Decadal Climate Warming." *Nature* 406 (2000): 978–981.

Oerlemans, J. "Extracting a Climate Signal from 169 Glacier Records." *Science* 308 (2005): 675–677.

Orr, J. C., V. J. Fabry, O. Aumont, L. Bopp, S. C. Doney, R. A. Feely, A. Gnanadeskian, N. Gruber, A. Ishida, F. Joos, R. M. Key, K. Lindsay, E. Maier-Reimer, R. Matear, P. Monfray, A. Mouchet, R. G. Najjar, G.-K. Plattner, K. B. Rogers, C. L. Sabine, J. L. Sarmiento, R. Schlitzer, R. D. Slater, I. J. Totterdell, M.-F. Weirig, Y. Yamanaka, and A. Yool. "Anthropogenic Ocean Acidification over the Twenty-first Century and Its Impact on Calcifying Organisms." *Nature* 437 (2005): 681–686.

Otto-Bliesner, B. L., S. J. Marshall, J. T. Overpeck, G. H. Miller, A. Hu, and CAPE Last Interglacial Project Members. "Simulating Arctic Climate Warmth and Icefield Retreat in the Last Interglacial." *Science* 311 (2006): 1751–1753.

Overpeck, J. T., B. L. Otto-Bliesner, G. H. Miller, D. R. Muhs, R. B. Alley, and J. T. Kiehl. "Paleoclimatic Evidence for Future Ice-Sheet Instability and Rapid Sea-Level Rise." *Science* 311 (2006): 1747–1750.

Pacala, S., and R. Socolow. "Stabilization Wedges: Solving the Climate Problem for the Next 50 Years with Current Technologies." *Science* 305 (2004): 968–972.

Parsons, B., M. Milligan, J. C. Smith, E. DeMeo, B. Oakleaf, K. Wolf, M. Schuerger, R. Zavadil, M. Ahlstrom, and D. Yen Nakafuji. *Grid Impacts of Wind Power Variability: Recent Assessments from a Variety of Utilities in the United States.* Conference Paper no. NREL/CP-500–39955. Washington, D.C.: Department of Energy, 2006. Available at http://www.osti.gov/bridge.

Patz, J. A., D. Campbell-Lendrum, T. Holloway, and J. A. Foley. "Impact of Regional Climate Change on Human Health." *Nature* 438 (2005): 310–317.

Peroich, D. K., B. Light, H. Eicken, K. F. Jones, K. Runciman, and S. V. Nghiem. "Increasing Solar Heating of the Arctic Ocean and Adjacent Seas, 1979–2005: Attribution and Role in the Ice-Albedo Feedback." *Geophysical Research Letters* 34 (2007), doi:10.1029/2007GL031480.

Peterson, B. J., J. McClelland, R. Curry, R. M. Holmes, J. E. Walsh, and K. Aagaard. "Trajectory Shifts in the Arctic and Subarctic Freshwater Cycle." *Science* 313 (2006): 1061–1066.

Petit, J. R., J. Jouzel, D. Raynaud, N. I. Barkov, J.-M. Barnola, I. Basile, M. Bender, J. Chappellaz, M. Davis, G. Delaygue, M. Delmotte, V. M. Kotlyakov, M. Legrand, V. Y. Lipenkov, C. Lorius, L. Pépin, C. Ritz, E. Saltzman, and M. Stievenard. "Climate and Atmospheric History of the Past 420,000 Years from the Vostok Ice Core, Antarctica." *Nature* 399 (1999): 429–436.

Pfirman, S., W. F. Haxby, R. Colony, and I. Rigor. "Variability in Arctic Sea Ice Drift." *Geophysical Research Letters* 31 (2004), doi:10.1029/2004GL020063.

Pierce, D. W., T. P. Barnett, K. M. AchutaRao, P. J. Gleckler, J. M. Gregory, and W. M. Washington. "Anthropogenic Warming of the Oceans: Observations and Model Results." *Journal of Climate* 19 (2006): 1873–1900.

Pinker, R. T., B. Zhang, and E. G. Dutton. "Do Satellites Detect Trends in Surface Solar Radiation?" *Science* 308 (2005): 850–854.

Plant, J. A., P. R. Simpson, B. Smith, and B. Windley. "Uranium Ore Deposits—Products of the Radioactive Earth." In *Uranium: Mineralogy, Geochemistry, and the Environment*, edited by P. C. Burns and R. Finch, 255–319. Reviews in Mineralogy, no. 38. Chantilly, Va.: Mineralogical Society of America, 1999.

Pollack, H. N., and S. Huang. "Climate Reconstruction from Subsurface Temperatures." *Annual Reviews of Earth and Planetary Sciences* 28 (2000): 339–365.

Proshutinsky, A. Y., and M. A. Johnson. "Two Circulation Regimes of the Wind-Driven Arctic Ocean." *Journal of Geophysical Research* 102 (1997): 12493–12514.

Rahmstorf, S. "A Semi-empirical Approach to Projecting Future Sea-Level Rise." *Science* 315 (2007): 368–370.

Ramachandran, S., V. Ramaswamy, G. L. Stenchikov, and A. Robock. "Radiative Impact of the Mt. Pinatubo Volcanic Eruption: Lower Stratospheric Response." *Journal of Geophysical Research* 105 (2000): 24409–24429.

Ramanathan, V., M. V. Ramana, G. Roberts, D. Kim, C. Corrigan, C. Chung, and D. Winker. "Warming Trends in Asia Amplified by Brown Cloud Solar Absorption." *Nature* 448 (2007): 575–578.

Ramaswamy, V., O. Boucher, J. Haigh, D. Hauglustaine, J. Haywood, G. Myhre, T. Nakajima, G. Y. Shi, S. Solomon, R. Betts, R. Charlson, C. Chuang, J. S. Daniel, A. Del Genio, R. van Dorland, J. Feichter, J. Fuglestvedt, P. M. de F. Forster, S. J. Ghan, A. Jones, J. T. Kiehl, D. Koch, C. Land, J. Lean, U. Lohmann, K. Minschwaner, J. E. Penner, D. L. Roberts, H. Rodhe, G. J. Roelofs, L. D. Rotstayn, T. L. Schneider, U. Schumann, S. E. Schwartz, M. D. Schwarzkopf, K. P. Shine, S. Smith, D. S. Stevenson, F. Stordal, I. Tegen, and Y. Zhang. "Radiative Forcing of Climate Change." In *Climate Change 2001: The Scientific Basis. Contribution of Working Group I to the Third Assessment Report of the Intergovernmental Panel on Climate Change*, edited by J. T. Houghton, Y. Ding, D. J. Griggs, M. Noguer, P. J. Van Der Linden, X. Dai, K. Maskell, and C. A. Johnson, 349–416. Cambridge: Cambridge University Press, 2001.

Ramaswamy, V., M. D. Schwarzkopf, W. J. Randel, B. D. Santer, B. J. Soden, and G. L. Stenchikov. "Anthropogenic and Natural Influences in the Evolution of Lower Stratospheric Cooling." *Science* 311 (2006): 1138–1141.

Randall, D. A., R. A. Wood, S. Bony, R. Colman, T. Fichefet, J. Fyfe, V. Kattsov, A. Pitman, J. Shukla, J. Srinivasan, R. J. Stouffer, A. Sumi, and K. E. Taylor. "Climate Models and Their Evaluation." In *Climate Change 2007: The Physical Science Basis. Contribution of Working Group I to the Fourth Assessment Report of the Intergovernmental Panel on Climate Change*, edited by S. Solomon, D. Qin, M. Manning, Z. Chen, M. Marquis, K. B. Averyt, M. Tignor, and H. L. Miller, 589–662. Cambridge: Cambridge University Press, 2007.

Raymo, M. E., L. E. Lisiecki, and K. H. Nisancioglu. "Plio-Pleistocene Ice Volume, Antarctic Climate, and the Global ^{18}O Record." *Science* 313 (2006): 492–495.

Reich, P. B., S. E. Hobbie, T. Lee, D. S. Ellsworth, J. B. West, D. Tilman, J. M. H. Knops, S. Naeem, and J. Trost. "Nitrogen Limitation Constrains Sustainability of Ecosystem Response to CO_2." *Nature* 440 (2006): 922–925.

Richter-Menge, J., J. Overland, A. Proshutinsky, V. Romanovsky, L. Bengtsson, L. Brigham, M. Dyurgerov, J. C. Gascard, S. Gerland, R. Graversen, C. Haas, M. Karcher, P. Kuhry, J. Maslanik, H. Melling, W. Maslowski, J. Morison, D. Perovich, R, Przyblak, V. Rachold, I. Rigor, A. Shiklomanov, J. Stroeve, D. Walker, and J. Walsh. *State of the Arctic Report*. NOAA/OAR Special Report. Seattle: National Oceanic and Atmospheric Administration (NOAA), Office of Oceanic and Atmospheric Research (OAR), and Pacific Marine Environmental Laboratory (PMEL), 2006.

Riebesell, U., K.G. Schulz, R.G.J. Bellerby, M. Botros, P. Fritsche, M. Meyerhöfer, C. Neill, G. Nondal, A. Oschlies, J. Wohlers, and E. Zöllner. "Enhanced Biological Carbon Consumption in a High CO_2 Ocean." *Nature* 450 (2007): 545–548.

Rignot, E., G. Casassa, P. Gogineni, W. Krabill, A. Rivera, and R. Thomas. "Accelerated Ice Discharge from the Antarctic Peninsula Following the Collapse of Larsen B Ice Shelf." *Geophysical Research Letters* 31 (2004), doi:10.1029/2004GL020697.

Rignot, E., and P. Kanagaratnam. "Changes in the Velocity Structure of the Greenland Ice Sheet." *Science* 311 (2006): 986–990.

Rigor, I., and J.M. Wallace. "Variations in the Age of Arctic Sea-Ice and Summer Sea-Ice Extent." *Geophysical Research Letters* 31 (2004), doi:10.1029/2004GL019492.

Rigor, I.G., J.M. Wallace, and R.L. Colony. "Response of Sea Ice to the Arctic Oscillation." *Journal of Climate* 15 (2002): 2648–2663.

Robock, A. "Volcanic Eruptions and Climate." *Reviews of Geophysics* 38 (2000): 191–219.

Rogner, H.-H., D. Zhou, R. Bradley. P. Crabbé, O. Edenhofer, B. Hare, L. Kuijpers, and M. Yamaguchi. "Introduction." In *Climate Change 2007: Mitigation of Climate Change. Contribution of Working Group III to the Fourth Assessment Report of the Intergovernmental Panel on Climate Change*, edited by B. Metz, O.R. Davidson, P.R. Bosch, R. Dave, and L.A. Meyer, 95–116. Cambridge: Cambridge University Press, 2007.

Romanovsky, V. E., T.S. Sazonova, V.T. Balobaev, N.I. Shender, and D.O. Sergueev. "Past and Recent Changes in Air and Permafrost Temperatures in Eastern Siberia." *Global and Planetary Change* 56 (2007): 399–413.

Rosenfeld, D., J. Dai, X. Yu, A. Yao, X. Xu, X. Yang, and C. Du. "Inverse Relations Between Amounts of Air Pollution and Orographic Precipitation." *Science* 315 (2007): 1396–1998.

Royal Society. *Ocean Acidification Due to Increasing Atmospheric Carbon Dioxide*. Policy Document no. 12/05. London: Royal Society, 2005.

Sabine, C.L., R.A. Feely, N. Gruber, R.M. Key, K. Lee, J.L. Bullister, R. Wanninkhof, C.S. Wong, D.W.R. Wallace, B. Tilbrook, F.J. Millero, T.-H. Peng, A. Kozyr, T. Ono, and A.F. Rios. "The Oceanic Sink for Anthropogenic CO_2." *Science* 305 (2004): 367–371.

Santer, B.D., T.M.L. Wigley, C. Doutriaux, J.S. Boyle, J.E. Hansen, P.D. Jones, G.A. Meehl, E. Roeckner, S. Sengupta, and K.E. Taylor. "Accounting for the Effects of Volcanoes and ENSO in Comparisons of Modeled and Observed Temperature Trends." *Journal of Geophysical Research* 106 (2001): 28033–28059.

Saunders, M.A., and A.S. Lea. "Large Contribution of Sea Surface Warming to Recent Increase in Atlantic Hurricane Activity." *Nature* 451 (2008): 557–560.

Scaife, A., C. Folland, and J. Mitchell. "A Model Approach to Climate Change." *Physics World* 20 (2007): 20–25.

Schär, C., P.L. Vidale, D. Lüthi, C. Frei, C. Häberli, M.A. Liniger, and C. Appenzeller. "The Role of Increasing Temperature Variability in European Summer Heatwaves." *Nature* 427 (2004): 332–337.

Schmidt, G.A. "The Physics of Climate Modeling, 2007." *Physics Today* 60 (2007): 72–73.

Schubert, S.D., M.J. Suarez, P.J. Pegion, R.D. Koster, and J.T. Bachmeister. "Causes of Long-Term Drought in the U.S. Great Plains." *Journal of Climate* 17 (2004): 485–503.

———. "On the 1930s Dust Bowl." *Science* 303 (2004): 1855–1859.

Seager, R., M. Ting, I. Held, Y. Kushnir, J. Lu, G. Vecchi, H.-P. Huang, N. Harnik, A. Leetmaa, N.-C. Lau, C. Li, J. Velez, and N. Naik. "Model Projections of an Imminent Transition to a More Arid Climate in Southwestern North America." *Science* 316 (2007): 1181–1184.

Serreze, M. C., M. M. Holland, and J. Stroeve. "Perspectives on the Arctic's Shrinking Sea-Ice Cover." *Science* 315 (2007): 1533–1536.

Severinghaus, J. P., and E. J. Brook. "Abrupt Climate Change at the End of the Last Glacial Period Inferred from Trapped Air in Polar Ice." *Science* 286 (1999): 930–934.

Shepherd, A., and D. Wingham. "Recent Sea-Level Contribution of the Antarctic and Greenland Ice Sheets." *Science* 315 (2007): 1529–1532.

Shepherd, A., D. Wingham, and E. Rignot. "Warm Ocean Is Eroding West Antarctic Ice Sheet." *Geophysical Research Letters* 31 (2004), doi:10.1029/2004GL021106.

Siegenthaler, U., T. F. Stocker, E. Monnin, D. Lüthi, J. Schwander, B. Stauffer, D. Raynaud, J.-M. Barnola, H. Fischer, V. Masson-Delmotte, and J. Jouzel. "Stable Carbon Cycle–Climate Relationships During the Late Pleistocene." *Science* 310 (2005): 1313–1317.

Sigurdsson, H., and S. Carey. "The Eruption of Tambora in 1815: Environmental Effects and Eruption Dynamics." In *The Year Without a Summer? World Climate in 1816*, edited by C. R. Harington, 16–45. Ottawa: Canadian Museum of Nature, 1992.

Sluijs, A., H. Brinkhuis, S. Schouten, S. M. Bohaty, C. M. John, J. C. Zachos, G.-J. Reichart, J. S. Sinninghe Damsté, E. M. Crouch, and G. R. Dickens. "Environmental Precursors to Rapid Light Carbon Injection at the Paloeocene/Eocene Boundary." *Nature* 450 (2007): 1218–1221.

Smil, V. *Energy at the Crossroads: Global Perspectives and Uncertainties.* Cambridge, Mass.: MIT Press, 2003.

Smith, L. C., Y. Sheng, G. M. MacDonald, and L. D. Hinzman. "Disappearing Arctic Lakes." *Science* 308 (2005): 1429.

Solomon, S. "Stratospheric Ozone Depletion: A Review of Concepts and History." *Reviews of Geophysics* 37 (1999): 275–316.

Solomon, S., D. Qin, M. Manning, R. B. Alley, T. Berntsen, N. L. Bindoff, Z. Chen, A. Chidthaisong, J. M. Gregory, G. C. Hegerl, M. Heimann, B. Hewitson, B. J. Hoskins, F. Joos, J. Jouzel, V. Kattsov, U. Lohmann, T. Matsuno, M. Molina, N. Nicholls, J. Overpeck, G. Raga, V. Ramaswamy, J. Ren, M. Rusticucci, R. Somerville, T. F. Stocker, P. Whetton, R. A. Wood, and D. Wratt. "Technical Summary." In *Climate Change 2007: The Physical Science Basis. Contribution of Working Group I to the Fourth Assessment Report of the Intergovernmental Panel on Climate Change*, edited by S. Solomon, D. Qin, M. Manning, Z. Chen, M. Marquis, K. B. Averyt, M. Tignor, and H. L. Miller, 19–92. Cambridge: Cambridge University Press, 2007.

Spahni, R., J. Chappellaz, T. F. Stocker, L. Loulergue, G. Hausammann, K. Kawamura, J. Flückiger, J. Schwander, D. Raynaud, V. Masson-Delmotte, and J. Jouzel. "Atmospheric Methane and Nitrous Oxide of the Late Pleistocene from the Antarctic Ice Cores." *Science* 310 (2005): 1317–1321.

Sridhar, V., D. B. Loopoe, J. B. Swinehart, J. A. Mason, R. J. Oglesby, and C. M. Rowe. "Large Wind Shift on the Great Plains During the Medieval Warm Period." *Science* 313 (2006): 345–347.

Stine, S. "Extreme and Persistent Drought in California and Patagonia During Mediaeval Time." *Nature* 369 (1994): 546–549.

Stocker, T. F., and S. J. Johnsen. "A Minimum Thermodynamic Model for the Bipolar Seesaw." *Paleoceanography* 18 (2003), doi:1010.1029/2003PA000920.

Stocker, T. F., D. G. Wright, and W. S. Broecker. "The Influence of High-Latitude Surface Forcing on the Global Thermocline Circulation." *Paleoceanography* 7 (1992): 529–541.

Stone, R. S., E. G. Dutton, J. M. Harris, and D. Longenecker. "Earlier Spring Snowmelt in Northern Alaska as an Indicator of Climate Change." *Journal of Geophysical Research* 107 (2002), doi:10.1029/2000JD000286.

Stothers, R. B. "The Great Tambora Eruption in 1815 and Its Aftermath." *Science* 224 (1984): 1191–1198.

Stott, P. A., D. A. Stone, and M. R. Allen. "Human Contribution to the European Heatwave of 2003." *Nature* 432 (2004): 610–614.

Streets, D. G., Y. Wu, and M. Chin. "Two-Decadal Aerosol Trends as a Likely Explanation of the Global Dimming/Brightening Transition." *Geophysical Research Letters* 33 (2006), doi:10.1029/2006GL026471.

Takahashi, T., R. A. Feely, R. F. Weiss, R. H. Wanninkhof, D. W. Chipman, S. C. Sutherland, and T. T. Takahashi. "Global Air–Sea Flux of CO_2: An Estimate Based on Measurements of Sea–Air pCO_2 Difference." *Proceedings of the National Academy of Science* 94 (1997): 8292–8299.

Thompson, D. W. J., and J. M. Wallace. "The Arctic Oscillation Signature in the Wintertime Geopotential Height and Temperature Fields." *Geophysical Research Letters* 25 (1998): 1297–1300.

Toggweiler, J. R., and J. Russell. "Ocean Circulation in a Warming Climate." *Nature* 451 (2008): 286–288.

Trenberth, K. E., P. D. Jones, P. Ambenje, R. Bojariu, D. Easterling, A. Klein Tank, D. Parker, F. Rahimzadeh, J. A. Renwick, M. Rusticucci, B. Soden, and P. Zhai. "Observations: Surface and Atmospheric Climate Change." In *Climate Change 2007: The Physical Science Basis. Contribution of Working Group I to the Fourth Assessment Report of the Intergovernmental Panel on Climate Change*, edited by S. Solomon, D. Qin, M. Manning, Z. Chen, M. Marquis, K. B. Averyt, M. Tignor, and H. L. Miller, 235–336. Cambridge: Cambridge University Press, 2007.

Trenberth, K. E., and D. J. Shea. "Relationships Between Precipitation and Surface Temperature." *Geophysical Research Letters* 32 (2005), doi:10.1029/2005GL022760.

Vecchi, G. A., and B. J. Soden. "Effect of Remote Sea Surface Temperature Change on Tropical Cyclone Potential Intensity." *Nature* 450 (2007): 1066–1070.

Velicogna, I., and J. Wahr. "Acceleration of Greenland Ice Mass Loss in Spring 2004." *Nature* 443 (2006): 329–331.

———. "Measurements of Time-Variable Gravity Show Mass Loss in Antarctica." *Science* 311 (2006): 1754–1756.

Vose, R. S., D. R. Easterling, and B. Gleason. "Maximum and Minimum Temperature Trends for the Globe: An Update Through 2004." *Geophysical Research Letters* 32 (2005), doi:10.1029/2004GL024379.

Waibel, A. E., T. Peter, K. S. Carslaw, H. Oelhaf, G. Wetzel, P. J. Crutzen, U. Pöschl, A. Tsias, E. Reimer, and H. Fischer. "Arctic Ozone Loss Due to Denitrification." *Science* 283 (1999): 2064–2069.

Walter, K. M., S. A. Zimov, J. P. Chanton, D. Verbyla, and F. S. Chapin III. "Methane Bubbling from Siberian Thaw Lakes as a Positive Feedback to Climate Warming." *Nature* 443 (2006): 71–75.

Walther, G.-R., E. Post, P. Convey, A. Menzel, C. Parmesan, T. J. C Beebee, J.-M. Fromentin, O. Hoegh-Guldberg, and F. Bairlein. "Ecological Responses to Recent Climate Change." *Nature* 416 (2002): 389–395.

Wang, Y., H. Cheng, R. L. Edwards, X. Kong, X. Shao, S. Chen, J. Wu, X. Jiang, X. Wang, and Z. An. "Millennial- and Orbital-Scale Changes in the East Asian Monsoon over the Past 224,000 Years." *Nature* 451 (2008): 1090–1093.

Watson, R. T., and the Core Writing Team, eds. *Climate Change 2001: Synthesis Report: Contribution of Working Groups I, II, and III to the Third Assessment Report of the Intergovernmental Panel on Climate Change.* Cambridge: Cambridge University Press, 2001.

Webster, P. J. "Large-Scale Structure of the Tropical Atmosphere." In *Large-Scale Dynamical Processes in the Atmosphere*, edited by B. J. Hoskins and R. F. Pearce, 235–275. London: Academic Press, 1983.

Weldeab, S., D. W. Lea, R. R. Schneider, and N. Andersen. "155,000 Years of West African Monsoon and Ocean Thermal Evolution." *Science* 316 (2007): 1303–1307.

Wentz, F. J., L. Ricciardulli, K. Hilburn, and C. Mears. "How Much More Rain Will Global Warming Bring?" *Science* 317 (2007): 233–235.

Westerling, A. L., H. G. Hidalgo, D. R. Cayan, and T. W. Swetnam. "Warming and Earlier Spring Increase Western U.S. Forest Wildfire Activity." *Science* 313 (2006): 940–943.

Wiedinmyer, C., and J. C. Neff. "Estimates of CO_2 from Fires in the United States: Implications for Carbon Management." *Carbon Balance and Management* 2 (2007), doi:10.1186/1750-0680-2-10.

Wigley, T. M. L. "The Climate Change Commitment." *Science* 307 (2005): 1766–1769.

Wild, M., H. Gilgen, A. Roesch, A. Ohmura, C. N. Long, E. G. Dutton, B. Forgan, A. Kallis, V. Russak, and A. Tsvetkov. "From Dimming to Brightening: Decadal Changes in Solar Radiation at Earth's Surface." *Science* 308 (2005): 847–850.

Willis, K. J., A. Kleczkowski, K. M. Briggs, and C. A. Gilligan. "The Role of Sub-Milankovitch Climatic Forcing in the Initiation of the Northern Hemisphere Glaciation." *Science* 285, (1999): 568–571.

Wing, S. L., G. J. Harrington, F. A. Smith, J. I. Bloch, D. M. Boyer, and K. H. Freeman. "Transient Floral Change and Rapid Global Warming at the Paleocene–Eocene Boundary." *Science* 310 (2005): 993–996.

Woodhouse, C. A., and J. T. Overpeck. "2000 Years of Drought Variability in the Central United States." *Bulletin of the American Meteorological Society* 79 (1998): 2693–2714.

Woods, A., K. D. Coates, and A. Hamann. "Is an Unprecedented Dothistroma Needle Blight Epidemic Related to Climate Change?" *BioScience* 55 (2005): 761–769.

Yokoyama, Y., K. Lambeck, P. De Deckker, P. Johnston, and L. K. Fifield. "Timing of the Last Glacial Maximum from Observed Sea-Level Minima." *Nature* 406 (2000): 713–716.

Zachos, J. C., U. Röhl, S. A. Schellenberg, A. Sluijs, D. A. Hodell, D. C. Kelly, E. Thomas, M. Nicolo, I. Raffi, L. J. Lourens, H. McCarren, and D. Kroon. "Acidification of the Ocean During the Paleocene–Eocene Thermal Maximum." *Science* 308 (2005): 1611–1615.

Zdanowicz, C. M., G. A. Zielinski., and M. S. Germani. "Mount Mazama Eruption: Calendrical Age Verified and Atmospheric Impact Assessed." *Geology* 27 (1999): 621–624.

Zeman, F. S., and K. S. Lackner. "Capturing Carbon Dioxide Directly from the Atmosphere." *World Resource Review* 16 (2004): 157–172.

Zhang, S., J. E. Walsh, J. Zhang, U. S. Bhatt, and M. Ikeda. "Climatology and Interannual Variability of Arctic Cyclone Activity: 1948–2002." *Journal of Climate* 17 (2004): 2300–2317.

Zhang, T., O. W. Frauenfeld, M. C. Serreze, A. Etringer, C. Oelke, J. McCreight, R. G. Barry, D. Gilichinsky, D. Yang, H. Ye, F. Ling, and S. Chudinova. "Spatial and Temporal Variability in Active Layer Thickness over the Russian Arctic Drainage Basin." *Journal of Geophysical Research* 110 (2005), doi:10.1029/2004JD005642.

Zielinski, G. A., and M. S. Germani. "New Ice-Core Evidence Challenges the 1620s B.C. Age for the Santorini (Minoan) Eruption." *Journal of Archaeological Science* 25 (1998): 279–289.

Zielinski, G. A., P. A. Mayewski, L. D. Meeker, S. I. Whitlow, M. S. Twickler, M. C. Morrison, D. Meese, R. Alley, and A. J. Gow. "Record of Volcanism Since 7000 B.C. from the GISP2 Greenland Ice Core and Implications for the Volcano-Climate System." *Science* 264 (1994): 948–952.

Zimov, S. A., E. A. G. Schuur, and F. S. Chapin III. "Permafrost and the Global Carbon Budget." *Science* 312 (2006): 1612–1613.

Zweibel, K., J. Mason, and V. Fthenakis. "A Solar Grand Plan." *Scientific American*, January 2008, 64–73.

STUDENT COMPANION

1 CLIMATE IN CONTEXT

Overview

One way to think about the difference between *weather* and *climate* is to consider the fact that we often dress for the weather but build houses in accordance with the climate. In other words, weather often dictates our day-to-day decisions, whereas our houses are built to withstand the climate in the region where they are located over many seasons and years. This example illustrates the important distinction between weather and climate: the former is the condition of the atmosphere at a given location and instant in time; the latter is the long-term average of the atmosphere within a region (ranging from global to local). Climate, the principal focus of this book, changes in time and space through the dynamic interactions of all parts of the Earth system: the *atmosphere*, *hydrosphere*, *cryosphere*, *biosphere*, and *lithosphere*. All these parts are responsible for moving energy and moisture horizontally and vertically about the globe, thus establishing the familiar climate patterns of our planet. The interactions among the various parts of Earth's *climate system* vary over many different timescales and can do so in complex ways with important consequences. For example, they are responsible for cycling important chemicals such as carbon dioxide and methane between the atmosphere, ocean, and land surface by processes that collectively make up the *carbon cycle*. They also produce processes known as *feedbacks*, or phenomena that enhance or diminish warming or cooling of the planet. Feedbacks have played an important role in changing climates of the past and will play an important role in the future climate as our planet continues to warm.

Understanding climate and its changes requires that we separate *facts* from *fears*. We can think of the facts as what we know about how the climate system works—the details of the climate system's interactions, as outlined in the previous paragraph—and how and why climate changes have occurred in the past. The basic facts that underlie climate change during the past 150 years are that Earth is warming at an accelerating rate, *greenhouse gases* in the atmosphere have risen exponentially due to human activity, and the link between these two observations is unequivocal. But what about the future? Projecting how changes will occur and what decisions societies will make in light of these changes is an endeavor that is inherently uncertain. Fears regarding climate change therefore can be thought of as what we expect in the future based on how we understand the climate system and what decisions we expect our societies to make. Three examples of fears for the twenty-first century are the magnitude of sea-level rise, the potential for megadroughts over large regions of the globe, and increases in extreme weather events such as heat waves and floods. The extent to which each of these fears is realized ultimately depends on the decisions our societies will make during the twenty-first century. Thus the facts of climate change suggest that our fears are a possibility, but they also imply that many of the direst consequences can be avoided by intelligent action and innovation, which places the future directly in our own hands.

Key Concepts

1. Weather is the condition of the atmosphere at any given instant in time, whereas climate is the average state of the atmosphere in a given region over years.

2. The climate system is composed of many interacting parts, generally divided into the atmosphere, the hydrosphere, the cryosphere, the biosphere, and the lithosphere.

3. Feedbacks are processes that enhance or diminish temperature changes within the climate system. These feedbacks can play important roles in *tipping points* within the climate system, or abrupt shifts in climate in response to factors that gradually cause climate to change.

4. Climate-change facts are what we know about how the climate system works and how climate has changed in the past. Climate-change fears are possible changes in future climate that we project based on what we know about the climate, but that also depend on societies' future decisions.

Key Terms

atmosphere
biosphere
carbon cycle
chaotic
climate
climate system
cryosphere
feedbacks
greenhouse gases
hydrosphere
inertia
lithosphere
tipping points
weather

Discussion Questions

1. The overview contrasts two decisions we make based on either weather or climate: the way we dress on a given day is often dictated by the weather, whereas the type of house we live in is likely related to the local climate. What other aspects of our daily lives demonstrate how either weather or climate impacts our decisions or living environment?

2. The climate changed dramatically in the past due to natural variations, but human-caused global warming is reason for great concern. Chapter 1 points out that the reasons for this concern are that projected changes may be substantial and that organized human societies did not exist during many of the large changes of the past. Are these arguments convincing? Are there other reasons why contemporary climate change should be cause for concern in light of past changes? Are there reasons why it should not?

3. Projecting how climate will change in the future is inherently uncertain, even though scientists know a great deal about the way in which the climate system works. There are certainly open science questions about the details of climate change, but these questions are not the biggest causes of uncertainty in future projections. Rather, human societies' decisions are the biggest source of uncertainty underlying how the future climate will unfold. What kinds of societal decisions might be the biggest sources of future uncertainty, and what role do these decisions play in affecting climate change?

2 THE CHARACTER OF THE ATMOSPHERE

Overview

Earth science teachers often begin discussions about Earth's atmosphere by asking students to estimate the scaled thickness of the atmosphere on a common geographic globe. Answers vary widely, but most students are surprised to learn that the thickness would amount to about that of a piece of tissue paper. Indeed, the atmosphere is a mere three-hundredths of the mass of the ocean and one-millionth of the mass of solid Earth. Despite this modest stature, however, the atmosphere is essential for moving energy and moisture within the climate system and for maintaining the comfortable conditions necessary for life as we know it on our planet.

The principal constituents of the atmosphere are nitrogen, oxygen, and argon, which together account for 99.96 percent of it. Some of the remaining gases are water vapor, carbon dioxide, methane, ozone, and nitrous oxide. Each of these five gases makes up a tiny fraction of the atmosphere, but plays an important role in trapping, absorbing, and reflecting incoming and outgoing radiation on Earth. The atmosphere extends to about 500 kilometers (300 miles) above Earth's surface and is divided into layers that are determined by temperature. The lowest layer is the *troposphere*, which decreases in temperature from the surface to the *tropopause*, a relatively constant temperature region that divides the troposphere from the overlying *stratosphere*. Temperatures in the stratosphere rise with height because ozone is concentrated in this layer and absorbs ultraviolet (UV) radiation from the Sun and infrared (IR) radiation emitted from Earth's surface. The final two layers are the *mesosphere*, which decreases in temperature with height, and the outermost *thermosphere*, which again increases in temperature with height. Perhaps the most important consequence of the atmosphere's stable thermal structure is that it prevents the planet from drying out: most of the water vapor in the atmosphere condenses in the troposphere and falls back to the surface as rain or snow, thereby preventing the disassociation of water molecules and the loss of hydrogen to space.

Due to its spherical shape, Earth receives more solar energy per square meter at the equator than it does at the poles. This difference in the distribution of incoming energy drives the circulation of the atmosphere, which redistributes energy from the equatorial regions to the poles. The general circulation of the atmosphere comprises a series of connected circulation cells known as the *Hadley*, *Polar*, and *Ferrel cells*. The dominant Hadley cells are established by rising, moist air in the *Intertropical Convergence Zone* (ITCZ) and descending, dry air in the subtropics. The Polar cells are the result of descending dry air in the polar regions and rising air in the extratropics. Between these two are the Ferrel cells, which are made up of descending, dry air in the subtropics and rising, moist air in the extratropics. These cells generally describe the vertical and latitudinal movement of atmospheric circulation, but the rotation of Earth also has an important impact on the circulation through the phenomenon known as the *Coriolis effect*. This effect causes the moving air in the circulation cells to be deflected, resulting in dominant wind patterns that we know at the surface as the *trade winds* and the *westerlies* and in the upper troposphere as the *jet streams*. In addition to general circulation patterns in the atmosphere, regional circulation patterns are also very important. Such patterns include *monsoons* and the *North Atlantic Oscillation* (NAO), both of which vary on annual and decadal timescales and can have considerable impacts on regional precipitation and temperature.

The story of the *ozone layer* and the Montreal Protocol, an international treaty on the depletion of the ozone layer signed and entered into force in the late 1980s, is important to consider in the context of contemporary climate change. *Ozone* is an important gas that can be harmful if it resides close to the surface in the troposphere or helpful if it resides in the stratosphere, where it absorbs UV radiation. Most of the ozone in the atmosphere is in the stratosphere, but ozone levels in the stratosphere have been reduced during the past several decades by the human manufacture of *halocarbons*. These compounds have interrupted the natural balance of ozone creation and destruction in the stratosphere, causing regions of significant ozone depletion, primarily over Antarctica and to a lesser extent over the Arctic. Although some outstanding questions remain, such as the connection between ozone depletion and climate change, international agreements established through the Montreal Protocol target the phasing out of halocarbon production and a return to pre-1980 levels of stratospheric ozone by the latter part of the twenty-first century.

Key Concepts

1. The atmosphere composes a tiny fraction of Earth's mass, but it is an essential part of Earth's climate system because it is responsible for redistributing energy and moisture about the globe and maintaining the comfortable conditions necessary for life as we know it on the planet.

2. The atmosphere is divided into several stable layers determined by the change in temperature with height above Earth's surface.

3. Because of its spherical shape, Earth receives more solar radiation per square meter near the equator than at the poles. This unevenness requires the redistribution of energy within the climate system, which is accomplished predominantly by the general circulation of the atmosphere by means of Hadley, Ferrel, and Polar cells.

4. The Coriolis effect is caused by the rotation of Earth and is the principal cause of large-scale longitudinal movement of air in the atmosphere. The trade winds and westerlies are due in part to the Coriolis effect, as are the existence of the jet streams.

5. Monsoons and the NAO are examples of regional climate phenomena that have large impacts on specific regions of the globe. These phenomena vary naturally over decades, but are also likely to be affected by global warming.

Key Terms

convective circulations
Coriolis effect
Ferrel cells
Hadley cells
halocarbons
Intertropical Convergence Zone (ITCZ)
jet streams
mesosphere
monsoon
North Atlantic Oscillation (NAO)
ozone
ozone layer
Polar cells
stratosphere
subpolar lows
thermosphere
trade winds
tropopause
troposphere
westerlies

Discussion Questions

1. Consider the circulation patterns described by the Hadley, Ferrel, and Polar cells. These cells are established in regions where warm, moist air is rising or cold, dry air is descending. Given the latitudinal regions of these rising or descending air masses discussed in chapter 2, how does the general circulation of the atmosphere explain many of the dominant climatic regions at Earth's surface?

2. As discussed in chapter 2, the Coriolis effect is responsible for deflecting the movement of air as it travels latitudinally in the large convection cells of the atmosphere. Considering the fact that the rotation of Earth is from west to east and that objects near the equator have larger west-to-east velocities than do objects near the poles, what is the direction of the deflection due to the Coriolis force in the Northern Hemisphere? In the Southern Hemisphere? To help your conceptualization, consider traveling first from the equator to the pole in each of the hemispheres and then from the poles to the equator.

3. The Montreal Protocol is an international agreement that targets phasing out the use and production of ozone-depleting chemicals. This protocol and ozone depletion are worth considering in the context of greenhouse-gas emissions and climate change in that avoiding dangerous depletion of the ozone layer required an international effort to eliminate the use of ubiquitous industrial chemicals. In what additional ways does the problem of ozone depletion compare to the contemporary problem of climate change? How might the two problems be quite different?

3 *THE WORLD OCEAN*

Overview

As molecules go, water is weird. This simple configuration of one oxygen and two hydrogen atoms has some very extreme properties. And it is a good thing that it does, because many of these unusual properties are responsible for why our climate system works the way that it does. Water's *thermal capacity* is one of the highest of all materials, and the energies required to change water from a solid to a liquid or a liquid to a vapor are again some of the highest among all substances. The density of ice is also less than the density of liquid water, causing ice to float when submerged in water. Each of these properties is related to the fact that water is a strongly *dipolar molecule*, and each has important consequences for the climate system's behavior.

The vast majority of water on our planet is, of course, in the ocean. Covering more than 70 percent of Earth's surface and extending to maximum depths of more than 4,000 meters (13,000 feet), the ocean plays a fundamental role in the moderation of climate by storing, releasing, and redistributing heat and chemicals such as carbon dioxide. But the ocean is not only water. More than 3 percent of it is dissolved salts, the vast majority of which is derived from sediments deposited by rivers. On average, a liter of ocean water contains about 35 grams of salt, but the exact level of salt content is ultimately affected by processes such as evaporation, precipitation, melting or freezing of sea ice, the influx of fresh river water, and ocean circulation. These processes combine to establish geographic and vertical patterns of *salinity* in the ocean.

The ocean is stratified into three major layers: the *mixed layer* (extending about 20 to 200 meters [66 to 660 feet] below the surface), the *thermocline* (extending about 500 to 900 meters [1,600 to 3,000 feet] below the mixed layer), and the *deep zone* (extending below the thermocline to the ocean's bottom). The ocean's surface currents occur predominantly in the mixed layer and are driven by the dominant wind patterns established by the general circulation of the atmosphere. These wind patterns, along with the positions of the continents, establish the large circular current patterns known as *subtropical gyres* in the major ocean basins. *Ekman transport*, which causes water to travel at 45- to 90-degree angles relative to the dominant wind direction, is an important feature of wind-driven gyres; it is also responsible for driving upwelling and downwelling of water along coastlines.

Water also circulates among all ocean basins over much longer timescales due to density changes in the ocean. This phenomenon is known as *thermohaline circulation* because the differences in water density that drive the circulation depend on temperature and salinity. The thermo-haline circulation is driven by cold, saline water that originates in two principal areas: the North Atlantic Ocean and the Southern Ocean. The cold, dense water formed in these two regions creates the bottom-water current that flows through all the major ocean basins before eventually mixing with warmer intermediate and surface waters in the middle to northern Pacific Ocean. The amount of water flowing in the global ocean conveyor system is massive, amounting to about 15 times that flowing in all the rivers of the world. The mixing is very slow compared with surface mixing, however: it takes about 1,000 years for the deep water to mix through all ocean basins.

The ocean and atmosphere are coupled, meaning that changes in one cause changes in the other. The important implications of this coupling are perhaps nowhere more evident than in the *El Niño–Southern Oscillation* (ENSO) phenomenon. El Niño is an ocean phenomenon, but it is intimately connected to atmospheric swings in barometric pressure between the eastern and western parts of the equatorial Pacific Ocean known as the Southern Oscillation. Under normal conditions in the equatorial Pacific, easterly trade winds maintain a tilted thermocline that comes close to the surface in the east and is driven to greater depths in the west. These conditions help drive the *Walker circulation*, which includes the rising of warm moist air in the east and the return of dry air in the west, further strengthening the easterly trade winds. These conditions collectively establish a strong, positive feedback between the ocean and the atmosphere. During El Niño years, this coupled ocean–atmosphere circulation pattern collapses, causing severe drought in Australia, Indonesia, and southern Africa; a weak Asian monsoon; mild but stormier winter weather in North America; and reduced hurricane activity in the North Atlantic. Overall, El Niño's effects on ecosystems and agriculture are dramatic.

Key Concepts

1. Water is a dipolar molecule, meaning that it has electrically positive and negative sides. This quality gives water unique properties that have important consequences for the climate system's behavior.

2. The ocean is salty due to the input of dissolved materials from rivers. Salinity in the ocean is relatively constant because the input from rivers is balanced by the removal of salts from the ocean through biological uptake and precipitation into ocean sediments.

3. The ocean is stratified into three main layers (the mixed layer, the thermocline, and the deep layer), which are established principally because of density differences caused by changes in temperature and salinity.

4. Winds drive the ocean's surface currents.

5. The ocean plays a fundamental role in the uptake, release, and redistribution of energy and chemicals in the climate system.

6. Thermohaline circulation is driven by temperature and salinity contrasts in the ocean and is a principal mechanism for regulating heat and carbon dioxide within the climate system over thousands of years.

7. ENSO is an important example of ocean–atmosphere coupling. Under "normal" conditions, strong easterly winds drive ocean responses, which in turn enhance the strength of the winds. During El Niño years, this feedback collapses, with serious consequences for many regions of the globe.

Key Terms

abyssal plains
continental shelves
continental slope
deep zone
dipolar molecule
Ekman transport
El Niño–Southern Oscillation (ENSO)
evapotranspiration
ion
meridional overturning circulation
mixed layer
salinity
specific heat
subtropical gyres
thermal capacity
thermocline
thermohaline circulation
Walker circulation
western boundary currents

Discussion Questions

1. Imagine that water is not a strongly dipolar molecule and therefore has more common properties—for example, that its density as a solid is higher than its density as a liquid and its thermal capacity is not as large. Speculate on how these more common properties would affect the way the climate system works. Would there be sea ice? How might maritime climates be different? Would the ocean circulation change, or would it circulate at all?

2. The thermohaline circulation is tied to the ocean's temperature and salinity, particularly in the North Atlantic and in the Southern Ocean. In a warming world, how might you expect the thermohaline circulation to be affected? What processes affecting the circulation would be most important?

3. The ENSO is the quintessential example of coupling between the ocean and the atmosphere. This coupling is established by numerous factors, including the strength of the equatorial winds and the temperature of the ocean waters. Speculate on how this coupling might be affected during the twenty-first century as the climate warms. Would you expect there to be more or fewer El Niño years? What might determine this outcome? Although there may not be obvious answers to these questions, think through how the various parts of the ENSO system will be affected by warming temperatures.

4 THE CARBON CYCLE AND HOW IT INFLUENCES CLIMATE

Overview

As if Earth were drawing a great breath, the carbon dioxide (CO_2) content of the atmosphere is drawn down each year in the spring and early summer as plants in the Northern Hemisphere begin to grow rapidly and take up CO_2. This process continues into the early autumn when CO_2 levels reach a minimum. With every intake of breath, however, an exhale must follow, and the CO_2 content in the atmosphere begins to rise again in the autumn when plants start to become dormant or die and return CO_2 to the atmosphere. This seasonal variation is a striking feature of the *Keeling curve*, the now famous series of CO_2 measurements begun by Charles Keeling in 1958, and a dramatic example of how carbon cycles through multiple carbon *reservoirs* in Earth's system. Reservoirs such as the soils, ocean, and living biomass modulate CO_2 concentrations in the atmosphere on seasonal to millennial timescales, but contain less than one-tenth of a percent of the total carbon near Earth's surface. The vast majority of the carbon, more than 50 million billion metric tons, is stored in the *rock reservoir*. The cycling of carbon through these various reservoirs is collectively defined as the *carbon cycle*.

The carbon cycle can be thought of as two cycles that operate over short and long timescales. The *long-term carbon cycle* involves the movement of carbon between the rock reservoir and the ocean and atmosphere, and it occurs on timescales of hundreds of thousands to millions of years and more. When rocks are weathered, CO_2 is removed from the atmosphere because it reacts with surface water and with the silicate and carbonate minerals of rocks. Calcium and bicarbonate ions are dissolved in surface waters and then transported by rivers to the ocean, where organisms use them to form shells. These organisms eventually die, and their carbonate shells sink to the ocean floor, where they eventually become limestone; the carbon is trapped in these rocks until the rocks are eventually broken down, at which point the carbon returns to the ocean or atmosphere. A variant of this process occurs on land, where CO_2 is removed from the atmosphere by plants, then buried as dead plant material, and ultimately stored as coal or carbonaceous shale until broken down. Overall, the long-term cycle has maintained a balance of carbon between the rock and *surface reservoirs* and is likely to have acted as a natural thermostat on Earth.

The *short-term carbon cycle* occurs on timescales of months to millennia and does not involve the rock reservoir. On land, plants remove CO_2 from the atmosphere through photosynthesis, and then about half of it is returned to the atmosphere through the respiration of plants and animals.

The other half accumulates in soils as dead debris, where it is returned to the atmosphere as microbes break down the organic matter and release CO_2 or methane. Although these land-based exchanges are important over months and years, the primary regulator of CO_2 in the atmosphere over decades to millennia is the ocean. Three sets of processes, known as the *biological*, *carbonate*, and *solubility pumps*, draw CO_2 from the atmosphere. The biological pump is driven by phytoplankton in the *photic zone* of the ocean. The carbonate pump involves the removal of carbon from the ocean when organisms build calcium carbonate shells. The solubility pump is driven by the cooling and sinking of CO_2-rich surface water at high latitudes, thus removing CO_2 from the atmosphere and transporting it to the deep ocean. All these pumps enrich the deep ocean with CO_2, and there it may be stored for many decades to centuries until the deep waters rise to the surface, warm up, and release CO_2 to the atmosphere.

Perhaps even more striking than the seasonal variation of CO_2 in the Keeling curve is the steady rise in atmospheric CO_2 over the duration of the record. The many possible consequences of this rise will depend in part on how the carbon cycle adjusts to *anthropogenic emissions* of carbon. One important consequence is the *acidification* of the ocean. Because the atmosphere and the ocean are in chemical equilibrium, the CO_2 content of the ocean will increase with that of the atmosphere. This increase will reduce the pH of ocean water and ultimately decrease the stability of carbonate minerals that constitute the skeletons and shells of calcifying organisms. In general, many areas of uncertainty remain in our understanding of the global carbon cycle and how it will operate in a warmer, more CO_2-rich world. It is clear, however, that natural processes are incapable of absorbing all 36 gigatons of anthropogenic CO_2 now being injected into the atmosphere every year, and the trend apparent in the famous Keeling curve will continue its rapid rise as long as such emissions continue.

Key Concepts

1. The Keeling curve has provided direct measurements of atmospheric CO_2 in the atmosphere since 1958. These measurements show a seasonal cycle in the atmosphere's CO_2 content and a steady rise in CO_2 levels over the entire period of observation.

2. The cycling of carbon through surface and rock reservoirs is collectively defined as the carbon cycle.

3. The long-term carbon cycle involves the movement of carbon between the rock reservoir and the ocean and atmosphere, and it occurs on timescales of hundreds of thousands to millions of years and more.

4. The short-term carbon cycle occurs on timescales from months to millennia and involves the cycling of carbon between the surface reservoirs.

5. Ocean acidification is occurring as the ocean takes up more CO_2 from the atmosphere, thereby reducing the ocean's pH. This reduction in pH reduces the stability of the carbonate minerals that constitute the skeletons and shells of calcifying organisms and will have significant impacts on ocean ecosystems.

6. The atmosphere contains more CO_2 now (385 parts per million [ppm]) than at any time in at least the past 800,000 years. CO_2 emissions will increase during the twenty-first century, and how the carbon cycle responds will be an important determinant of the changes in climate we experience.

Key Terms

anthropogenic emissions
biological pump
carbon cycle, long term and short term
carbonate compensation depth
carbonate pump
Keeling curve
ocean acidification
photic zone
reservoir
rock reservoir
solubility pump
surface reservoir

Discussion Questions

1. The long-term carbon cycle is tied to the erosion of rocks. Over the course of Earth's history, mountain belts have formed at different periods by the thrusting of huge and newly exposed rock upward from the surface of the continents. These new mountain ranges are subject to increased and rapid erosion, and thus to increased chemical weathering. Speculate on how these processes might impact the long-term carbon cycle. Would you expect more carbon to be sequestered in the rock reservoir or less? What other geologic processes during these periods of mountain formation might also have important impacts on the long-term carbon cycle?

2. Increases in atmospheric CO_2 are often discussed in terms of their warming effects on the planet, but these increases will have many consequences that are not necessarily tied to warmer surface temperatures. Ocean acidification is one example of such a consequence. How might these additional effects change our discussion of carbon emissions? Consider, for instance, that ocean acidification is tied directly to atmospheric CO_2 content.

3. The carbon cycle is an important source of feedbacks on climate change. For example, large amounts of carbon are stored in the permafrost regions of the Northern Hemisphere. What will happen to this carbon as the climate system warms? What important processes should be considered when answering this question? Are there reasons to believe that climate change will cause permafrost zones to yield a net increase or decrease of carbon to the atmosphere? How does this example illustrate the transfer of carbon between different reservoirs?

5 A SCIENTIFIC FRAMEWORK FOR THINKING ABOUT CLIMATE CHANGE

Overview

The climate system is an enormous heat engine, and the Sun is its principal energy source. The Sun's radiation is not distributed evenly, however, and excess energy at the equator drives winds and ocean currents that move energy toward the poles. Indeed, many of the climate system's common features are simply the result of processes that work to move energy from one part of the planet to another: the midlatitude storms that occur during boreal winters, hurricanes that emerge in late summer and early fall of the Northern Hemisphere, and the steady Gulf Stream are all parts of the heat engine that works to redistribute the Sun's energy.

Understanding the energy accounting within the climate system requires first starting with the energy received from the Sun. Incoming solar radiation has a peak in the visible range of the electromagnetic spectrum and is known as *short-wave radiation*. Earth receives about 342 *watts* per square meter of solar radiation averaged over the surface of the outer atmosphere. Not all of this radiation reaches Earth's surface: the atmosphere absorbs a small fraction and reflects some of the radiation back to space. Of the radiation that does make it to Earth's surface, much is absorbed, and some is reflected back to space by a fraction determined by the surface's *albedo*. Earth's average albedo is about 0.3, which means that approximately 30 percent of the Sun's radiation is reflected back to space by the surface or atmosphere. The radiation that is absorbed warms the surface and is in turn reemitted as *infrared radiation* (IR), also termed *long-wave radiation*. A small amount of this long-wave radiation reaches space, but most of it is radiated back and forth between the surface and the atmosphere. These processes of transmission, reflection, and absorption collectively dictate Earth's radiation budget and indicate one of the principal axioms of the climate system: the energy received by Earth must be equal to the energy it emits. In order to achieve this balance, Earth either heats up or cools down when changes occur in the amount of energy it receives from the Sun or in the amount of *greenhouse gases* in the atmosphere.

Various factors can change Earth's energy balance, and the concept of *radiative forcing* is used to quantify and assess the relative importance of these factors. Greenhouse gases collectively are the most important forcing factor and include carbon dioxide (CO_2), methane, nitrous oxides, and halocarbons. The relative importance of the different greenhouse gases depends on their abundances and on how much they absorb specific wavelengths of energy within spectral absorption bands. Carbon dioxide is the most important greenhouse gas because of its abundance and absorption characteristics. *Aerosols* are the second most important forcing factor and tend to cool the surface, although the magnitude of aerosol forcing is highly uncertain. Land-use changes are an additional forcing factor, but their effect on climate is also quite uncertain.

Changes in these forcing factors are associated with human activity, but purely natural forcing factors also affect climate. Volcanoes influence climate when especially large, explosive eruptions inject sulfur dioxide into the stratosphere, causing warming in the stratosphere and cooling at Earth's surface. Because stratospheric aerosols take several years to settle out, the impact of such an eruption on climate lasts well beyond the time it occurred. The Sun's irradiance also varies on different timescales and can cause warming or cooling depending on the direction of the change.

Feedbacks are important dynamic responses that can influence the climate system's sensitivity to changes in forcing factors. Feedbacks that amplify the effect of a process are known as *positive feedbacks*, whereas feedbacks that dampen responses are known as *negative feedbacks*. Both play important roles in thresholds within the climate system, otherwise known as *tipping points* or forcing limits, that when crossed cause the climate to transition to a new state very rapidly. Tipping points represent significant risks associated with contemporary climate change; they are difficult to predict, but we do know that they have been reached in the past and have caused dramatic climatic shifts. Finally, *inertia* is an important characteristic of the climate system that represents the time required for the system to reach a new balance in response to a change in a forcing factor. Based on current estimates, even if atmospheric CO_2 content and other forcing factors were held constant at 2003 levels, Earth will still unavoidably be committed to 0.6°C (1.1°F) of warming over the course of the next several decades.

Key Concepts

1. Earth's energy budget must ultimately be balanced. When this condition is not met, the climate system will adjust until a balance is achieved.

2. Albedo is a measure of an object or region's reflectivity. Light-colored objects or regions (for example, snow or ice) have high albedos, whereas dark objects or regions (for example, ocean or asphalt) have low albedos.

3. A surplus of incoming radiation in the equatorial regions and a deficit at the poles drive a net transport of energy toward the poles.

4. The *greenhouse effect* refers to the absorption of IR by the atmosphere, which traps heat close to Earth's surface.

5. Radiative forcing is a measure of the amount that a given factor changes Earth's radiation balance. These factors can be divided into natural contributions (for example, volcanoes and irradiance changes) or anthropogenic contributions (for example, greenhouse-gas emissions, aerosols, and land use).

6. Feedbacks influence the climate system's sensitivity to forcing factors and complicate its response to them.

7. Tipping points are forcing limits that when crossed cause the climate to transition to a new state at a rate that is much faster than the process causing the change. Inertia represents the time required for the climate system to reach a new balance in response to a change in a forcing factor.

Key Terms

aerosol
albedo
black body radiation
feedback, positive and negative
forcing factors
greenhouse effect
greenhouse gas
inertia
infrared (IR) radiation
insolation
latent heat
Milankovitch cycle
radiative forcing
sensible heat
short-wave radiation
solar constant
thermal infrared (IR) radiation
tipping point
watt
zenith

Discussion Questions

1. Discussions of declining Arctic sea ice often focus on the amount and extent of ice during the summer months of the Northern Hemisphere. Considering specifically Earth's radiation balance, why would you expect the amount of Arctic summer ice to be more important than the amount of winter ice?

2. During the past several decades, satellites have observed cooling trends in the middle to upper stratosphere and warming trends in the troposphere (much like the surface warming measured in observational records). How is this behavior consistent with the idea that greenhouse gases are a principal cause of contemporary global warming?

3. Large volcanic eruptions eject massive amounts of sulfur dioxide into the stratosphere, causing a short-term (several years) cooling in the climate system. Some scientists have proposed that we mimic this effect in order to mitigate global warming by intentionally ejecting tons of sulfur dioxide into the stratosphere. Based on what you have studied about the effect of volcanoes on climate, what would likely be some of the challenges and negative effects of this plan? What might be some of the potential benefits?

6 LEARNING FROM CLIMATES PAST

Overview

The guiding premise of *paleoclimatology* is that the secrets of the past are locked away in the world around us: ancient trees, coral skeletons, cave formations, peat bogs, deep-sea sediments—to name only a few—all have a story to tell about climates past. These stories are a precious archive of Earth history and are essential both for putting current climate change in perspective and for studying how the climate can change under many different scenarios. Some of these scenarios may have parallels to the near future and offer insights into what we might expect. One such epoch of change occurred during the late Paleocene, about 55 million years ago, when a sudden and enormous mass of carbon flooded the ocean and atmosphere. The carbon increase was accompanied by an average surface temperature rise of 5 to 9°C (9 to 16°F) and widespread ocean acidification. This period of time, known at the *Paleocene–Eocene Thermal Maximum* (PETM), lasted for about 120,000 years before Earth returned to cooler conditions. Where all the carbon came from is still a mystery, but the sudden release seems to have been preceded by gradual warming, suggesting that a warming climate has the potential to trigger much more dramatic shifts.

Another important period of Earth's climate history occurred about 3 million years ago. Beginning in the early to middle *Pliocene* (5.3 million to 1.8 million years ago), Earth was quite warm. At the start of this period, ice covered eastern Antarctica, but there was no ice cap in the Northern Hemisphere. Around 3 million years ago, however, the Southern Hemisphere ice cap gradually expanded, and a Northern Hemisphere ice cap was established for the first time in nearly 200 million years. These changes were accompanied by a gradual cooling that was punctuated by a continuous cycling between glacial and interglacial intervals as ice sheets periodically advanced and retreated. This periodicity is still at work.

Paleoclimatologists use *Milankovitch theory* to explain the recent glacial–interglacial cycles. This theory suggests that the cycles are driven by cyclic changes in Earth's three orbital characteristics: *obliquity*, *eccentricity*, and *precession*. The interplay of these three characteristics is timed with the advance and retreat of the *Pleistocene* ice sheets, but the *insolation* changes caused by *orbital forcings* are not sufficient to explain the magnitude of glacial–interglacial cycles. This fact has given rise to multiple theories about the existence of an amplifier internal to the climate system that enhances the effect of the orbital forcings.

Ice cores from Antarctica and Greenland give us high-resolution records that extend back several hundred thousand years and over multiple glacial–interglacial cycles. Ice cores from the East Antarctic Ice Sheet provide a continuous record of the atmospheric concentrations of carbon dioxide (CO_2) and other greenhouse gases for the past 800,000 years. The fluctuations in CO_2 content display a very close correspondence to variations in local temperature, implying a strong and stable coupling between the two. Greenland ice cores have also provided very important records of temperature, dust, and atmospheric composition over the past 100,000 years. An important observation in the Greenland cores involves the *Younger Dryas*, an interval of abrupt climate shifts that were far greater and more rapid than any climate change experienced since the rise of organized human society. It is hypothesized that the Younger Dryas was caused by the disintegration of an ice dam, which released the waters of glacial Lake Agassiz into the North Atlantic and shut down the thermohaline circulation of the global ocean.

The termination of the Younger Dryas is considered the start of the *Holocene* epoch. Climate has fluctuated during the Holocene, but these changes have been much smaller than those associated with glacial–interglacial cycles. The best-characterized part of the Holocene is the Common Era, two millennia for which proxy records are relatively abundant and highly resolved, therefore allowing spatial variations in climate to be inferred. Climate reconstructions for this period have demonstrated that the twentieth century was warmer than any other time during the previous 400 years. The data also suggest, albeit with some uncertainty, that the late twentieth century was likely warmer than any other time during the previous 1,000 years. Two important periods during the Common Era are the putative *Medieval Warm Period* and the *Little Ice Age*. Both of these periods are well documented in Europe and elsewhere around the North Atlantic, but are more difficult to characterize globally because they appear to have occurred at different times in different places, if at all. Nevertheless, these relatively small climate perturbations appear to have had significant impacts on some of the organized societies of their time, again pointing to the importance of using the past as a guide to the future.

Key Concepts

1. Earth's climate history is replete with rapid changes and cyclic variability.

2. Carbon dioxide levels have at times in the past been much higher than they are today, but not for at least 800,000 years and likely many more.

3. The PETM demonstrates that the injection of large quantities of carbon into the atmosphere can significantly heat the planet and that it takes tens of thousands of years for the climate system to recover to its former state.

4. The past million years have been dominated by 100,000-year glacial–interglacial cycles driven by changes in the obliquity, eccentricity, and precession of Earth's orbit, as described by Milankovitch theory.

5. The Holocene has been a relatively stable period of climatic history, but has nevertheless contained climate swings with considerable consequences for organized societies.

6. Paleoclimate records provide evidence of tipping points that clearly illustrate the climate system's capacity to change abruptly.

Key Terms

bipolar seesaw
Dansgaard-Oeschger (D-O) events
eccentricity
Holocene
Ice Age
ice drainage divide
insolation
Little Ice Age
Medieval Warm Period
Milankovitch theory
obliquity
orbital forcing
Paleocene–Eocene Thermal Maximum (PETM)
paleoclimatology
Pleistocene
Pliocene
precession
speleotherm
tree-ring chronology
varves
Younger Dryas

Discussion Questions

1. This chapter discusses the PETM as a period of time that has implications for contemporary climate change. Similar to the present situation, the PETM is associated with a sudden and massive release of carbon into the atmosphere and the subsequent warming of Earth's surface. Are other aspects of the PETM similar to what is happening now? What characteristics of the climate during the PETM might not make it a perfect analog to today's changing climate?

2. Review figure 6.5, which shows the temperature and CO_2 record estimated from an Antarctic ice core. Can you estimate the approximate changes in CO_2 and temperature from the most recent glacial maximum to the Holocene? What about the rates of change? How do the magnitude of these changes and their rates of change compare with global changes during the twentieth century (temperature is estimated to have changed about 0.6°C [1.1°F] and atmospheric CO_2 by about 100 parts per million [ppm])?

3. Again looking at figure 6.5, note the strong relationship between CO_2 and temperature. Using the parlance of statistics, we would say that these two time series are strongly positively correlated. A strong correlation, however, is not a sufficient demonstration of causation—in other words, that a change in CO_2 or temperature is physically causing the change in the other. If you were a paleoclimatologist, what other things would you like to know to determine causation? What other relationships might you try to determine from the CO_2 and temperature time series? What other measurements would you like to have?

A CENTURY OF WARMING AND SOME CONSEQUENCES

Overview

Is Earth warming, and if so, what is the cause? These two fundamental questions underlie much of the discussion about contemporary climate change and its connections to human activity. Modern observations fortunately provide much of the information needed to address these questions.

With regard to warming, the data are unequivocal: all parts of Earth's climate system—from the atmosphere to the oceans, from the continental subsurface to the cryosphere—warmed during the twentieth century. Direct observations of temperature are available on a global scale beginning about the mid-nineteenth century and indicate a global mean surface air temperature increase of about 1°C (1.8°F) since that time. Surface air temperature measurements are not the only records we have, however, and observations of many other elements of the system independently reflect warming at Earth's surface. Glaciers in most parts of the world have been retreating at rates that indicate a mean global temperature increase commensurate with that of the *instrumental record*. Temperatures measured in terrestrial boreholes also indicate long-term warming of the global land surface over the past 500 years, with more than half of the warming occurring during the past century alone. Still more measurements from weather balloons and satellites during the latter part of the twentieth century demonstrate that the lower and middle sections of the troposphere have also been warming. The oceans too have been warming. Global sea-surface temperatures increased during the twentieth century, and the warming has been observed to depths of about 3,000 meters (9,800 feet), although at slower rates. All these measurements and many more indicate a global and persistent warming during the twentieth century that is substantially greater than can be accounted for by natural variations in the climate system.

With regard to what is causing the observed warming, it is known that multiple factors influence climate change, but the human-induced buildup of greenhouse gases is quite clearly tied to the warming during much of the twentieth century. This conclusion is supported by the fact that significant increases in atmospheric greenhouse gases have been observed and can be traced to the burning of fossil fuels (for example, the isotopic composition of carbon in the atmosphere is changing in accordance with the lower carbon-13/carbon-12 [$^{13}C/^{12}C$] ratios in fossil fuels). It is well known that these greenhouse gases absorb infrared radiation (IR), and warming therefore is an expected physical consequence of increases in their concentrations. Furthermore, all observations of warming and other planetary changes are consistent with increased greenhouse-gas forcing. For example, the troposphere is warming, but the stratosphere is cooling, which is consistent with the fact that greenhouse gases are trapping more heat near Earth's surface. Finally, no other known changes in forcing factors can account for the warming we have observed in the climate over the past several decades. Solar irradiance, for example, has remained essentially unchanged since it has been continuously measured by satellite. Each of these observations and arguments points strongly to the role that increasing greenhouse-gas emissions and therefore human activity are playing in the changing climate.

The consequences of the observed warming are and will be pervasive. Changes in the hydrologic cycle are linked to the warming climate and are just beginning to be manifest. Patterns of precipitation and drought are changing and will have direct and potentially severe impacts on agriculture, water supplies, and ecosystems. Historical occurrences such as the *Dust Bowl* of the 1930s have demonstrated the dire consequences of drought, and paleoclimate examples from the past millennium suggest the possibility of droughts even more prolonged and severe than those of the early twentieth century. Water supplies for many sensitive areas are also being depleted. Dwindling snow and ice in areas such as the Himalayas threaten to reduce the principal source of water for hundreds of millions of people. A warming climate can also increase the probability of *extreme events*, leaving open the possibility that we will experience more frequent droughts, heat waves, and heavy storms. In addition, ecosystems are being affected in a variety of ways. Warming can directly impact photosynthesis and respiration, increase plants' susceptibility to pests and disease, force the redistribution of species, reduce biodiversity, and modify *phenology*. Warming and drought also increase the incidence of wildfires, again with serious consequences for ecosystems and human infrastructure. Some of these effects are only just beginning to be detected by modern observations and felt around the world, and their projections into the future are associated with some uncertainty. Nevertheless, they represent very real and dire possibilities for the future.

Key Concepts

1. Observations across all parts of Earth's climate system indicate a global and persistent warming during most of the twentieth century.

2. Increases in carbon dioxide (CO_2) in the atmosphere have been observed since measurements began in 1958 and can be clearly traced to human activity.

3. Increases in human-induced greenhouse gases in the atmosphere are very likely the cause of observed warming since the mid-twentieth century.

4. Changes in precipitation and drought have been observed and will likely continue to be important and serious consequences of contemporary climate change.

5. Changes in the occurrence of extreme events have been observed and will be another serious consequence of contemporary climate change.

6. Ecosystem changes due to warming and hydrological changes—such as species migration, biodiversity losses, and alterations in phenology—have been observed.

Key Terms

$^{13}C/^{12}C$ (isotopic ratio)
Dust Bowl

extreme event
instrumental temperature record
megadrought
Palmer Drought Severity Index
phenology
radiosonde
stratospheric cooling
tropospheric warming

Discussion Questions

1. As discussed in this chapter, temperature observations beginning in about the mid-nineteenth century have allowed scientists to make estimates of hemispheric and global temperature changes that have occurred since that time. These measurements have a degree of uncertainty, however, that scientists must adjust for. What kinds of variations in technology and knowledge over time might be important for scientists to think about when using temperature measurements from a given observational station to estimate climate changes? For example, think about how the instruments for measuring temperature have likely changed over time. What other things may have changed, and how might scientists account for them?

2. It is impossible to point to one event, such as the 2003 heat wave in Europe or Hurricane Katrina in 2005, and say that it was caused by global warming. Nevertheless, the frequency of extreme events will increase in a warming climate. What might be some of the consequences of an increase in extreme events, such as drought or heavy storms, for our organized societies? How might such an increase impact our planning and infrastructure?

3. This chapter discusses several observations consistent with the idea that increases in greenhouse gases are causing contemporary climate change—for example, the observation that the troposphere is warming, whereas the stratosphere is cooling. Pretend that you are a climate scientist and that you would like to explore further the association between greenhouse-gas increases and climate change. What other observations might you like to have, and what would you look for in those observations? Do you think you would be able to "prove" the connection between greenhouse-gas increases and climate change? What does it mean to conclude that observations are consistent with the idea that greenhouse gases and climate change are connected? What might you look for in the observations to determine if effects are inconsistent with the connection?

8 THE SENSITIVE ARCTIC AND SEA-LEVEL RISE

Overview

The polar regions are special places on our planet. They are home to many unique species and have captured the imagination of explorers throughout much of human history. But these regions are also important within the climate system and play a unique role in global circulation. For many reasons, they are the "canary in the coal mine," and the observed rapid polar changes are likely harbingers for much greater changes to come on the planet as a whole.

In the Arctic, temperatures have been rising at a rate that is more than double that observed for the globe. A potential feedback associated with this warming involves reductions in *permafrost* extent and thickness. Permafrost currently underlies about 25 percent of the land area in the Northern Hemisphere. Although projections are still somewhat uncertain, most predict considerable reductions in permafrost extent during the twenty-first century. These reductions may cause significant increases in carbon emissions that will likely act as a feedback on the warming induced by greenhouse gases. Warming and melting of permafrost also will affect local hydrology through the drying out of soils, the disappearance of lakes, and increased river runoff. These changes will in turn have an impact on local infrastructure and ecology and will present considerable adaptive challenges to northern societies and ecosystems.

Sea ice in the Arctic has also been declining steadily during the period of recorded observation. During the twentieth century, ice cover in the Arctic Ocean typically reached a minimum of about 6 million square kilometers (2.3 million square miles) in the early fall. In 2005, a new low of 5.6 million square kilometers (2.2 million square miles) of ice cover was reached, and in 2007 it dwindled to a mere 4.1 million square kilometers (1.6 million square miles). Some of the changes in sea-ice extent are tied to natural climate variations in the *Arctic Oscillation*. The minimum in 2007 was caused in part by an unusual combination of a strong *Transpolar Drift Stream* (a rapid flow of ice through the central Arctic Ocean, across the pole, and out through the Fram Strait) and an influx of warm water flowing into the Arctic Ocean, both tied to the positive phase of the Arctic Oscillation. Perhaps more important was the paucity of old, thick ice leading into the summer of 2007, again caused by predominantly positive phases of the Arctic Oscillation since the late 1980s. Sea ice would likely build in age and thickness again with a prolonged negative phase of the Arctic Oscillation, but human-induced warming will likely cause the proportion of old, thick ice to continue to dwindle. The consequences of these continued reductions in Arctic sea ice include impacts on ocean circulation, enhanced albedo feedback, reduced habitat for polar bears and other Arctic species, loss of hunting grounds for indigenous peoples, and increased coastal erosion.

Sea-level rise is another important impact tied to warming of the poles. The two largest *ice sheets* of the world are in Greenland and Antarctica, and together they represent nearly 80 percent of the freshwater on the planet and many meters of sea-level rise should they melt. Although it is thought that the *Greenland Ice Sheet* would take several centuries to melt under business-as-usual scenarios, there are several reasons for concern now. Greenland's climate is particularly sensitive to global warming due to albedo feedback. Much of the ice sheet is also at high elevation, but as the ice sheet shrinks and elevation decreases, the surface will become warmer and enhance melting. The Greenland Ice Sheet is also losing mass by the flow of its marginal glaciers into the ocean. This process is poorly understood, complicating estimates for the rate of ice loss. On the other end of the world, the fate of the *West Antarctic Ice Sheet* is also the subject of much uncertainty. Unlike the base of the *East Antarctic Ice Sheet*, which is mostly above sea level, much of the base of the West Antarctic Ice Sheet is below sea level. This condition makes it particularly sensitive to rising ocean temperatures in the region. The Antarctic Peninsula of West Antarctica also extends much farther north than any other part of the continent, where air temperatures are warmer. These two factors combine to make the West Antarctic Ice Sheet particularly sensitive to fluctuations in climate and open to the possibility of losing considerable mass as the climate warms.

The prospect of sea-level rise on the order of many meters is very unlikely within the twenty-first century. Nevertheless, a great amount of uncertainty associated with the magnitude and rate of sea-level rise remains, particularly regarding connections to changes in the great ice sheets at the poles. The not too distant past certainly witnessed periods in which sea level was 4 to 6 meters (13 to 20 feet) above present levels. Furthermore, it is possible that tipping points are looming in the near future that, once crossed, will render the ice sheets unstable. Should such tipping points be reached, significant changes in sea level will become irreversible regardless of future decisions. Such unknowns are therefore reason for concern and deserve considered attention by scientists and policy makers alike.

Key Concepts

1. The Arctic Oscillation defines two dominant and distinct regimes in Arctic climate.

2. The Arctic has been warming at twice the rate observed for the globe.

3. Arctic sea ice has been declining steadily since direct observations began in the late 1970s.

4. Ice losses in the Arctic have been caused by a combination of natural variations in the Arctic Oscillation and the prolonged warming that has occurred there.

5. Sea-ice losses cause significant positive feedbacks on warming due to albedo changes.

6. Large-scale reductions in permafrost are occurring in the Arctic, causing ecological and hydrological changes as well as potentially increased carbon emissions.

7. Sea level is projected to rise by several tens of centimeters during the twenty-first century, but these projections do not include uncertainties surrounding the role of ice sheet dynamics in the stability of the large ice sheets.

Key Terms

active layer
anticyclonic
Arctic Oscillation
cyclonic
East Antarctic Ice Sheet
Greenland Ice Sheet
ice sheet
ice shelves
loess
permafrost
sea ice
thermokarst
Transpolar Drift Stream
West Antarctic Ice Sheet
yedoma

Discussion Questions

1. The melting of sea ice does not affect sea-level rise, but the melting of ice sheets does. What is the reason for this difference? To help you understand the process, think of adding ice to a glass of water. What happens to the water level in the glass when you first add the ice? What happens to the water level once the ice has melted?

2. This chapter discusses observed reductions in permafrost and the greening of Arctic regions. The former activity will potentially cause net releases of carbon into the atmosphere, whereas the latter indicates enhanced vegetation growth and therefore the likelihood of carbon uptake. How might these two effects compete in the net uptake or release of carbon in the Arctic? How might some of the other changes in the Arctic—such as reductions in snow cover and soil moisture—potentially determine which of these effects is dominant?

3. An ice-free Arctic in the summer will have many climatic implications, but it will also carry many political and economic ramifications. Most obvious among the latter will be the prospect of new and shorter trade routes and additional oil prospecting. What are some of the likely positive and negative impacts of these activities in the Arctic? How might these activities have specific relevance to the ecology of the Arctic and the problem of climate change globally?

9 CLIMATE MODELS AND THE FUTURE

Overview

One can think of *climate models* as giant accounting tools: they divide space into a series of boxes (three-dimensional grid points) and calculate the exchange of energy, moisture, and momentum among these boxes over a period of specified time steps. All the processes related to these exchanges are represented by mathematical equations, such that each grid point represents a certain volume and specified set of properties that dictate the rate, character, and amount of the exchanges. As the properties of one point change, so do those of all neighboring points. In the presence of time-varying external forcings, such as the amount of solar energy received, the models calculate how the properties of each point change with time. Good spatial resolution is necessary for simulating certain features of the climate, and the length of the time steps can determine whether realistic solutions are achieved or not. All models also contain approximations of physical or chemical processes that are known as *parameterizations*. Different climate models have different parameterizations, which is one reason why they can yield different results. When a climate model is being built, decisions on each of these features can dictate how well the model simulates realistic features of the climate system.

One important evaluation of models is whether they produce *emergent behavior*—in other words, whether large-scale features develop within model simulations that have not been built into the model structure. For example, the Intertropical Convergence Zone (ITCZ) appears in model simulations and results from only the forces acting on each grid point, not because it has been explicitly represented in the model. Models also can be evaluated in terms of their ability to reproduce past climates when driven with historical data of greenhouse-gas emissions and other known forcings. A general measure of climate is the global mean surface temperature, which climate models can reproduce faithfully for the late nineteenth century and the twentieth century. An important finding related to modeling global mean surface temperatures involves the causes of warming. When the models are run with both *anthropogenic* and *natural forcings*, they closely replicate observed twentieth-century warming. If, however, the models are run with only natural forcings, the simulations cannot reproduce the observed warming after about 1960. These experiments thus strongly indicate that anthropogenic greenhouse-gas emissions are the likely cause of warming in the latter part of the twentieth century.

Despite the growing confidence in climate models, interpretation of their results is subject to some uncertainty. For instance, some poorly understood natural phenomena, such as water vapor and cloud feedbacks, are not included in climate models. Most models also do not include feedbacks involving the terrestrial biosphere. Each of these scientific factors undoubtedly infuses uncertainty into a given model's *projections* of the future. Aside from these scientific unknowns, however, the greatest uncertainty in model projections is associated with hard-to-predict societal decisions that will affect future greenhouse-gas emissions. These decisions cannot be accurately known, but it is possible to establish a number of *emission scenarios* that reflect different possible paths for the future. These scenarios can then be used as the basis for different simulations that reflect how the future climate might unfold. The Intergovernmental Panel on Climate Change (IPCC) used this approach to describe the possible outcomes of a range of high-, medium-, and low-emission scenarios. Given these scenarios, the IPCC projects that global mean surface temperatures during the twenty-first century will change, relative to the 1980–1999 average, by 3.6°C, 2.8°C, and 1.8°C (6.5°F, 5.0°F, and 3.2°F) for the high-, medium-, and low-emission scenarios, respectively.

In addition to temperature projections, many important insights into the future climate have come from modeling studies. Significant changes in precipitation patterns as well as more frequent and intense heat waves are expected, and it is anticipated that precipitation occurrences will be concentrated in more intense but less frequent events. These findings reflect the general prediction that extreme events will increase in a warmer climate. The oceans are also projected to continue to warm. The warming will initially be restricted to the mixed layer, but will later extend to the deep ocean. Climate feedbacks are also expected to be important:

1. The ocean and living biota are projected to become progressively less efficient at removing carbon dioxide (CO_2) from the atmosphere.
2. Warming permafrost is expected to yield increased emissions of CO_2 and methane.
3. Sea ice will continue to melt and reduce the albedo of the poles.

High-emission scenarios indicate that by the latter part of the twenty-first century, the Arctic will become ice free in the summer, and current observations are outpacing these projections. Sea level is also expected to rise by several tens of centimeters during the twenty-first century, but this projection does not consider dynamic changes in the Greenland and West Antarctic ice sheets, which may work to exceed these projections. All of these model projections indicate that the world during the twenty-first century may be very different than the one to which we have become accustomed.

Key Concepts

1. Global climate models are discrete representations of the physical and chemical processes that transfer energy, mass, and momentum within the climate system.
2. Climate model simulations are estimates of reality and include approximations of poorly understood processes, whereas some processes are not represented at all.
3. Despite approximations and known deficiencies, models reproduce well many features of the observed climate.

4. Representative model simulations of observed twentieth-century warming are possible only if anthropogenic increases in greenhouse gases are included as forcings.

5. Future societal decisions that will impact emission scenarios represent the greatest uncertainty associated with twenty-first-century climate model projections.

Key Terms

anthropogenic forcings
atmosphere–ocean general circulation models
chaotic
climate models
emergent behavior
emission scenarios
ensemble
initial conditions
natural forcings
parameterization
projections

Discussion Questions

1. Model projections are not future weather forecasts. Common experience tells us that we cannot predict the weather more than several days, let alone several decades, into the future. How then are we to interpret model projections? What is the difference between predicting whether it will be raining on a given day in June ten years from now and suggesting that increases in extreme precipitation events are likely to occur in a warming world? Why is a predicted increase in global mean temperatures likely a robust future projection, but temperature forecasts beyond several days in your local region most likely useless? What do these kinds of questions tell us about the type of information we can expect from climate model projections?

2. Most scientific disciplines are well suited for multiple experiments that allow scientists to study the influence of one variable at a time. In climate science, however, it is difficult to establish a control experiment. The observed climate represents only one possible scenario under which the system can evolve, and it is very difficult to separate the influence of many different changing variables on the climate's state. How do climate models help circumvent this difficulty? Do they completely solve the problem? How might models complicate our ability to establish control experiments regarding the climate system?

3. Pretend you are a policy maker for the twenty-first century and are addressing the prospects of global climate change. What kinds of model simulations would you like to perform to help you in your challenge? What kinds of questions would you ask, and what kinds of scenarios would you establish? In working to establish policy recommendations based on the model projections, how would you address uncertainties in those recommendations?

10 ENERGY AND THE FUTURE

Overview

Decreasing carbon dioxide (CO_2) emissions while meeting the world's growing energy needs is essential if we are to avoid potentially catastrophic climate change. The knowledge and nearly all the technology are at hand, but the practical implementations needed to achieve this goal remain a daunting challenge. Central to this issue is the production of clean electricity. About one-third of present-day CO_2 emissions is from energy generation, and these emissions are growing far more rapidly in that sector than in others.

Coal is cheap and abundant, and energy production from coal is already supported by a large existing infrastructure. Coal will therefore continue to provide a large portion of the world's energy in the decades to come, but the burning of coal presents significant problems. It produces more CO_2 per unit energy than do other *fossil fuels*; it produces multiple toxic pollutants; and the mining and transport of coal are themselves energy intensive. Part of the production of cleaner electricity therefore is tied to more efficient burning of coal in combination with CO_2 capture and sequestration. The efficiency and emissions associated with coal are largely tied to how it is burned. The burning of *subcritical pulverized coal* is most common in the United States; it yields maximum generating efficiencies of about 34 percent. More efficient approaches use *supercritical* to *ultrasupercritical pulverized coal* (PC) technologies with maximum generating efficiencies of up to 46 percent. Coal also may be burned with limestone, which does not significantly increase efficiency but eliminates most pollutants.

It is possible to capture the emissions associated with the burning of coal, but at significant decreases in generating efficiency. The common capturing process involves chemical absorption of flue gas CO_2. Once CO_2 is captured, however, the question becomes what to do with it. *Carbon sequestration* in underground rock reservoirs is likely the best technical and economical means of storing CO_2, but the CO_2 must be sequestered in liquid form and stored in rocks that have high porosity and permeability. These kinds of rocks are found in sedimentary basins that fortunately are common in many settings. There are, however, uncertainties associated with such large-scale carbon sequestration. To date, the process has not been demonstrated on a large enough scale to suggest that the technology is a viable means of offsetting emissions. Site-specific considerations such as leaks and local reactions between CO_2, fluids, and minerals also have to be evaluated in the context of long-term storage. There is also the question of whether power plants can be located near sequestration sites in order to avoid substantial transportation costs. In spite of these challenges, however, existing sequestration projects have succeeded as planned, and the technical and economic evaluations of these projects have in general been positive.

Some energy alternatives do not produce CO_2 emissions. *Nuclear power* is one such alternative that has the potential to reduce carbon emissions significantly. Large-scale implementation of nuclear power would be a massive undertaking, requiring construction of sixteen 1,000-megawatt plants per year for the next 40 years. An expansion of this magnitude would require the availability of sufficient uranium reserves; current estimates suggest known and inferred uranium deposits would be enough to meet 20 to 25 percent of the world's electricity needs with nuclear power during the twenty-first century. Nevertheless, the implementation of large-scale nuclear power production ultimately hinges on overcoming four challenging but not insurmountable hurdles: cost, operational safety, waste storage, and control of weapons proliferation.

Of all the clean and renewable energy sources, *wind* and *solar power* are the most promising. They do not produce greenhouse gases, and they have the potential of producing massive and sustained amounts of electricity. Collectively, wind and solar power are likely to provide a significant and growing proportion of the world's energy during the twenty-first century.

Wind power production is expanding rapidly as its costs approach the cost of producing electricity from coal-burning plants; readily accessible wind resources are also enormous. Since 1996, globally installed wind power capacity has increased an average of more than 28 percent per year. The expansion is being driven by the world's ever-growing demand for energy, which is ratcheting up the costs of fossil fuels and thus making wind power increasingly more cost competitive. Government policies have also encouraged the development of wind power as well as increasing investment in the sector by large oil, utility, and manufacturing entities. Nevertheless, several hurdles to the expansion of wind power in the United States remain. The most important is that the existing transmission infrastructure is in places not capable of transporting large amounts of power from potential generation sites to population centers. Limitations in energy storage also reduce the proportion of power that wind can contribute to a local *power grid*.

Although the large-scale expansion of solar power lags behind that of wind power, solar power represents perhaps the greatest hope for meeting future needs while driving down CO_2 emissions. The amount of available solar power also dwarfs any other renewable energy source. In general, the cost of generating solar power is much more than the cost of generating power from fossil fuel sources. Great efforts therefore are being directed at producing solar power with better efficiency, cheaper manufacturing costs, and more efficient means of storage. In spite of some of the challenges, the use of solar power is now expanding at an increasingly rapid rate, often as a result of public-incentive programs. The development of a network of high-voltage transmission lines is also important to the large-scale expansion of solar power because efficiency dictates that solar power plants be located in sunny parts of the world. In the United States, large installations in the Southwest would cost an estimated $400 billion over the next 40 years, but would potentially provide 70 percent of electricity and more than 33 percent of total energy requirements by 2050.

Overall, the shift to clean electricity will take decades. Rapid implementation of energy conservation and efficiency measures must also be a part of these efforts. We have no panacea for getting rid of the problem of greenhouse-gas emissions, so efforts on multiple fronts are necessary. What is clear, however, is that there is no shortage of ideas on how to meet the energy and climate challenges of a new century. Only a lack of will and conviction to make these ideas a reality would prevent us from moving forward.

Key Concepts

1. Coal is a cheap and abundant resource that will likely continue to contribute a significant portion of the world's energy supply during the twenty-first century.

2. Burning coal presents significant problems: it produces the highest CO_2 output per unit energy of any fossil fuel; it generates significant pollutants; and the mining and transport of coal are energy intensive.

3. Capture of CO_2 from coal-burning power plants is technically viable, but significantly reduces generating efficiency.

4. Carbon sequestration in underground rock reservoirs is a potentially important climate-change mitigation strategy, but the implementation of the technology is still subject to challenges.

5. Nuclear energy is a viable energy source that does not produce CO_2. The four hurdles to expansion of nuclear power, however, are cost, safety, storage, and proliferation of nuclear weapons.

6. Wind and solar power are the most promising renewable energy sources because of their potential competitive pricing and large-scale capacity.

7. In addition to high costs, storage and transmission of wind and solar power represent the biggest hurdles to their expansion.

Key Terms

carbon sequestration
closed fuel cycle
feed-in tariff
fission
fossil fuels
high-voltage alternating current (HVAC) transmission
high-voltage direct current (HVDC) transmission
integrated gas combined cycle (IGCC)
joule
nuclear power
open fuel cycle
oxy-fuel pulverized coal (PC) combustion
photoelectric effect
photovoltaic
power grid
pulverized coal (PC)
saline formations
solar power
stabilization triangle
Stirling engine
turbine
watt
wind power

Discussion Questions

1. Coal is an abundant and cheap energy source that likely will continue to contribute to a large fraction of the world's energy production during the twenty-first century. Reducing the amount of CO_2 emitted to the atmosphere as a result of burning coal will be an important component of any climate-change mitigation effort. How will attempts to reduce the emissions from coal, such as carbon capture and sequestration, also reduce the efficiency of coal burning and potentially contribute some additional carbon emissions? Should these additional efforts be considered in the total price of energy production from coal, and how would such pricing be enacted?

2. This chapter discusses the *feed-in tariff* policies that Germany enacted to encourage the production of energy from solar power. What other policy decisions might encourage transitions to renewable energy sources? Divide your thinking into two areas: *incentives* (such as providing tax reductions for companies wishing to develop renewable energy sources) and *penalties* (such as cap-and-trade policies that impose limits on emissions of greenhouse gases and fines for companies that exceed them).

3. Transitioning to renewable energy sources such as wind, solar, biomass, hydro-, and geo-thermal power is an essential step to avoiding potentially catastrophic climate change. Regardless of these efforts, however, we have already committed ourselves to some degree of climate change during the twenty-first century, which in turn will impact the energy potential of some renewable energy sources. What are some of these impacts, and will they generally be positive or negative? Which renewable energy sources will not be affected by climate change?

INDEX